Selected Topics in Vibrational Mechanics

SERIES ON STABILITY, VIBRATION AND CONTROL OF SYSTEMS

Founder and Editor: Ardéshir Guran
Co-Editors: C. Christov, M. Cloud, F. Pichler & W. B. Zimmerman

About the Series

Rapid developments in system dynamics and control, areas related to many other topics in applied mathematics, call for comprehensive presentations of current topics. This series contains textbooks, monographs, treatises, conference proceedings and a collection of thematically organized research or pedagogical articles addressing dynamical systems and control.

The material is ideal for a general scientific and engineering readership, and is also mathematically precise enough to be a useful reference for research specialists in mechanics and control, nonlinear dynamics, and in applied mathematics and physics.

Selected Volumes in Series B

Proceedings of the First International Congress on Dynamics and Control of Systems, Chateau Laurier, Ottawa, Canada, 5–7 August 1999
    Editors: A. Guran, S. Biswas, L. Cacetta, C. Robach, K. Teo, and T. Vincent

Selected Volumes in Series A

Vol. 2 Stability of Gyroscopic Systems
       Authors: A. Guran, A. Bajaj, Y. Ishida, G. D'Eleuterio, N. Perkins, and C. Pierre

Vol. 3 Vibration Analysis of Plates by the Superposition Method
       Author: Daniel J. Gorman

Vol. 4 Asymptotic Methods in Buckling Theory of Elastic Shells
       Authors: P. E. Tovstik and A. L. Smirinov

Vol. 5 Generalized Point Models in Structural Mechanics
       Author: I. V. Andronov

Vol. 6 Mathematical Problems of Control Theory: An Introduction
       Author: G. A. Leonov

Vol. 7 Analytical and Numerical Methods for Wave Propagation in Fluid Media
       Author: K. Murawski

Vol. 8 Wave Processes in Solids with Microstructure
       Author: V. I. Erofeyev

Vol. 9 Amplification of Nonlinear Strain Waves in Solids
       Author: A. V. Porubov

Vol. 10 Spatial Control of Vibration: Theory and Experiments
       Authors: S. O. Reza Moheimani, D. Halim, and A. J. Fleming

Vol. 11 Selected Topics in Vibrational Mechanics
       Editor: I. Blekhman

SERIES ON STABILITY, VIBRATION AND CONTROL OF SYSTEMS

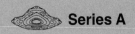 Series A   Volume 11

Founder and Editor: **Ardéshir Guran**
Co-Editors: **C. Christov, M. Cloud,
F. Pichler & W. B. Zimmerman**

Selected Topics in Vibrational Mechanics

Edited by

Ilya Blekhman
Russian Academy of Science, Russia

Contributing Authors:

I. I. Blekhman   P. Rodionov
L. I. Blekhman   E. Shishkina
H. Dresig   S. V. Sorokin
H. Duckstein   L. Sperling
O. A. Ershova   D. M. Tcherniak
A. Fidlin   J. J. Thomsen
A. L. Fradkov   L. A. Vaisberg
S. V. Grishina   V. B. Vasilkov
D. A. Indeitsev   K. S. Yakimova
J. S. Jensen   N. P. Yaroshevich
P. S. Landa

NEW JERSEY • LONDON • SINGAPORE • SHANGHAI • HONG KONG • TAIPEI • CHENNAI

Published by

World Scientific Publishing Co. Pte. Ltd.
5 Toh Tuck Link, Singapore 596224
USA office: Suite 202, 1060 Main Street, River Edge, NJ 07661
UK office: 57 Shelton Street, Covent Garden, London WC2H 9HE

British Library Cataloguing-in-Publication Data
A catalogue record for this book is available from the British Library.

SELECTED TOPICS IN VIBRATIONAL MECHANICS

Copyright © 2004 by World Scientific Publishing Co. Pte. Ltd.

All rights reserved. This book, or parts thereof, may not be reproduced in any form or by any means, electronic or mechanical, including photocopying, recording or any information storage and retrieval system now known or to be invented, without written permission from the Publisher.

For photocopying of material in this volume, please pay a copying fee through the Copyright Clearance Center, Inc., 222 Rosewood Drive, Danvers, MA 01923, USA. In this case permission to photocopy is not required from the publisher.

ISBN 981-238-055-8

Foreword

This volume of World Scientific's SVCS series covers many recent developments in Vibrational Mechanics. The latter is a young and fast-growing branch of nonlinear dynamics with applications not only in technology (e.g. building new vibrating machines, processing granular materials, eliminating harmful vibrations, selfbalancing) but also in celestial mechanics, physics, chemistry, biology, physiology, medicine—even economics and sociology. The term vibrational mechanics was introduced by professor Blekhman, the leading authority in the fields of applied mathematics and mechanics. His work has evoked much interest—in the US, Japan, Denmark, Germany, Bulgaria, Iran, Russia, and Canada—because of its potential applications and theoretical implications. It has also served as an inspiration for many other researchers. The present volume contains contributions from his collaborators in Russia, Denmark, and Germany. Other researchers will undoubtedly make further contributions to this fascinating field.

The book "Selected Topics in Vibrational Mechanics" represents an excellent contribution to the literature on nonlinear dynamics and oscillation theory.

Professor Blekhman is to be congratulated for organizing, editing, and co-authoring the volume. Furthermore, I would like to call attention to the forthcoming volume of this same series, entitled "What Vibration Can Do" also written by Professor Blekhman. It will be of great interest to all those—physicists, mathematicians, engineers, and of course to specialists in the science of mechanics—who sooner or later will deal with such phenomena.

Wien, Austria   Ardeshir Guran
September 2003

Preface

Vibrational Mechanics is a new, intensively developing section of nonlinear dynamics and of the theory of nonlinear oscillations. It presents a general approach to the study of the effect of vibration on nonlinear systems. This approach is characterized by the simplicity of application and by physical clearness.

The impetus to the development of vibrational mechanics was given by the work of P.L.Kapitsa "Dynamic stability of the pendulum when the point of suspension is oscillating" [Journ. of Exp. and Theor. Phys., 21, 5, 1951 (in Russian)]. I learned about that work when I was just beginning my scientific researches. I was delighted by the fact how simple and physically transparent—by means of direct separation of slow and fast motions—a bright result was achieved: was established the stability of the upper position of the pendulum and was found the corresponding condition of the stability.

In those years many outstanding scientists, mostly mathematicians, were of the opinion that this result was obtained in a non-strict way and does not have a sufficiently general significance. Therefore in my theses, devoted to the theory of self-synchronization of rotating bodies and to the theory of vibrational displacement, I preferred to use the Poincare-Lyapunov methods and, wherever possible, exact methods.

How surprised I was when it turned out that the results, found by means of a rather complicated investigation can be easily obtained by means of the method (though not grounded mathematically!), generalizing the method of P.L.Kapitsa for more complicated systems. It became obvious that the success of that method was not accidental. And I made it my task to extend the method of Kapitsa to the sufficiently general nonlinear systems and to prove it mathematically, as far as possible, in the lines of asymptotic methods of nonlinear mechanics. This brought me, first, to the solution of a number of problems, important for applications, about the action of vibration on nonlinear systems—problems in the range from the theory of mechanisms and machines to celestial mechanics. Secondly, the apparatus of the method was formalized mathematically. And, eventually, thirdly, conceptions were formulated which were called "vibrational mechanics" and "vibrorheology". They were a special case of a more general conception—of the mechanics of the systems with the hidden (ignored) motions.

It is the enumerated results that made the subject of the author's book "Vibrational Mechanics", published in 1994 in the Russian language and then in an extended version, in 2000 in English (World Scientific, Singapore, New Jersey, London, Hong Kong, 2000).

The method of direct separation of motions as well as the conceptions of vibrational mechanics and the mechanics of systems with hidden motions are comparatively universal (within the frames, of course, of certain natural suppositions). That, however, cannot be said about the methods of an approximate solution of equations of fast motions, used by P.L.Kapitsa and his followers. The establishing of the region of the applicability of those methods (when in the equations of fast motions the slow variables are considered to be "frozen", and also when they are limited by the so-called purely inertial approximation) remains the task which has not yet been completely solved. Elaborating efficient means of solving problems beyond the region of the above mentioned approximate methods is also an important problem.

In recent years considerable progress has been made in solving the above problems. Besides, a number of interesting applied problems have been solved.

This book reflects these results in the ingenious presentation of their authors—well-known scientists of Germany, Denmark and Russia.

A certain number of interesting publications which refer to the problem, but belong to other investigators, are reflected in the book in the form of short reviews.

The book consists of six parts.

Part I, for the convenience of the readers, presents the foundations of vibrational mechanics and gives its mathematical apparatus. As compared to what was given in the corresponding section of the book "Vibrational Mechanics", the presentation is shortened but another chapter (chapter 4) has been added. The classical problem of a pendulum with a vibrating axis is solved there by five main methods of the theory of nonlinear oscillations and by the method of direct separation of motions. Emphasizing the simplicity and the physical clearness of the latter, the authors hope that the reader will form his own idea about the advantages and disadvantages of these methods.

Parts II–V are devoted to the use of the approach of vibrational mechanics and of the method of direct separation of motions when solving modern important applied problems of mechanics.

Part II considers the pendulum and the pendulum systems under a high frequency excitation

Atom is said to be inexhaustible. The statement that pendulum is inexhaustible is valid as well. This refers to the pendulum as to a physical object and as to the corresponding mathematical model presenting the basic fundamental model of oscillatory phenomena. It will not be an exaggeration to say that the overwhelming majority of the oscillatory systems, both discrete and distributed, can be presented as a set of oscillators, i.e. objects which are "generalized" pendulums.

I hope this part is a clear illustration of what has been said. The approach of vibrational mechanics, used in every chapter, makes it possible to display the peculiar, as a rule unknown before, specific features of the behavior of the pendulum and of the pendulum systems under vibration.

Part III is devoted to the problems of the synchronization of dynamic objects. The theory of synchronization, which was developed first as a conception of frequency ("Huygens") synchronization, is now flourishing as a conception of the coordinated behaviour of several objects or processes in time in the most general sense. This conception finds important applications not only in technology but also in celestial mechanics, in physics, in chemistry, in biology, in physiology and medicine, and even in economy and sociology. In this connection the first (the eleventh) chapter of part III is devoted to the general definition of synchronization. One of its aims is to be an aid in making the achievements in one branch of the theory be used more widely in other branches. It should be noted that this definition specifies and generalizes the one given in the English edition of "Vibrational Mechanics".

Other chapters of part III are devoted to developing the theory and to new investigations of the frequency synchronization of vibro-exciters, i.e unbalanced rotors, installed on mobile bases, driven by asynchronous engines.

Part IV gives an outline of the idea of the new method of producing materials with peculiar dynamic properties—the so-called dynamic materials. It also considers one of the possible ways of realizing this idea.

Part V contains an analogue of Bernoulli's classical equation for the slow flows of fluid, subjected to the action of a high frequency vibration.

In these equations, like in other equations of vibrational mechanics, apart from the ordinary slow forces (in this case—pressure difference) there also appear some additional (vibrational) forces. This part also gives a rather simple theory of two remarkable phenomena—of the so-called vibro-jet effect and a phenomenon of vibro-injection of gas into the fluid.

The final part—part VI is devoted to two mathematical supplements to the theory, stated in part I.

As a Supplement are given the contents of the English edition of the book "Vibrational Mechanics".

I express my deep gratitude to Prof. Ardeshir Guran for his idea to publish this book. I am also thankful to my colleagues and friends—authors of the sections—for their consent to present the results of their investigations.

I am thankful to doctor M.M.Perelman who has translated some chapters of the book, to L.G.Titova for the trouble she has taken in preparing the manuscript and also to Professor M.E.Levinshtein for his useful advice.

I greatly appreciate the contact with those colleagues which gave me much pleasure while working at this book.

St. Petersburg   I. Blekhman
October 2002

Contents

Foreword v

Preface vii

Part I. The Basis of Vibrational Mechanics

Chapter 1. On Some Nonlinear Oscillatory Effects. The Main
Idea of Vibrational Mechanics 3
I. I. Blekhman (Translated by M. Perelman)

1.1 On the Effects Caused by the Action of Vibration in
Nonlinear Oscillatory Systems 3
1.2 The Main Idea of Vibrational Mechanics.
Observer O and Observer V 5

Chapter 2. The Main Mathematical Apparatus of Vibrational
Mechanics and of the Method of Direct Separation
of Motions 9
I. I. Blekhman (Translated by M. Perelman)

2.1 Preliminary Remarks 9
2.2 The Initial Equation and Its Reduction to a System of
Integro-differential Equations 10
2.3 The Case When a Separate Equation for the Slow
Component is Obtained 12
2.4 The Main Assumption of Vibrational Mechanics, Its
Formalization and Conditions of its Fulfillment 14
2.5 The Main Equation of Vibrational Mechanics. Vibrational
Forces, Observers O and V 18

2.6	Method of an Approximate Derivation of the Expression of Vibrational Forces and of Composing the Main Equation of Vibrational Mechanics	19
2.7	Important Special Case	20
2.8	On the Case of a Mechanical Systems with Constraints	24
2.9	On the Simplifications of Solving Equations for the Fast Component of Motion. Purely Inertial Approximation	27
2.10	Additional Remarks, Certain Generalizations	29
2.11	Summary: On the Procedure of the Practical Use of the Method	32

Chapter 3. On Other Methods of Obtaining Expressions for the Vibrational Forces and the Main Equations of Vibrational Mechanics 35
I. I. Blekhman (Translated by M. Perelman)

3.1	Well Known Methods	35
3.2	Two Other Methods	36

Chapter 4. A Simplest Example: Solving the Problem about a Pendulum with a Vibrating Axis of Suspension by Different Methods of the Theory of Nonlinear Oscillations 39
I. I. Blekhman, D. A. Indeitsev
(Translated by M. Perelman)

4.1	Preliminary Remarks	39
4.2	Equation of Motion	40
4.3	The Poincaré-Lyapunov Method of Small Parameter	41
4.4	The Use of Floquet-Lyapunov's Theory and of Ince-Strutt's Diagram	43
4.5	Asymptotic Method	45
4.6	Method of Multiple Scales	46
4.7	Methods of Harmonic Balance and of Bubnov–Galerkin	48
4.8	Method of Direct Separation of Motions	50

4.9   Discussion 53

Chapter 5. Conclusion: On the Main Peculiarities and Advantages of the Approaches of Vibrational Mechanics and of the Method of Direct Separation of Motions as Compared to Other Methods of Nonlinear Mechanics 55
I. I. Blekhman (Translated by M. Perelman)

5.1   Peculiarities and Limitations 55
5.2   Advantages 56
5.3   Final Remarks 58

References to Part 1 60

Part II. Pendulum and Pendulum Systems under High-Frequency Excitation — Non-Trivial Effects

Chapter 6. Quasi-equilibrium Positions and Stationary Rotations of the Pendulums with a Periodically Vibrating Axis 65
I. I. Blekhman, H. Dresig, P. Rodionov
(Translated by M. Perelman)

6.1   Preliminary Remarks, Equation of Motion 65
6.2   Regimes of Quasi-Equilibrium 66
6.3   Regimes of Rotation 69
References 72

Chapter 7. Non-Trivial Effects of High-Frequency Excitation for Pendulum Systems 73
J. S. Jensen, J. J. Thomsen, D. M. Tcherniak

7.1   Preliminary Remarks 73
7.2   Chelomei's Pendulum — Resolving a Paradox 74
       7.2.1   Introduction 74

	7.2.2	History of the Problem and the Controversy 75
	7.2.3	Experimental Observations 76
	7.2.4	A Theoretical Model 80
	7.2.5	Predicting Stationary Solutions 83
	7.2.6	Analysing the Results 88
	7.2.7	Conclusions 96
7.3	Nonlinear Dynamics of the Follower-Loaded Double Pendulum with Added Support-Excitation 97	
	7.3.1	The Model and Model Equations 97
	7.3.2	Direct Partition of Motion 99
	7.3.3	Linear Stability of the Upright Pendulum Position 101
	7.3.4	Global Dynamic Behavior 110
	7.3.5	Effects of Bi-Directional Support-Excitation 113
	7.3.6	Conclusions 119
7.4	Articulated Pipes Conveying Fluid Pulsating with High Frequency 120	
	7.4.1	The Model and Model Equations 120
	7.4.2	Autonomous Model Equations 122
	7.4.3	Linear Stability of the Hanging Position 124
	7.4.4	Periodic Motion of the Autonomous System 127
	7.4.5	Conclusions 132
References 134		

Chapter 8. On the Theory of the Indian Magic Rope 139
I. I. Blekhman, H. Dresig, E. Shishkina
(Translated by M. Perelman)

8.1	Preliminary Remarks 139	
8.2	Equation of Oscillations and Its Consideration 141	
8.3	Solving the Problem by Method of Direct Separation of Motions 142	
8.4	Analysis of the Result. Physical Explanation of the Effect of the Indian Rope 144	
References 149		

Chapter 9. Conjugate Resonances and Bifurcations of
Pendulums under Biharmonical Excitation 151
I. I. Blekhman, P. S. Landa
(Translated by M. Perelman)

9.1 Preliminary Remarks 151
9.2 Equation of Motion 152
9.3 Equation of Slow Motions 153
9.4 The Case of "Ordinary" Resonance. Conjugate
    Resonances and Bifurcations 155
9.5 The Case of Parametric Resonances 159
9.6 The Biharmonic Effect on the System, Described by
    Duffing's Equation 160
9.7 Some Remarks on the Application of the Results 163
References 165

Chapter 10. On the Investigations of the Electromechanical
Systems. On the Behavior of the Conductivity
Bodies of Pendulum Types in High-Frequency
Magnetic Fields 167
I. I. Blekhman (Translated by M. Perelman)

10.1 On Some New Results of the Theory of
    Electro-Mechanical Systems 167
10.2 The Problem about a Passive (Resonant) Electrostatic
    Suspension 168
10.3 The Problem about the Motion of the Pendulum with the
    Closed Circuit of High-Current in Frequency Magnetic Field ... 172
References 176

Part III. Problems of the Theory of Selfsynchronization

Chapter 11. On General Definitions of Synchronization 179
I. I. Blekhman, A. L. Fradkov

11.1	Preliminary Remarks	179
11.2	Evolution of the Synchronization Concept	180
11.3	General Definition of Synchronization	181
11.4	Examples	182
11.5	Discussion	185
References		187

Chapter 12. A Guide to Solving Certain Self-Synchronization Problems 189
L. Sperling, H. Duckstein

12.1	Preliminary Remarks	189
12.2	Unbalanced Rotors on an Oscillatory System	190
12.3	Application of the Method of Direct Separation of Motion	194
12.4	Harmonic Influence Coefficients and Vibrational Moments	195
12.5	Conditions for the Existence of Synchronous Motions	199
12.6	Conditions of Stability	201
12.7	Two Rotors	202
12.8	Example 204	
12.9	Summary	207
References		208

Chapter 13. The Setting Up of the Self-Synchronization Problem of the Dynamic Objects with Inner Degrees of Freedom and Methods of Its Solution 209
I. I. Blekhman, L. Sperling
(Translated by M. Perelman)

13.1	Preliminary Remarks	209
13.2	Statement of the Problem	210
13.3	Structure of the Kinetic and Potential Energy of the System. The Generalized Forces	213
13.4	Integral Criterion (Extreme Property) of the Stability of Synchronous Motions	215
	13.4.1 Introduction	215

13.4.2 Systems with almost uniform rotations, non-quasiconservative idealization of rotation 215
13.4.3 Systems with the objects, idealized as quasiconservative 218
13.4.4 Comparison of the results at different types of idealization. Some conclusions 220
13.5 Method of Direct Separation of Motions. Methods of Small Parameter 221
13.6 Self-Synchronization of Two Identical Vibro-Exciters with the Internal Degrees of Freedom, Whose Axes Pass through the Center of Gravity of a Solid Body (Plane Motion) 222
    13.6.1 Description of the system 222
    13.6.2 Synchronization of the system in the absence of inner degrees of freedom of the exciters 224
    13.6.3 Equations of motion of the system 224
    13.6.4 Stationary solutions and their stability 226
    13.6.5 Discussion of the results of the section 232
References 234

Chapter 14. On the Expansion of the Field of Applicability of the Integral Signs (Extreme Property) of Stability in Problems on the Synchronization of the Dynamic Objects with Almost Uniform Rotations 235
I. I. Blekhman, N. P. Yaroshevich
(Translated by M. Perelman)

14.1 Preliminary Remarks 235
14.2 On the Integral Sign (Extreme Property) of the Stability of Synchronous Motions 236
14.3 The Statement of the Problem on the Synchronization of Objects with Almost Uniform Rotations 237
14.4 Solution of the Problem by Method of Direct Separation of Motions 238

14.5 Extended Formulation of Integral Signs 241
14.6 Examples, Comparison to the Results, Obtained by Other
Methods 243
    14.6.1 Double Synchronization of Two Vibro-Exciters on
    a Platform with One Degree of Freedom 243
    14.6.2 Double Synchronization of Three Unbalanced
    Vibro-Exciters, Located Symmetrically on a Softly
    Vibro-Isolated Flatly Oscillating Solid Body. 246
References 251

Part IV. Problems of Creating Dynamic Materials

Chapter 15. On Dynamic Materials 255
    I. I. Blekhman (Translated by M. Perelman)
15.1 Briefly on the Idea of Dynamic Materials 255
15.2 On the Development of the Idea of Creating Dynamic
Materials 256
References 257

Chapter 16. The Active Control of Vibrations of Composite
Beams by Parametric Stiffness Modulation 259
    S. V. Grishina, O. A. Ershova, S. V. Sorokin
16.1 Preliminary Remarks 259
16.2 The Governing Equations 261
16.3 Modal Analysis of Vibrations 263
16.4 Direct Partition of Motions 267
16.5 The Eigenvalue Problem for a Beam with the Resonant
Parametric Stiffness Modulation 272
16.6 Forced Vibrations. An Influence of Internal Damping 277
16.7 Vibrations of a Beam with Parametrically Modulated
Stiffness in Heavy Fluid Loading Conditions 280
16.8 Discussion of the Parametric Stiffness Modulation 284

16.9 A Modal Formulation of the Control Problem for a
Sandwich Beam.. 285
16.10 Analysis of Vibration Control for a Model Two-Degrees
of Freedom Mechanical System... 290
  16.10.1 Analysis by the method of multiple scales................... 291
  16.10.2 Analysis by the method of direct partition of
    motions... 302
  16.10.3 'Standard' dynamical absorber 305
16.11 Conclusions.. 307
References 309

Part V. Vibrational Hydrodynamics and Hydraulics

Chapter 17. On Vibrational Hydrodynamics and Hydraulics 313
  I. I. Blekhman (Translated by M. Perelman)

17.1 Reinolds' Equation as an Equation of Vibrational
  Mechanics 313
17.2 The Analog of Bernoulli's Equation for the Flows,
  Subjected to Vibration... 314
References 315

Chapter 18. On the Vibro-Jet Effect and on the Phenomena of
  Vibrational Injection of the Gas into Fluid.................... 317
  *I. I. Blekhman, L. I. Blekhman., L. A. Vaisberg,
  V. B. Vasilkov, K. S. Yakimova*
  (Translated by M. Perelman)

18.1 On Phenomenon under Consideration.. 317
18.2 Common Expression for the Gas or Fluid Discharge
  through a Hole in Vibrating Vessel... 319
18.3 On the Theory of Vibro-jet Effect... 324
18.4 On the Theory of the Vibrational Injection of Gas into the
  Fluid 325

18.5 Results of the Experiments.. 326
References 330

Part VI. Some Mathematical Supplements and Generalization

Chapter 19. On Asymptotic Analysis of Systems with Fast
                Excitation.. 333
                A. Fidlin

19.1 Introduction. Classification of Systems with Fast
      Excitation. Weakly Excited Systems 333
19.2 Systems with Strong Excitation. General Analysis................... 335
19.3 Systems with Very Strong Excitation in a Special Case of
      Fast Oscillating Inertial Coefficients... 340
19.4 Two Mathematical Examples of Systems with Strong
      Excitation 343
19.5 Response of a One Degree of Freedom System to Strong
      and Very Strong, High Frequency External and Parametric
      Excitation 347
19.6 Conclusions 354
References 355

Chapter 20. On the Averaging of Discontinuous Systems................ 357
                A. Fidlin

20.1 Introduction. Types of Discontinuities in Oscillating
      Systems. Short Review of the Investigations 357
20.2 Averaging of Constant Order Discontinuous Systems.............. 363
20.3 Averaging of Variable Order Discontinuous Systems 367
20.4 On the Averaging of Discontinuous Systems in the
      Vicinity of a Strong Nonlinear Resonance................................ 370
20.5 Constant Order Discontinuous Averaging in Systems with
      Quasi-Elastic Collisions. 374

20.6 Resonant Motions in a Quasi-Elastic Colliding Oscillator
  Excited by an Inertial Source of Limited Power 379
  20.6.1 The problem statement 379
  20.6.2 Dimensionless formulation. Small parameters 381
  20.6.3 Discontinuous variables transformation. Main
         resonance 382
  20.6.4 First step of the hierarchical averaging.
         Stationary resonance solution. 383
  20.6.5 Second step of the hierarchical averaging.
         Attraction area of the resonant solution 385
  20.6.6 Concluding remarks 390
20.7 Variable Order Discontinuous Averaging in Systems with
  Inelastic Collisions 390
20.8 Conclusions 394
20.9 Concluding Remarks of the Editor of the Book 394
References 397

Index 403

Appendix. Table of Contents of the Book *Vibrational Mechanics*
(*Nonlinear Dynamic Effects, General Approach, Applications*)
Iliya I. Blekhman, World Scientific, 2000

Part I

The Basis of Vibrational Mechanics

Chapter 1
On Some Nonlinear Oscillatorz Effects. The Main Idea of Vibrational Mechanics

I. I. Blekhman

Institute for Problems of Mechanical Engineering,
Russian Academy of Sciences
and
Mekhanobr-Tekhnika Corp.
22 Liniya 3, V.O., 199106, St. Petersburg, Russia

1.1 On the Effects Caused by the Action of Vibration in Nonlinear Oscillatory Systems

The action of vibration in nonlinear mechanical systems often leads to peculiar and sometimes quite unexpected results. These effects, on the one hand, can be used in technology, the principles of action of quite a number of most efficient machines being based on them, on the other hand, the same effects may be the cause of undesirable and even disastrous situations. Here we have mainly to do with the following four groups of effects.

1. *The change in the behaviour of oscillatory systems and mechanisms under the action of vibration.* This group of effects comprises such cases as the disappearance of the former equilibrium positions and former types of motions of the system and the appearance of the new positions of equilibrium and new types of motion; the change in the character of the equilibrium positions (i.e. of their stability or instability); the disappearance of the former points of bifurcation and the appearance of the new ones: the change in frequencies of free small oscillations near the positions of stable equilibrium; effects of a vibrational link, in particular the self-synchronization of unbalanced rotors (vibro-exciters); the effect of vibrational maintenance and retarding of rotation of the unbalanced rotors, a peculiar behavior of the so-called "oscillatory systems with limited excitation" and some others.

2. *Effects of vibrational displacement and drift.* This group comprises effects of vibrational transportation of solid bodies, granular materials and more complex media; effects of vibro-dipping and vibro-extracting of grooves, piles and shells; separation of particles of the material according to their properties on the vibrating surfaces and in the oscillating vessels with granular material or with liquid; the appearance of slow flows of granular materials or streams of liquid in vibrating vessels (particularly a phenomenon of vibro-bunkering – filling up the bunker with granular materials from below upwards); an essential change in the velocity and sometimes in the character of leaking the fluid and the granular material from the holes in the vibrating vessels, (say, vibro-jet effect and the phenomenon of vibrational injection);the drift and location of particles in heterogeneous vibrational fields; mutual microshifts and wear of contiguous parts in the nominally fixed joints; the drift of pointers of the devices and of the axes of gyroscopes under the action of vibration of the foundation they are installed on.

3. *Vibrorheological effects.* These effects denote the change, caused by vibration, in the rheological properties of bodies with regard to slow actions, or, as it is sometimes said, they denote the seeming change in the rheological properties of bodies under vibration. We mean here the seeming transformation of dry friction into viscous friction (pseudo-liquefaction) which occurs under vibration; the reduction of dry friction coefficients, the seeming change in the viscosity coefficient (the classical example is the transfer from the laminar flow of liquid to the turbulent stream); the effect of vibro-creep; the increase of the stiffness of elastic systems under vibration and many others.

4. *The appearance of an intensive mechanical interaction between particles and volumes in the multicomponent systems.* This group of effects comprises the loosening of granular materials in the vibrating pan chutes and vessels – the formation of the so-called vibro-boiling layer; the appearance of intensive relative oscilations of solid particles, different in size and density, in the vibrating liquid or in the granular material, etc. It is natural that such effects facilitate the intensification of chemical reactions, being a basis for the use of vibration for the fine grinding of the material and also for the abrasive processing of articles.

Many of the effects stated above will be considered in this book.

It should be noted that this classification is to a certain extent arbitrary. It is easy to notice that some effects can belong to two groups. The effects of the fourth group can sometimes be interpreted by means of linear models. Should we add to the listed effects the fundamental phenomena of resonance and of self-excited oscillations which had already been mentioned by us, we will obtain an almost exhaustive list of the effects, connected with the action of vibration on mechanical systems.

A man who first comes across these effects is greatly impressed by them. One cannot very well stay indifferent, seeing that as a result of but a slightly appreciable vibration the upper position of the pendulum becomes stable, a heavy metal ball "flows up" in a layer of a sand, a pile quite easily goes down into the ground under the effect of its own weight, a heavy body or a layer of granular material moves upward along a slope, the rotation of a rotor is maintained steadily with the electric engine being switched off, etc. Very often and, as we will see later, not without a ground, we gain an impression that a gravitational force has changed its direction, and that a well-known statement that it is impossible to accelerate or retard the motion of the center of masses of the system only at the expense of its inner forces has become invalid, that the law of mechanics of action and reaction has lost its validity, that the essentially non-conservative system behaves as if it were conservative, etc.

No wonder that the above effects time and again gave rise to delusions, including the "overthrow" of the laws of mechanics. Though, to do justice it should be mentioned that those "subverters" sometimes quite accidentally made useful intricate inventions, stimulating interesting investigations (just like their predecessors – inventors of the "perpetuum mobile").

1.2 The Main Idea of Vibrational Mechanics. Observer O and Observer V

Most of the enumerated effects are characterized by the fact that the motion which appears in the system under vibration can be presented as a sum of two components – the "fast", "vibrational" component and the "slow" component which changes very little in one period of vibration, and it is the slow motion that is of special interest in the overwhelming majority of cases. Let us imagine that there is an observer who does not notice (or does not want to notice) either those fast motions or fast forces. This observer is either wearing special glasses which do not let him see the fast motions of the system, or he may be watching the motion in the stroboscopic (i.e. interrupted) light, the frequency of flashes being equal to that of vibration. This observer **V**, unlike the ordinary observer **O** who "sees everything", will notice only the slow component of motion, and lest he should contradict the laws of mechanics, he will have to explain all those paradoxical effects by the appearance of certain additional slow forces or moments acting together with the ordinary slow forces. We will call them after Kapitsa [14, 15] "*vibrational forces*". From the point of view of that "biassed" observer it is those forces that cause the above effects on which the technical use of vibration is based.

In terms of differential equations it looks like this. Let the motion of the system be described by the equation (We consider here a system with one degree of freedom, the generalization of a system with many degrees of freedom presenting no special difficulty, see below):

$$m\ddot{x} = F(\dot{x},x,t) + \Phi(\dot{x},x,t,\omega t) \qquad (1.1)$$

where m is the mass, x is the coordinate, F is the "slow" force, and Φ is the "fast" force. Differentiation with respect to the "ordinary", "slow" time t is designated by a dot. The force unlike F depends not only on t, but also on the "fast" time ωt proportional Φ to a "large parameter" ω that is the frequency of vibration. In the simplest case the function Φ is a periodic function ωt with a period 2π. Let the motion be presented as

$$x = X(t) + \psi(t, \omega t), \qquad (1.2)$$

where X is the "slow" component, and ψ is the "fast" component of the motion (not necessarily small compared to X). Then to the observer **V** who does not see the fast motions ψ and the fast force Φ it will seem that the slow motion can be described by the equation

$$m\ddot{X} = F(\dot{X},X,t) + V(\dot{X},X,t), \qquad (1.3)$$

where V is the vibrational force.

Chapter 2

The Main Mathematical Apparatus of Vibrational Mechanics and of the Method of Direct Separation of Motions

I. I. Blekhman

Institute for Problems of Mechanical Engineering,
Russian Academy of Sciences
and
Mekhanobr-Tekhnika Corp.
22 Liniya 3, V.O., 199106, St. Petersburg, Russia

2.1 Preliminary Remarks

In this chapter for the convenience of the readers the mathematical apparatus of vibrational mechanics are stated briefly as reference material. A more detailed presentation can be found in the books [1, 2]. The books also contain the history of the method of direct separation of motions whose origin lies in the above cited work by P.L.Kapitsa, about the pendulum with a vibrating axis of suspension [14, 15]. The books [1, 2] also considers the connection of vibrational mechanics with a more general conception – with the mechanics of systems with hidden motion.

Along with that, the chapter contains some revision as compared to the analogous chapter of the books [1, 2], and also some extensions, in particular, the extension to the case of mechanical systems with constraints (see section 2.8).

2.2 The Initial Equation and Its Reduction to a System of Integro-differential Equations

With rather general suppositions, the differential equations of motion of the systems under consideration can be presented in the following form [1]:

$$m\ddot{\mathbf{x}} = \mathbf{F}(\dot{\mathbf{x}},\mathbf{x},t) + \mathbf{\Phi}(\dot{\mathbf{x}},\mathbf{x},t,\omega t) \qquad (2.1)$$

where $\mathbf{x}$ is the n-dimensional vector of the generalized coordinates, m is a positive constant ("mass"), ω is the positive parameter (in future the "large" parameter), $\mathbf{F}$ and $\mathbf{\Phi}$ are the n-dimensional vectors of forces, with $\mathbf{\Phi}$ being an almost periodical function of the argument $\tau = \omega t$ (in particular—the periodic function τ with a period of 2π); in future the time t will be called *slow time*, $\mathbf{F}$ will be called *slow force*, while τ will be called *fast time*, and $\mathbf{\Phi}$ *fast force*. As for the smoothness of the functions $\mathbf{F}$ and $\mathbf{\Phi}$, the usual conditions are supposed to be satisfied, which provides the existence of all the solutions of differential equations which are considered below. The conditions of the existence of the average values, introduced below, are also believed to be satisfied.

In accordance with what has been said in 1.2, we will assume that

$$\mathbf{x} = \mathbf{x}(t,\tau) = \mathbf{X}(t) + \mathbf{\Psi}(t,\tau) \qquad (2.2)$$

where $\mathbf{X}$ is the "slow", and $\mathbf{\Psi}$ is the "fast" component of the vector of the generalized coordinates. Here we will assume that $\mathbf{\Psi}$ is an almost periodical (in particular—periodical) function of τ, assuming for certainty that

$$\langle \mathbf{\Psi}(t,\tau) \rangle = 0 \qquad (2.3)$$

The angular brackets here and below indicate the averaging with respect to argument $\tau = \omega t$ which can enter into the expression which is being averaged either explicitly or via the function $\mathbf{\Psi}$, with $\langle ... \rangle = \dfrac{1}{T}\displaystyle\int_0^T ... d\tau$ in

---

[1] All the variable and constant values, involved in the equations, whenever necessary should be considered dimensionless. So in the inequality $\omega \gg 1$ the frequency ω is believed to be dimensionless: $\omega = \omega_* / \omega_0$ where ω_* is the dimensional frequency and ω_0 is a certain frequency characteristic of the system, say, the largest frequency of free oscillations of the linearized system.

the case of an almost periodic function (T is the so-called averaging period) and $\langle ... \rangle = \dfrac{1}{2\pi} \int\limits_0^{2\pi} ... d\tau$ in the case of 2π- periodic function of τ.

The equality (2.3) means that we will assume that the average value of the fast component with respect to τ with the fixed ("frozen") slow time t is equal to zero. According to (2.3) the slow component **X** in accordance with (2.2) is the corresponding average value of the coordinate **x**:

$$\mathbf{X}(t) = \langle \mathbf{x}(t,\tau) \rangle. \qquad (2.4)$$

Substituting expressions (2.2) into the differential equations (2.1), and then adding and subtracting in their right sides the expression

$$\langle \widetilde{\mathbf{F}}(\dot{\mathbf{X}}, \mathbf{X}, \dot{\mathbf{\Psi}}, \mathbf{\Psi}, t) \rangle - \langle \mathbf{\Phi}(\dot{\mathbf{X}} + \dot{\mathbf{\Psi}}, \mathbf{X} + \mathbf{\Psi}, t, \tau) \rangle \qquad (2.5)$$

where

$$\widetilde{\mathbf{F}}(\dot{\mathbf{X}}, \mathbf{X}, \dot{\mathbf{\Psi}}, \mathbf{\Psi}, t) = \mathbf{F}(\dot{\mathbf{X}} + \dot{\mathbf{\Psi}}, \mathbf{X} + \mathbf{\Psi}, t) - \mathbf{F}(\dot{\mathbf{X}}, \mathbf{X}, t) \qquad (2.6)$$

denotes the function which becomes zero when $\dot{\mathbf{\Psi}} = 0$, $\mathbf{\Psi} = 0$.

Now, having the right to choose (instead of one unknown function x two functions **X** and **Ψ** have been introduced), we will demand that a certain group of terms of the relation that has been obtained should also become zero. And, namely, we will demand that the equation

$$m\ddot{\mathbf{\Psi}} = \widetilde{\mathbf{F}}(\dot{\mathbf{X}}, \mathbf{X}, \dot{\mathbf{\Psi}}, \mathbf{\Psi}, t) + \mathbf{\Phi}(\dot{\mathbf{X}} + \dot{\mathbf{\Psi}}, \mathbf{X} + \mathbf{\Psi}, t, \tau) \\ - \langle \widetilde{\mathbf{F}}(\dot{\mathbf{X}}, \mathbf{X}, \dot{\mathbf{\Psi}}, \mathbf{\Psi}, t) \rangle - \langle \mathbf{\Phi}(\dot{\mathbf{X}} + \dot{\mathbf{\Psi}}, \mathbf{X} + \mathbf{\Psi}, t, \tau) \rangle \qquad (2.7)$$

should be satisfied. Then the equation

$$m\ddot{\mathbf{X}} = \mathbf{F}(\dot{\mathbf{X}}, \mathbf{X}, t) + \langle \widetilde{\mathbf{F}}(\dot{\mathbf{X}}, \mathbf{X}, \dot{\mathbf{\Psi}}, \mathbf{\Psi}, t) \rangle + \langle \mathbf{\Phi}(\dot{\mathbf{X}} + \dot{\mathbf{\Psi}}, \mathbf{X} + \mathbf{\Psi}, t, \tau) \rangle \quad (2.8)$$

should also be satisfied, its right side being, as can be easily seen, the result of averaging the right side of the initial equation (2.1) with respect to τ. The argument in favour of such a "splitting" of equation (2.1) will be given below. Here we will mark that the system of integro-differential equations (2.7), (2.8) that has been obtained, is equivalent to the initial equation (2.1), at least in the sense that if there is a certain solution **X**, **Ψ** of this system, the function $\mathbf{x} = \mathbf{X} + \mathbf{\Psi}$ will be the solution of equation

(2.1). In other words, for the existence of the solution of the equation (2.1) of type (2.2) it is sufficient that there should be the corresponding solutions **X**, **Ψ** of the system (2.7), (2.8).

The system of equations (2.7), (2.8) can also be arrived at in the following way. Let us substitute expression (2.2) into the initial equation (2.1) and average both sides of it with respect to the time τ which enters it both explicitly and via the function **Ψ**. Then, after specifying the function $\mathbf{F}(\dot{\mathbf{X}}, \mathbf{X}, t)$, we arrive at the equation (2.8). Equation (2.7) is obtained by means of subtracting equation (2.8) from the initial equation (2.1).

Of special interest for applications are cases when part of the components of the vector in equalities (2.2) is identically equal to zero. To make it concrete let us assume that

$$x_1 = X_1 + \psi_1, ..., \quad x_k = X_k + \psi_k; \quad x_{k+1} = \overline{\psi}_{k+1}, ..., \quad x_n = \overline{\psi}_n \quad (2.9)$$

For the coordinates $x_{k+1}, ..., x_n$ the corresponding components of the functions **F** and of the average $\langle \Phi \rangle$ are equal to zero. As a result, equations (2.7) for those coordinates with due regard for equalities (2.2) coincide with the initial equations (2.1). Then the number n of equations (2.7) for the components **Ψ** remains the same, and the number of equations (2.8) for the components **X** which are to be considered appears to be equal to k, i.e. it is reduced by the number $n-k$ of the fast generalized coordinates $x_{k+1}, ..., x_n$.

2.3 The Case when a Separate Equation for the Slow Component is Obtained

As was mentioned, solutions of the system (2.7), (2.8) in which **Ψ** are periodical or almost periodical with respect to $\tau = \omega t$ are of a considerable interest. Let us assume that we have managed to find such solutions

$$\Psi = \Psi^*(\dot{\mathbf{X}}, \mathbf{X}, t, \tau) \quad (2.10)$$

which are isolated at the given **X** and $\dot{\mathbf{X}}$. Let them also be asymptotically stable with respect to the generalized coordinates $\overline{\Psi}$ (i.e. to the coordinates which do not contain the components of **X**), and also

2.3. The Case when a Separate Equation for the Slow Components obtained

with respect to the corresponding generalized velocities $\dot{\overline{\Psi}}$ all over the region of the change of other generalized coordinates and velocities which is of interest to us. Then for every such solution $\Psi = \Psi^*$ a certain additional force can be obtained:

$$V(\dot{X},X,t) = \langle \widetilde{F}(\dot{X},X,\dot{\Psi}^*,\Psi^*,t) \rangle + \langle \Phi(\dot{X}+\dot{\Psi}^*, X+\Psi^*,t,\tau) \rangle, \quad (2.11)$$

and equation (2.8) can be written as

$$m\ddot{X} = F(\dot{X},X,t) + V(\dot{X},X,t) \quad (2.12)$$

The fulfillment of this condition of asymptotic stability provides, under rather general assumptions, the definiteness of the expression for the additional force within a certain range of changing the initial conditions for the coordinates $\overline{\Psi}$ at sufficiently large values of t. It also provides a mutual conformity of properties of the stability of motions of the initial system (2.1) with respect to $\dot{x}, x$ and of system (2.12) with respect to $\dot{X}, X$. And namely, if the transformation of the variables $\dot{x}, x$ to the variables $\dot{X}, X, \dot{\overline{\Psi}}, \overline{\Psi}$, defined by the relations

$$x_s = X_s(t) + \psi_s(\dot{X},X,t,\omega t) \quad (s=1,...,k);$$
$$x_s = \overline{\psi}_s(\dot{X},X,t,\omega t) \quad (s=k+1,...,k+r=n),$$

and also the inverse transformation possess the properties of uniqueness and continuity in the vicinity of the non-perturbed motions under consideration, then to the stable (asymptotically stable), with respect to $\dot{X}$ and X, solutions of (2.12) correspond the stable (asymptotically stable) solutions $x = X + \Psi$ of the initial system (2.1). And vice versa, to the stable (asymptotically stable) solutions x of the initial system (2.1) correspond the stable (asymptotically stable) solutions X of the system (2.12).

This statement can also be formulated in the following way: if we "do not want to notice" either the component of the motion Ψ or the force Φ, which depend on the argument $\tau = \omega t$, then at the assumed version of "splitting" the initial equation (2.1), we must first find from equation (2.7) the component Ψ, possessing the above-mentioned properties, and, secondly, we must determine the component X from (2.12) in which the force Φ is not given in its direct form and a certain additional force $\mathbf{V}$ is added to the force $\mathbf{F}$.

It should be noted that the present section is directly connected with a more general conception—that of the mechanics of the systems with the hidden motions [1, 2].

So far we have not yet made use of the assumption about the value of the parameter ω and, accordingly, about the rate of the change of the forces $\mathbf{F}$ and Φ as well as components $\mathbf{X}$ and $\mathbf{\Psi}$. At the same time without that assumption relations (2.7)—(2.12) have a formal rather than constructive meaning, since the solution of system (2.7), (2.8) is, in the general case, not simpler than that of the initial equation (2.1).

2.4 The Main Assumption of Vibrational Mechanics, its Formalization and Conditions of its Fulfillment

Let us now make the *main assumption of vibrational mechanics*, which implies that the initial equation (2.1) may have solutions of type (2.2) or that system (2.7), (2.8) may have adequate solutions $\mathbf{X}$, $\mathbf{\Psi}$ with the component $\mathbf{X}$ in those solutions being "actually slow" and the component $\mathbf{\Psi}$ being "actually fast". First, however, we should define the mathematical meaning of the notions "slow" and "fast".

For this purpose we will use the symbols X_0 and ψ_0 to denote the scale of the components X and ψ defined in such a way that the values $X_* = X/X_0$ and $\psi_* = \psi/\psi_0$ should be of the order of unity

$$X_* = X/X_0 \sim 1, \quad \psi_* = \psi/\psi_0 \sim 1, \qquad (2.13)$$

and let T be the shortest period of time t during which the variable ψ undergoes changes of the order of ψ_0. Here, as before, the components of the corresponding vectors, coordinates and forces are printed in the ordinary (not bold) type. However, the indices, indicating the number of the component are omitted, which makes the notation simpler. We will assume that the component X is changing very slowly as compared to ψ (or for brevity sake in comparison with ψ it is slow), and, accordingly, ψ is fast in comparison with X, provided the following relationship is satisfied:

2.4. The Main Assumption of Vibrational Mechanics

$$\frac{X|_{t+T} - X|_t}{X_0 T} : \frac{\psi|_{t+T} - \psi|_t}{\psi_0 T} \approx \frac{\dot X}{X_0 T} : \frac{\dot\psi}{\psi_0 T} \sim \varepsilon$$

In this relationship, and also in (2.14)—(2.19) the differences $X|_{t+T} - X|_t$, $\psi|_{t+T} - \psi|_t$ and the derivatives of X and ψ imply the maxima of their absolute values.

That relationship expresses the demand that the relative speed of changing the component X should be a value of the order of ε as compared to the relative speed of changing ψ, with ε being a small parameter. Having calculated the derivative of the function $\psi(t,\omega t)$ with respect to t, we will write the relationship under consideration as

$$\frac{\dot X}{X_0 T} : \frac{\dot\psi}{\psi_0 T} = \frac{1}{X_0}\frac{dX}{dt} : \frac{1}{\psi_0}\left(\frac{\partial\psi}{\partial t} + \omega\frac{\partial\psi}{\partial \tau}\right) \sim \varepsilon \qquad (2.14)$$

Hence it can be seen that for the validity of the assumption about the rate of changing the components X and ψ, it is sufficient (though not necessary!) to identify the small parameter ε with the value $1/\omega$ and demand that the values $(1/\psi_0)(\partial\psi/\partial\tau)$ and $(1/X_0)(\partial X/\partial t)$ should be of the same order, and $(1/\psi_0)(\partial\psi/\partial t)$ be of the same or of a higher order with respect to $\varepsilon = 1/\omega$; in particular it may be $(1/\psi_0)(\partial\psi/\partial t) \equiv 0$. Then

$$\frac{\dot X}{\dot\psi}\frac{\psi_0}{X_0} \sim \varepsilon = \frac{1}{\omega} \qquad (2.15)$$

Thus we come to the following statement (denoting, as usual, the order of the value x by $O(x)$):

If there exist solutions of the initial differential equation (2.1) of type (2.2) or the adequate solutions X, ψ of system (2.7), (2.8), then the conditions

$$\omega = \frac{1}{\varepsilon} \gg 1,$$

$$O\left(\frac{1}{\psi_0}\frac{\partial\psi}{\partial\tau}\right) = O\left(\frac{1}{X_0}\frac{\partial X}{\partial t}\right), \quad O\left(\frac{1}{\psi_0}\frac{\partial\psi}{\partial t}\right) \geq O\left(\frac{1}{\psi_0}\frac{\partial\psi}{\partial\tau}\right) \qquad (2.16)$$

are sufficient for the validity of the main assumption of vibrational mechanics.

Notions of the fast and slow forces can also be formalized in a similar way, there is however no necessity in it, as will be shown later.

The simplest example of the pair of functions X and ψ, satisfying condition (2.16) is provided by $X = X_0 \sin t$ and $\psi = \psi_0 \sin \omega t$. In this connection we must emphasize that condition (2.16) and, consequently, the main assumption of vibrational mechanics, as it was formulated above, does not impose any restriction on the ratio of the absolute values of the components X and ψ: the value ψ_0 is not necessarily small in comparison to X_0, it may be comparable to X_0 and even larger than that. In other words, the amplitude of the vibration of high frequency ψ_0 can be of the same order or even much larger than the scale of change of the slow component. Such cases are met in some problems of celestial mechanics [1, 2].

In much the same way as in equality (2.14) we find

$$\frac{\ddot{X}}{X_0 T^2} : \frac{\ddot{\psi}}{\psi_0 T^2} = \frac{1}{X_0}\frac{d^2 X}{dt^2} \Big/ \frac{1}{\psi_0}\left(\frac{\partial^2 \psi}{\partial t^2} + 2\omega \frac{\partial^2 \psi}{\partial t \partial \tau} + \omega^2 \frac{\partial^2 \psi}{\partial^2 \tau}\right). \quad (2.17)$$

Hence we believe that the following conditions to be satisfied

$$O\left(\frac{1}{\psi_0}\frac{\partial^2 \psi}{\partial \tau^2}\right) = O\left(\frac{1}{X_0}\frac{d^2 X}{dt^2}\right), \quad O\left(\frac{1}{\psi_0}\frac{\partial^2 \psi}{\partial t^2}, \frac{\varepsilon}{\psi_0}\frac{\partial^2 \psi}{\partial t \partial \tau}\right) \geq O\left(\frac{1}{\psi_0}\frac{\partial^2 \psi}{\partial \tau^2}\right) \quad (2.18)$$

and we obtain

$$\frac{\ddot{X}}{\ddot{\psi}}\frac{\psi_0}{X_0} \sim \varepsilon^2 = \frac{1}{\omega^2} \quad (2.19)$$

From relations (2.15) and (2.19) it follows that if

$$\psi_0 / X_0 \sim \varepsilon^n \quad (n = ...,-1,0,1,2,...), \quad (2.20)$$

i.e. if the fast component ψ is the value of the order n as compared to the slow component X, then

$$\dot{X}/\dot{\psi} \sim \varepsilon^{1-n}, \quad \ddot{X}/\ddot{\psi} \sim \varepsilon^{2-n}. \quad (2.21)$$

2.4. The Main Assumption of Vibrational Mechanics

Particularly, in an important case when ψ is a small value of the first order as compared to X, we have

$$n = 1, \quad X/\psi \sim \frac{1}{\varepsilon}, \quad \dot{X}/\dot{\psi} \sim 1, \quad \ddot{X}/\ddot{\psi} \sim \varepsilon, \qquad (2.22)$$

and in case ψ and X are of the same order,

$$n = 0, \quad X/\psi \sim 1, \quad \dot{X}/\dot{\psi} \sim \varepsilon, \quad \ddot{X}/\ddot{\psi} \sim \varepsilon^2 \qquad (2.23)$$

Comparing relation (2.21) to equations (2.7) and (2.8), we conclude that for the validity of the main assumption it is necessary that the right side of equation (2.7) should be of the order ε^{n-2} if the order of the right side of equation (2.8) is taken to be unity, which, evidently, can always be done. In other words, it is necessary that these equations should be presented as

$$m\ddot{\mathbf{X}} = \mathbf{Q}, \qquad m\ddot{\mathbf{\Psi}} = \mathbf{P}/\varepsilon^{2-n} \qquad (2.24)$$

where

$$\mathbf{Q} = \mathbf{F} + \langle \tilde{\mathbf{F}} \rangle + \langle \mathbf{\Phi} \rangle, \quad \mathbf{P}/\varepsilon^{2-n} = \tilde{\mathbf{F}} - \langle \tilde{\mathbf{F}} \rangle + \mathbf{\Phi} - \langle \mathbf{\Phi} \rangle,$$

with $|\mathbf{P}|$ and $|\mathbf{Q}|$ being the values of the same order.

Along with that, should we write down equations (2.7), (2.8), using the relative variables $\mathbf{X}_* = \mathbf{X}/X_0$, $\mathbf{\Psi}_* = \mathbf{\Psi}/\psi_0$, then according to (2.18), the orders of their right sides must always differ by two units, i.e. these equations may be written as

$$m\ddot{\mathbf{X}}_* = \mathbf{Q}_*, \quad m\ddot{\mathbf{\Psi}}_* = \mathbf{P}_*/\varepsilon^2 \qquad (2.25)$$

where $|\mathbf{P}_*|$ and $|\mathbf{Q}_*|$ are values of the same order, connected with $|\mathbf{P}|$ and $|\mathbf{Q}|$ by the relationships

$$|\mathbf{P}_*| = |\mathbf{P}|\varepsilon^n/\psi_0, \quad |\mathbf{Q}_*| = |\mathbf{Q}|/X_0$$

It is necessary to bear in mind that when using the variables $\mathbf{X}_*$ and $\mathbf{\Psi}_*$ equality (2.2) is written as

$$\mathbf{x} = X_0 \mathbf{X}_*(t) + \Psi_0 \mathbf{\Psi}_*(t, \omega t) \qquad (2.26)$$

Thus, the main assumption under consideration leads to a certain demand to the relative order of the right sides of equations (2.7) and (2.8), depending on the relative order of the components $\mathbf{X}$ and $\mathbf{\Psi}$. Finally, as it should be expected, this imposes certain conditions on the functions $\mathbf{F}$ and $\mathbf{\Phi}$ in the right sides of the initial equation (2.1).

What has been said can be formulated as the following statement:

For the existence of the solutions of the initial differential equation (2.1) of type (2.2), or the corresponding solutions $\mathbf{X}$ and $\mathbf{\Psi}$, of system (2.7), (2.8), satisfying conditions (2.15), (2.19) and such that $\psi \sim \varepsilon^n X$ (n being either an integer or zero), it is necessary that system (2.7), (2.8) should allow the notation in form (2.24) or (2.25).

2.5 The Main Equation of Vibrational Mechanics. Vibrational Forces, Observers O and V

Let the main assumption of vibrational mechanics be satisfied, i. e. let there be a solution of equation (2.17) of type (2.2) or the corresponding solutions of system (2.7), (2.8) satisfying conditions (2.16). Let us assume further that the conditions of section 2.3 are satisfied. Then we will call equation (2.12), containing only the slow component of motion, the *main equation of vibrational mechanics*, or the *equation of slow motions*, and the expression $\mathbf{V}(\dot{\mathbf{X}}, \mathbf{X}, t)$ for the additional force—the *vibrational force*.

Equation (2.8) will then be called *equation of fast motions*. As has been mentioned in section 1.2, vibrational mechanics can be considered to be a mechanics in which the observer does not notice any fast forces and fast motions, and perceives only the slow forces and slow motions. Such an observer (the *observer* **V**) can be contrasted to the "ordinary" *observer* **O** who perceives both the slow and fast forces and motions. It is convenient to use these images, taking, as may be required, the positions of either the first or the second observer. In accordance with what has been said, the observer **V**, so as not to come in conflict with the laws of mechanics, must take into consideration not only the "ordinary" slow forces $\mathbf{F}$, but also the additional slow forces i.e. the vibrational forces $\mathbf{V}$.
Note that according to equations (2.6), (2.9) and (2.11), the vibrational force is obtained by means of averaging the eigenfast force $\mathbf{\Phi}$ and the fast contribution $\tilde{\mathbf{F}}$ which differs from the slow force $\mathbf{F}$. In accordance

with it, we will distinguish the *eigenvibrational force*

$$\mathbf{V}^{(s)} = \langle \Phi \rangle \tag{2.27}$$

and the *induced vibrational force*

$$\mathbf{V}^{(i)} = \langle \widetilde{\mathbf{F}} \rangle \tag{2.28}$$

Mark that the induced vibrational force may be different from zero even in the absence of a fast external perturbation due to fast self-excited oscillations which may appear in systems with slow forces.

One can see (and it will be demonstrated by numerous examples) that equation (2.12) at least in its adequate approximate version is much simpler than the initial equation (2.1); the slow component **X** whose change this equation describes being of utmost interest for researchers.

It should be once more emphasized that vibrational mechanics can be regarded as a special case of a more general conception—of the *mechanics of systems with hidden motions*.

2.6 Method of an Approximate Derivation of the Expression of Vibrational Forces and of Composing the Main Equation of Vibrational Mechanics

As before, we will consider valid both the main assumption of vibrational mechanics and the conditions of section 2.3. Then it seems natural to use the following method of an approximate finding of the vibrational force **V** and of composing the main equation (2.12). We will describe it here at an euristic level. We begin by solving the equation of fast motions (2.7). Since the change of the values $\dot{\mathbf{X}}$, **X** and t for the typical period of the fast motion $2\pi/\omega$ is relatively small, in solving Eq.(2.7) those values are being regarded as "frozen", i.e. as fixed parameters.

Let Eq. (2.7), with $\dot{\mathbf{X}}$, **X** and t being frozen, admit either one or several almost-periodical (particularly 2π-periodical) with respect to $\tau = \omega t$ solutions $\Psi = \Psi^*(\dot{\mathbf{X}}, \mathbf{X}, t, \tau)$ satisfying condition (2.3) and asymptotically stable with respect to all fast generalized coordinates and velocities, while all the other fast variables changing and $\dot{\mathbf{X}}$, **X** and t being frozen all over the region under consideration. This assumption is usually checked up very easily and it is really valid under the conditions of section 2.3. Besides, Eq.(2.7) is such that the necessary condition (2.3) of the existence of almost-periodical (particularly 2π-periodical with respect to τ) solutions is automatically fulfilled for it. To make sure of it, it is enough to average Eq.(2.7) with respect to $\tau = \omega t$ ($\dot{X}, X$ and t being frozen). What has been said holds the key to the way adopted by us of "splitting" the initial equation (2.1) into two equations (2.7) and (2.8).

Substituting the definite solution $\Psi = \Psi^*(\dot{X}, X, t, \tau)$ into expression (2.11), we will find an approximate expression for the vibrational force $V(\dot{X}, X, t)$ after which the main equation (2.12) can be composed. This equation will,of course, also be approximate and must be valid on the one hand with $t > t_0$ where t_0 is the time of achieving a steady state of the fast motions, and on the other hand—with $t < T_0$, where T_0 is the boundary of the interval of validity of the asymptotic approximation (it is naturally assumed that $T_0 \gg t_0$).

The grounds of the described approximate method for the case when the solution of type (2.22) are being sought are given in the books [1, 2] It is based on the scheme of averaging, suggested by V.M.Volosov [25]. For some other suppositions the grounds are given by O.Z.Malakhova [19] and for the case of systems with the constraints by V.I.Yudovich [27]; the results of their works being given below (sections 2.7 and 2.8). As for the grounds of the method for more a general case, see Part VI, written by A. Ja. Fidlin.

2.7 Important Special Case

Let us consider the case when the initial system (2.1) can be presented in the following way (to simplify our reasoning and notation we will first consider x to be a scalar):

$$m\ddot{x} = F(\dot{x}, x, t) + \omega \Phi_1(x, t, \tau), \quad \omega \gg 1 \qquad (2.29)$$

where the almost-periodic with respect to τ function Φ_1 has a zero average value with respect to this argument at the fixed x and t:

$$\langle \Phi_1(x, t, \tau) \rangle = 0 \qquad (2.30)$$

2.7. Important Special Case

Here we are searching for the solution of the type

$$x = X(t) + \varepsilon \psi_1(t,\tau), \quad \varepsilon = 1/\omega \ll 1 \tag{2.31}$$

In this case, considered by Malakhova [19], finding the function ψ_1 and the expression for the vibrational force becomes much more simple and the ground of the approximate method stated in section 2.6 is obtained directly by means of using the first and second theorems of Bogolyubov.

In this case the equation of fast motions (2.7) looks as

$$m\varepsilon\ddot{\psi}_1 = F(\dot{X}+\varepsilon\dot{\psi}_1, X+\varepsilon\psi_1, t) - \langle F(\dot{X}+\varepsilon\dot{\psi}_1, X+\varepsilon\psi_1, t) \rangle$$
$$+ \frac{1}{\varepsilon}[\Phi_1(X+\varepsilon\psi_1, t, \tau) - \langle \Phi_1(X+\varepsilon\psi_1, t, \tau) \rangle]$$

or, considering (2.30) and with accuracy to the terms of a higher order with respect to ε

$$m\ddot{\psi}_1 = \frac{1}{\varepsilon^2}\Phi_1(X,t,\tau) \tag{2.32}$$

Looking ahead, we will note that the latter equation corresponds to the so-called *purely inertial approximation* (see section 2.9).

Let $\psi_1^* = \psi_1^*(X,t,\tau)$ be a certain almost-periodic solution of equation (2.32), found at the frozen X and t. Then, according to the approximate method, stated in section 2.6, in the main equation of vibrational mechanics (2.12), i.e. in the equation

$$m\ddot{X} = F(\dot{X}, X, t) + V(\dot{X}, X, t), \tag{2.33}$$

written with the same accuracy with respect to ε as equation (2.32), the expression for the vibrational force (2.11) will be

$$V(\dot{X},X,t) = \left\langle F\left[\dot{X}+\frac{\partial \psi_1^*(X,t,\tau)}{\partial \tau}, X, t\right] - F(\dot{X},X,t) \right\rangle \tag{2.34}$$
$$+ \left\langle \frac{\partial \Phi_1(X,t,\tau)}{\partial X}\psi_1^*(X,t,\tau) \right\rangle$$

(Recall that according to what has been said in section 2.4, the values $\varepsilon\dot\psi_1$ and $\dot X$ are of the same order, while $\varepsilon\psi_1$ is supposed to be small as compared to X). We will show that equation (2.33), in which the vibrational force is determined according to equation (2.34), can be obtained by using the first theorem of Bogolyubov [8]. To this end we transform equation (2.29) into a standard form by means of changing the variables

$$x = X + \varepsilon\psi_1^*(X,t,\tau), \quad x' = \varepsilon Y + \frac{\partial \psi_1^*(X,t,\tau)}{\partial \tau}$$

where the full derivative with respect to τ is again marked with a prime. As a result, we obtain the system

$$\begin{aligned} X' &= \varepsilon[Y + \varepsilon A(Y,X,t,\tau,\varepsilon)], \quad Y' = \varepsilon\{\frac{1}{m}F[Y + \frac{\partial \psi_1^*(X,t,\tau)}{\partial \tau}, X, t] \\ &+ \frac{1}{m}\frac{\partial \Phi_1(X,t,\tau)}{\partial \tau}\psi_1^*(X,t,\tau) - \frac{\partial^2 \psi_1^*(X,t,\tau)}{\partial t \partial X}Y \frac{\partial^2 \psi_1^*(X,t,\tau)}{\partial \tau \partial t} \\ &+ \varepsilon B(Y,X,t,\tau,\varepsilon)\}, \quad t' = \varepsilon \end{aligned} \quad (2.35)$$

where A and B are certain functions of the indicated arguments. Let functions F and Φ_1 be such that the right sides of equations (2.35) should satisfy the conditions of Bogolyubov's first theorem. Then applying to these equations the principle of averaging, and retaining the same designations for the averaged variables, we will obtain the following equation of the first approximation

$$X' = \varepsilon Y, \quad Y' = \varepsilon R(Y,X,t), \quad t' = \varepsilon. \quad (2.36)$$

where

$$R(Y,X,t) = \frac{1}{m}\langle (F+V) \rangle$$

$$= \frac{1}{m}\left\langle F\left[Y + \frac{\partial \psi_1^*(X,t,\tau)}{\partial \tau}, X, t\right] + \frac{\partial \Phi_1(X,t,\tau)}{\partial X}\psi_1^*(X,t,\tau) \right\rangle$$

It is obvious that the latter equations lead to equation (2.33) in which $V(\dot X, X, t)$ are determined according to (2.34), which is precisely what

2.7. Important Special Case

was to be proved. It should be noted that the relations of type (2.33), (2.34) were obtained by S. V. Chelomey [12] in a different way.

Let us consider the question of stability of the positions of equilibrium $Y = 0$, $X = X_*$, which are determined by equations (2.35), i.e. the so-called positions of quasi-equilibrium for the initial equation (2.29). Here we will assume that functions F and Φ_1 do not depend explicitly on t and consequently $R = R(Y,X)$. The indicated positions of equilibrium, if they exist, are determined as solutions of the equation

$$R(0,X) \equiv \frac{1}{m}[F(0,X) + V(0,X)] = 0 \qquad (2.37)$$

Then, under more strict requirements to the right sides of equation (2.29) it is possible to apply Bogolyubov's second theorem [8]. If, according to this theorem, the roots of the equation

$$\begin{vmatrix} -\lambda & 1 \\ \frac{\partial R}{\partial X}\bigg|_{\substack{Y=0 \\ X=X_*}} & \frac{\partial R}{\partial Y}\bigg|_{\substack{Y=0 \\ X=X_*}} - \lambda \end{vmatrix} \qquad (2.38)$$

have negative real parts, then the stationary solutions $Y = 0$, $X = X_*$ of the equations of the first approximation will be answered, at the sufficiently small ε all over the interval $-\infty < \tau < \infty$ by the asymptotically stable almost-periodic solutions of the initial equations (2.29), i.e. by the quasi-equilibrium positions for this equation.

System (2.36) is equivalent to equation (2.33), and equation (2.38) is equivalent to the characteristic equation for the equation in variations, corresponding to the stationary solution of equation (2.33). From here it follows that the asymptotically stable stationary solutions $\dot{X} = 0$, $X = X_*$ of Eq.(2.33) are answered by the asymptotically stable almost-periodic solutions of the initial equations (2.29), i.e. by the positions of quasi-equilibrium for this equation.

What has been said can be formulated as the following statement:

Let us assume that the initial differential equation can be presented in form (2.29) where the function $\omega\Phi_1$ (the "fast force") satisfies equality (2.30) and the functions F, Φ_1 are such that the right sides of equations (2.35) satisfy the conditions of Bogolyubov's first theorem, and the solutions of type (2.31) are being considered. Then in the first *asymptotic approximation, the slow component* X *satisfies equation* (2.33) (*the main equation of vibrational mechanics*), *in which the fast component* ψ_1^* *is found as an almost-periodic solution of the differential equation* (2.32) *at the fixed* ("*frozen*") X *and t*.

If the functions F and Φ_1 do not explicitly depend on the time t, and the right sides of equation (2.35) satisfy the conditions of Bogolyubov's second theorem, then the asymptotically stable positions of the equilibrium $\dot{X} = 0$, $X = X_*$ of the system, defined by Eq. (2.33), correspond to the asymptotically stable positions of the quasi-equilibrium (2.31) of the system, defined by equation (2.29).

Note that in the case of the system with many degrees of freedom, when the functions F, φ, x, X, ψ_1 and V are the vectors $\mathbf{F}, \mathbf{\Phi}, \mathbf{x}, \mathbf{X}, \mathbf{\Psi}_1$ and $\mathbf{V}$, the expression in the last angular brackets of equation (2.34) for the vibrational force $\mathbf{V}$ should be understood as the product of the matrix $\partial \mathbf{\Phi}_1 / \partial \mathbf{X}$ multiplied by the vector $\mathbf{\Psi}_1^*$.

We will also note that another approach to the substantiation of the presented approximate method in the special case under consideration was suggested by Kirghetov on the basis of the concept of μ-stability introduced by him [16].

The specific procedure of averaging suggested by Fidline made it possible not only to obtain very accurately the results stated here, but also to consider strictly and under more general assumptions, the case when the function Φ_1 in equation (2.29) depends on $\dot{x}$ (see chapter 19).

2.8 On the Case of a Mechanical Systems with Constraints

A detailed strict consideration of this case was given by V.Yu.Yudovich.[27, 28]. Here we will state only the main result of that work, using somewhat different designations and terms, adopted in this book.

The system is believed to obey a certain number q of ideal holonomic constraints. The equation of motion of such a system is written as follows

$$m\ddot{\mathbf{x}} = \mathbf{F}(\dot{\mathbf{x}}, \mathbf{x}, t) + \omega \mathbf{\Phi}_1(\mathbf{x}, t, \tau) - \sum_{j=1}^{q} \Lambda_j \nabla S_j \qquad (2.39)$$

2.8. Case of Mechanical Systems with Constraints

and the equation of the constraints has the form

$$S_j(\mathbf{x},t,\tau) = 0 \quad (j=1,...,q) \tag{2.40}$$

Here $\mathbf{x}$ is a n-dimensional vector of the coordinates of the system, and, as before, $\mathbf{F}$ is a slow force and Φ_1 is a fast force, t is a slow time, and τ is a fast time, Λ_j are the Lagrange multipliers, and

$$\nabla = \sum_{s=1}^{n} \mathbf{e}_s \frac{\partial}{\partial x_s}$$ is the Hamilton operator where $\mathbf{e}_s$ is a unit coordinate vector.

Φ_1 and S_j are supposed to be the 2π–periodic functions of τ The following asymptotic expansion is valid for the functions S_j at $t \to \infty$:

$$S_j(\mathbf{x},t,\tau) = \langle S_j(\mathbf{x},t,\tau) \rangle + \frac{1}{\omega} s_j(\mathbf{x},t,\tau) + O\left(\frac{1}{\omega^2}\right) \tag{2.41}$$

The angular brackets denote here, as before, averaging with respect to τ. It is also believed that the equations

$$\langle S_j(\mathbf{x},t,\tau) \rangle \equiv \overline{S}_j(\mathbf{x},t) = 0 \quad (j=1,...,q) \tag{2.42}$$

determine for any t in the space $x_1,...,x_n$ the smooth $n-q$-dimensional subvariety Γ, with the vectors $\nabla \overline{S}_1(\mathbf{x},t),...,\nabla \overline{S}_q(\mathbf{x},t)$ for all $\mathbf{x}$, belonging to Γ being linearly independent.

The asymptotic solution of equations (2.39) (2.40) with ω being large for Cauchy's problem with the initial conditions independent of ω, is sought as

$$\mathbf{x}(t) = \mathbf{X}(t) + \frac{1}{\omega} \Psi_1(\mathbf{X},t,\tau) \tag{2.43}$$

$$\Lambda_j(\mathbf{x},t,\tau) = \overline{\Lambda}_j(\mathbf{X},t) + \omega \lambda_j(\mathbf{X},t,\tau) \tag{2.44}$$

As a result of the investigation, the author comes to the following equations for determining the slow component $\mathbf{X}$:

$$m\ddot{\mathbf{X}} = \overline{\mathbf{F}}(\dot{\mathbf{X}},\mathbf{X},t) - \sum_{j=1}^{q} \overline{\Lambda}_j \nabla \overline{S}_j + \mathbf{V}_1(\mathbf{X},t) \tag{2.45}$$

$$\overline{S}_j(\mathbf{X},t) = 0 \quad (j=1,...,q) \tag{2.46}$$

Here

$$\overline{\mathbf{F}}(\dot{\mathbf{X}},\mathbf{X},t) = \left\langle \mathbf{F}[\dot{\mathbf{X}} + \frac{1}{\omega}\dot{\psi}_1(\mathbf{X},t,\tau),t,\tau] \right\rangle \tag{2.47}$$

and $\mathbf{V}_1(\mathbf{X},t)$ is the force, called by V.I.Yudovich "*vibro-genous force*".

Equation (2.45) can also be written in the form corresponding to the main equation of vibrational mechanics

$$m\ddot{\mathbf{X}} = \overline{\mathbf{F}}(\dot{\mathbf{X}},\mathbf{X},t) - \sum_{j=1}^{q}\overline{\Lambda}_j \nabla S_j + \mathbf{V}(\mathbf{X},t) \tag{2.48}$$

where

$$\mathbf{V}(\dot{\mathbf{X}},\mathbf{X},t) = \mathbf{V}_1(\mathbf{X},t) + \overline{\mathbf{F}}(\dot{\mathbf{X}},\mathbf{X},t) - \mathbf{F}(\dot{\mathbf{X}},\mathbf{X},t) \tag{2.49}$$

is the force, called in this book *vibrational force*. It is easy to see that the vibro-genous force $\mathbf{V}_1(\mathbf{X},t)$ corresponds to the *eigenvibrational force* $\mathbf{V}^{(s)}$ while the difference $\overline{\mathbf{F}} - \mathbf{F}$ corresponds to the *induced vibrational force* $\mathbf{V}^{(i)}$, introduced in section 2.5 of this book.
Thus,

$$\begin{aligned} \mathbf{V} &= \mathbf{V}^{(s)} + \mathbf{V}^{(i)}, \quad \mathbf{V}^{(s)} = \mathbf{V}_1, \\ \mathbf{V}^{(i)} &= \overline{\mathbf{F}}(\dot{\mathbf{X}},\mathbf{X},t) - \mathbf{F}(\dot{\mathbf{X}},\mathbf{X},t) \end{aligned} \tag{2.50}$$

V.I.Yudovich, in the cited article, gives the formulas to determine the functions $\psi_1(\mathbf{X},t,\tau)$ and $\mathbf{V}_1(\mathbf{X},t)$. Besides he splits the force $\mathbf{V}$ into three summands

$$\mathbf{V}_1 = \mathbf{V}_{1\Phi} + \mathbf{V}_{1S} + \mathbf{V}_{1\Phi S}$$

where $\mathbf{V}_{1\Phi}$ is the component caused by the action of the force $\omega\Phi_1$, $\mathbf{V}_{1S}$ is the force caused by the vibration of the constraints, and $\mathbf{V}_{1\Phi S}$ is the force, formed by the interaction of the force $\omega\Phi_1$ with the constraint.

One can make sure that the results, obtained in the case of the absence of the constraints coincide with those given in section 2.7.

The work [27] also contains the condition of the potentiality on the average of the forces $V_{1\Phi}$ and $V_{1\Phi S}$ i.e. the condition of the *potentiality of the system on the average* (see chapter 4 of the book [2]). It should be noted that the force V_{1S} is always potential.

V.Yu.Yudovich has used the results obtained to solve the problems on the behavior of a material particle on a smooth vibrating surface in the field of the gravitational force [28]. Both cases of motion have been investigated – the general case of motion on a solid vibrating surface and that on the surface of an ellipsoid with pulsating axes. A quasi-equilibrium position of a particle on the surface has been found. Results of the investigation show a possibility to control by means of vibration the states of the particle on the surface.

2.9 On the Simplifications of Solving Equations for the Fast Component of Motion. Purely Inertial Approximation

Apart from these mentioned above, there are other expedient methods of an approximate solution of the equation for the fast component Ψ. From general positions, the rightfulness of such approximate methods follows from the fact that the information which is contained in the initial equation (2.1) and also in the system (2.7), (2.8) is excessive if we are interested only in the change of the slow component X. It is natural that then we can restrict ourselves to an approximate finding of the fast component Ψ. To be concrete, the possibility of an approximate finding of Ψ follows from the fact that this function appears in equation (2.11) for the vibrational force only under the integration sign.

One of the possible simplifications when solving equation (2.7) consists in finding Ψ in the form of a sum of a small number of harmonics or a small number of terms of a power series of a small parameter, which need not necessarily coincide with the parameter $\varepsilon = 1/\omega$. We must emphasize that here we mean an approximate calculation of Ψ when $\dot{X}, X$ and t are "frozen", i.e. within the frames of the method, given in section 2.6. Besides, as was already mentioned, Ψ is often considered to be small as compared to X and it is possible to linearize the expression for the forces F and Φ with respect to Ψ (and sometimes to $\dot{\Psi}$). In particular, in many cases it is possible to consider only the linear terms in the series of the function $\tilde{F}$ by the powers $\dot{\Psi}$ and Ψ, believing that according to (2.6)

$$\tilde{F} = \left(\frac{\partial F}{\partial \dot{x}}\right) \cdot \dot{\Psi} + \left(\frac{\partial F}{\partial x}\right) \cdot \Psi \qquad (2.51)$$

where the derivatives are calculated when $\dot{\Psi} = 0$, $\Psi = 0$. It is necessary however to bear in mind that due to (2.28) and (2.3) there exists only the eigenvibrational force $V^{(s)}$, while the induced vibrational force $V^{(i)}$ is absent. In other words, the induced component exists in the case when the slow force F is non-linear with respect to $\dot{x}$ and x.

On the other hand, even in case of the absence of the fast force Φ in the initial equation (2.1), the vibrational force can be different from zero on account of its induced component. Such a situation is characteristic of the autonomous systems in which fast oscillations appear in spite of the fact that there are only slow forces in the system.

When solving the equations of fast motions (2.7), the approximation which may be called *purely inertial approximation* is of special significance. It is based on the assumption that in this equation the "oscillating component" of the fast force is much greater than the "oscillating component" of the slow force, i.e.

$$\left|\Phi - \langle\Phi\rangle\right| \gg \left|\tilde{F} - \langle\tilde{F}\rangle\right|, \qquad (2.52)$$

and the second component can therefore be neglected as compared to the first one. In many cases relation (2.52) appears in a most natural way from the statement of the problem; it can be formalized by means of introducing a small parameter (the system considered in section 2.7 can serve as one of the examples).

Quite frequently, when $\Psi = \Phi - \langle\Phi\rangle$ does not depend on $\dot{\Psi}$ or Ψ, and is an almost-periodic function of τ, presented as a sum

$$\Psi = \sum_{s=1}^{p} \mathbf{f}_s(\dot{X}, X, t, \tau) \qquad (2.53)$$

where

$$\mathbf{f}_s(\dot{X}, X, t, \tau) = \mathbf{A}_s(\dot{X}, X, t)\cos v_s t + \mathbf{B}_s(\dot{X}, X, t)\sin v_s t$$

and $v_s > 0$ are certain numbers, the wanted approximate solution of equation (2.7) has the form

$$\Psi = -\frac{1}{m}\sum_{s=1}^{p}\frac{\mathbf{f}_s(\dot{\mathbf{X}},\mathbf{X},t)}{v_s^2} \qquad (2.54)$$

When solving equation (2.7) approximately, the method of multiple scales (see section 4.5) can be also used quite successfully.

2.10 Additional Remarks, Certain Generalizations

In connection with the method of an approximate obtaining of the expressions for the vibrational force and for the main equation of vibrational mechanics, stated in sections 2.6–2.9, we will make the following remarks and additions.

1. The main equation (2.12) will not change, that is there won't be any mistake if either all the slow forces or just some of them are referred to the fast forces.

2. If the initial equation is written in form of the Lagrange equation of the second type

$$\mathbf{E}(T) = \mathbf{Q} \qquad (2.55)$$

where

$$\mathbf{E}(\ ) = \frac{d}{dt}\frac{\partial}{\partial \dot{\mathbf{x}}} - \frac{\partial}{\partial \mathbf{x}} \qquad (2.56)$$

is the Euler operator, $\mathbf{Q}$ is the generalized force, and T is the kinetic energy of the system, then the equation of slow motions can be written in the following way

$$\langle \mathbf{E}(T_{\mathbf{X}+\Psi})\rangle = \langle \mathbf{Q}_{\mathbf{X}+\Psi}\rangle \qquad (2.57)$$

where $T_{\mathbf{X}+\Psi}$ and $Q_{\mathbf{X}+\Psi}$ denote the result of substituting the values $\mathbf{x} = \mathbf{X} + \Psi$ and $\dot{\mathbf{x}} = \dot{\mathbf{X}} + \dot{\Psi}$ into the corresponding expression. Then the equation of fast motions will be

$$\mathbf{E}(T_{\mathbf{X}+\Psi}) - \langle \mathbf{E}(T_{\mathbf{X}+\Psi})\rangle = \mathbf{Q}_{\mathbf{X}+\Psi} - \langle \mathbf{Q}_{\mathbf{X}+\Psi}\rangle \qquad (2.58)$$

In this case equation (2.57) can be always written in the form, corresponding to that of the main equation of vibrational mechanics (2.12):

$$\mathbf{E}(T_{\mathbf{X}}) = \mathbf{F} + \mathbf{V} \qquad (2.59)$$

Here the expression for the vibrational force $\mathbf{V} = \mathbf{V}(\dot{\mathbf{X}}, \mathbf{X}, t)$ can be easily obtained by means of comparing equations (2.57) and (2.59) and taking into account that the initial equation (2.55) is always soluble with respect to x; the "ordinary" slow force is designated by $\mathbf{F}(\dot{\mathbf{X}}, \mathbf{X}, t)$.

3. The approximate method presented here can also be used in the case when under the 2π-periodic function Φ with respect to τ the equation of fast motions (2.7) allows a periodic solution with respect to τ, its period being $T = 2\pi q / p$ (with q and p being relatively prime numbers), and also in the case of the autonomous initial equation (2.1). In the first case the corresponding expressions should be subject to $2\pi q / p$ averaging while in the second case the period of averaging T, unknown beforehand, is determined in the process of solving the equation. It should be also noted that in the case when the fast forces do not depend on τ, equations (2.7) or (2.58) admit the solutions $\Psi \equiv 0$.

Finally, the equation of fast motions can allow solutions of a stochastic character. In that case the necessary condition for the applicability of that method is the existence of the average

$$\langle ... \rangle = \frac{1}{T_*} \int_0^{T_*} (...) d\tau$$

where T_* is the so-called averaging period.

4. In section 2.4 the case in point is the condition, necessary for the existence of solutions of the initial differential equation (2.1), which can be presented in form (2.2), or for the corresponding solutions X and Ψ of system (2.7) and (2.8), satisfying the main assumption of vibrational mechanics about the rate of change of the components X and Ψ. The question about the real existence of such solutions can be solved efficiently by means of constructing them actually in every separate case and by checking a posteriori the assumption about the rate of change of the components X and Ψ. Such checking is most desirable since the motions, described by Eq. (2.12), may prove to be fast, despite the fact that the motions, described by the same equation at $\mathbf{V} \equiv 0$, were slow

2.10. Additional Remarks, Certain Generalizations

(see, e.g. [4] where special attention was paid to that circumstance in confirmity to the method of harmonic linearization). Usually it is sufficient that the typical period of changing the component $\mathbf{X}$ should be at least three times larger than that of the component Ψ.

It should be also noted that in case the forces $\mathbf{F}$ and Φ do not depend on the slow time t, which happens quite often, (hence $\mathbf{V}$ does not depend on t either), and when it is only the stable stationary solutions of the equations of slow motions $\dot{\mathbf{X}} = 0$, $\mathbf{X} = \mathbf{X}_*$ that are of interest, as a rule it is possible to state a priori that in a sufficiently close vicinity of such solutions the conditions we are discussing are fulfilled. Indeed, for such solutions the right sides of the equations of slow motions become zero, and in the indicated vicinity they can be arbitrarily small with a proper choice of the vicinity. At the same time the right sides of the equations of fast motions at $\dot{\mathbf{X}} = 0$, $\mathbf{X} = \mathbf{X}_*$ generally speaking do not become zero.

In case other regions of the change of $\dot{\mathbf{X}}, \mathbf{X}$ are of interest, it is necessary to compare the orders of the right sides of the equations for those regions, and then to establish the correlation between the orders of $\mathbf{X}$ and Ψ at which the conditions of the validity of this method will be fulfilled. So, for instance, if the right sides of equations (2.7) and (2.8) prove to be of the same order, it can be expected that the solutions found will be valid under the initial conditions for $\dot{\mathbf{X}}, \mathbf{X}$ and under the system parameters which provide the fulfillment of the condition $\Psi \sim \varepsilon^2 \mathbf{X}$. If the system (either discrete or continuous), describing the slow motions, is close to the linear system with the eigenfrequencies $\lambda_1, \lambda_2, \ldots$, then it is natural to expect this system to describe sufficiently well the motion with the frequencies $\lambda_i \ll \omega$ and to be not valid for the motions with higher frequencies. The consideration of a number of concrete problems shows that it is usually sufficient that the condition $3\lambda_i < \omega$ should be valid.

5. Some important classes of nonlinear systems have been established (those with synchronized objects, of quasi-conservative objects and also canonical systems, systems with quasi-cyclic coordinates, those with kinematic and dynamic excitation of vibration) in which vibrational forces, despite their being nonconservative, are the derivatives of a certain function $D = D(X_1, \ldots, X_k)$:

$$V_s = -\frac{\partial D}{\partial X_s} \quad (s = 1, \ldots, k) \tag{2.60}$$

The function D, which plays the role of additional potential energy, is called the *potential function* and the corresponding systems are called the *potential on the average systems*. For the stationary and quasi-stationary motions of such systems the statements analogous to the classical theorem of Lagrange-Dirichlet are valid: the stable motions of the initial system may correspond (and under certain conditions they do correspond) to the points of minimum of the function D.

The above results are described in detail in books [1, 2].

2.11 Summary: On the Procedure of the Practical Use of the Method

The use of any method, including even the strict mathematical method or result, when solving an applied problem comprises, as a rule, a number of stages which do not belong to those mathematically strict, but can be referred to the methods realized at the so-called rational level of rigour. This circumstance and some other specific features of applied mathematical research are discussed in detail in the book [5]. This certainly refers to the method of direct separation of motions as well. The procedure of using that method, at least in the form stated above, comprises a number of rational elements, including the heuristic ones. We will describe this procedure schematically. We will emphasize that it allows various modifications and improvements, in future however we will try to keep to it when solving concrete problems.

1. A preliminary conclusion based on the experimental data and/or on the euristic grounds, deals with the question whether the motion of the system under consideration presents an imposing of fast oscillations upon the slow motion. In other words, a prognosis is made about the fulfillment of the main assumption of vibrational mechanics.

2. A system of equations (2.7), (2.8) is composed, based on the initial equations of motion.

3. A supposition is made about the relative order of the value of the fast and slow components X and Ψ, i.e. the supposition about the n in the relation $\Psi \sim \varepsilon^n X$, and the values of the parameters and of groups of summands are estimated, and large and small parameters and groups of summands are selected. On account of that supposition and those estimations, due simplifications are made in the equations.

2.11. Summary

4. Almost-periodic (in particular 2π—periodic) with respect to $\tau = \omega t$ solutions $\Psi = \Psi^*(\dot{X}, X, t, \omega t)$ of the equation of fast motions at the frozen (fixed) $\dot{X}, X$ and t are sought as a rule approximately. Among them such solutions are selected which are asymptotically stable with respect to all the fast generalized coordinates and velocities all over the region of values of the variables which interests us. These solutions are used while finding approximate expressions for the vibrational force by expression (2.11), after which the corresponding main equations of vibrational mechanics (2.12) are composed.

This approximate method is well-grounded, at least for the case when $\Psi \sim \varepsilon^n X$, i.e. when $n = 1$. As that takes place, if the expression for the fast force Φ can be presented as (2.32), the finding of the function Φ and of the vibrational force is essentially simplified.

5. Solutions of equations (2.12) which are of interest are being studied.

6. In questionable cases the suppositions made are checked a posteriori (taking into account remark 4 in section 2.10).

Chapter 3

On Other Methods of Obtaining Expressions for the Vibrational Forces and the Main Equations of Vibrational Mechanics

I. I. Blekhman

Institute for Problems of Mechanical Engineering,
Russian Academy of Sciences
and
Mekhanobr-Tekhnika Corp.
22 Liniya 3, V.O., 199106, St. Petersburg, Russia

3.1 Well Known Methods

The method of direct separation of motions, presented in chapter 2, is just one of the possible methods of obtaining the expressions for vibrational forces and equations of vibrational mechanics. Relations, containing the vibrational forces or moments in the form of summands can in many cases be obtained solving problems by other methods, such as the Poincare method of a small parameter, asymptotic methods, multiple scales method, the method of a harmonic balance. It is natural that such relationships present in fact equations of vibrational mechanics. Examples of the use of the above-mentioned methods are given in chapter 4. We will also mention the method of successive approximations, proposed by Kolovsky for solving problems of the dynamics of machine aggregates, but which can also be used to solve problems of other types (see [17] and [2], section 6.4).

3.2 Two Other Methods

Two more methods, in a sense diametrically opposite, should be metioned, with regard to which the method of direct separation of motions takes an intermediate position. The first of the methods refers to those rare cases when the solution of the initial equation (2.1) – either the exact, or the approximate one – is known. Then the finding of the vibrational force **V** is reduced to selecting the summands **X** and **Ψ** in that solution taking into account equality (2.3), and to the averaging according to formula (2.11). Such calculation is not useless, particularly in connection with the fact that the expression obtained for **V** can be used as an element when solving more complicated problems in an approximate way (see [2], sections 9.4 and 15.1).

Expressions for the vibrational forces can also be obtained by computer, solving numerically the corresponding equations and selecting a proper approximating formula, based on the series of calculations that have been performed.

The second method refers to the opposite case, when the problem is so intricate that neither the initial equation (2.1), nor equation (2.7) for the fast component **Φ** can be solved (and sometimes even composed) in a satisfactory way. Then we must resort to the hypotheses about the type of the function $V(\dot{X}, X, t)$ based mostly on the experimental data and on the ideas, following from the theory of dimensionality and similarity. Examples of such an approach are the *semi-empirical theories of turbulence*, the first publications on the theory of vibratory pile driving, in which the ground, which is under the action of vibration, is regarded as a fluid whose viscosity coefficient depends on the parameters of vibration [2], and also a number of investigations on vibratory creep and those which study the effect of vibration upon the substances of the type of concrete mixtures [2]. In spite of a kind of dissatisfaction a theoretician begins to feel at such an approach, there is no denial that such investigations are of practical benefit and they should be considered acceptable, at least until the problem has been elaborated more thoroughly. It should be also noted that the general statements of the mechanics of systems with the hidden motions and of vibrational mechanics, given here, are in fact a basis of a possibility to present the "main" results of the action of vibration upon the system as those caused by the action of vibrational forces. So both approaches to find these forces – the phenomenological and the semi-empirical – can in principal be considered acceptable.

We will also emphasize that the "macroscopic" equation (2.12), describing only the slow component of motions, which is of utmost practical interest, requires for its composition far less information than the initial "micro-equation"(2.1) or the system that corresponds to it (2.7), (2.8).

Finally, in a number of cases the vibrational force can be found by direct or indirect experiments (see, say, [2], section 10.1).

Chapter 4

A Simplest Example: Solving the Problem about a Pendulum with a Vibrating Axis of Suspension by Different Methods of the Theory of Nonlinear Oscillations

I. I. Blekhman [1], D.A.Indeitsev [2]

[1]*Institute for Problems of Mechanical Engineering,*
Russian Academy of Sciences
and
Mekhanobr-Tekhnika Corp.
22 Liniya 3, V.O., 199106, St. Petersburg, Russia

[2]*Institute for Problems of Mechanical Engineering ,*
Russian Academy of Sciences
61 Bolshoy pr. V.O., 199178, St. Petersburg, Russia

4.1 Preliminary Remarks

Nowadays the investigators of the nonlinear oscillations have at their disposal quite a number of analytic methods. These methods differ essentially in their abilities, their labor-consuming character and the degree of their mathematical grounding. The aim of this chapter is to illustrate this circumstance by the example of solving a classical problem about the stability of the "overturned" pendulum with a vibrating axis of suspension in its simplest formulation.

4.2 Equation of Motion

The motion of the pendulum (Fig. 4.1) whose suspension axis performs vertical oscillations according to the law

$$y = G\cos\Omega t \tag{4.1}$$

is described by the equation

$$I\ddot\varphi + h\dot\varphi + ml(g + G\Omega^2 \cos\Omega t)\sin\varphi = 0 \tag{4.2}$$

Here φ is the pendulum angle of deviation from the lower position of equilibrium, m, I and l, are the mass, the moment of inertia and the distance from the pendulum center of gravity C to the axis of suspension O, g is the acceleration of the free fall, h is the viscous resistance coefficient, G and Ω are the frequency and the amplitude of vibration.

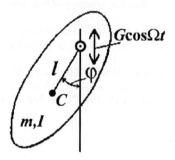

Figure 4.1. Pendulum with a vertically vibrating axis of suspension.

Designating

$$\varepsilon = \Omega t, \quad \frac{G}{l_0} = \varepsilon b, \quad \varepsilon = \frac{p_0}{\Omega}, \quad l_0 = \frac{I}{ml}, \quad p_0 = \sqrt{\frac{g}{l_0}} = \sqrt{\frac{mgl}{I}}, \quad \varepsilon h_0 = \frac{h}{I\Omega} \tag{4.3}$$

where p_0 is the small free oscillations frequency of the pendulum in the absence of oscillations, l_0 is the equivalent length of the pendulum, we will present equation (4.2) in the following dimensionless form

$$\varphi'' = -\varepsilon h_0 \varphi' - \varepsilon(\varepsilon + b\cos\tau)\sin\varphi = 0 \tag{4.4}$$

We consider the value $\varepsilon > 0$ to be the small parameter, i.e. we assume that the free oscillation frequency p_0 is small as compared with the frequency Ω, the oscillation amplitude G is small as compared with the equivalent length of the pendulum; the resistance coefficient is also considered to be small.

4.3 The Poincare-Lyapunov Method of Small Parameter

Following this classical method [20, 3], we seek the 2π-periodic over τ solution of equation (4.4) in the form of the series

$$\varphi = \varphi_0(\tau) + \varepsilon \varphi_1(\tau) + \varepsilon^2 \varphi_2(\tau) + \ldots \quad (4.5)$$

where $\varphi_0, \varphi_1, \varphi_2 \ldots$ are periodic functions of the given period. The so-called *generating equation*, corresponding to equation (4.4), is

$$\ddot{\varphi}_0 = 0, \quad (4.6)$$

i.e. the equation we get at $\varepsilon = 0$. This equation admits a periodic solution

$$\varphi_0 = \alpha, \quad (4.7)$$

where α is an arbitrary constant. Thus we are dealing with a special case in the theory of periodic solutions by Poincare [20]. In this case the constant α can be found from the conditions of periodicity of the following approximations.

Substituting series (4.5) into equation (4.4) and equating to the terms at ε and ε^2 in the right-hand and left-hand parts of the equality, we obtain the equations to find the following approximations

$$\varphi_1'' = -h_0 \varphi_0' - b \cos \tau \sin \alpha \quad (4.8)$$

$$\varphi_2'' = -h_0 \varphi_1' - \sin \alpha - b \varphi_1 \cos \tau \cos \alpha \quad (4.9)$$

From equation (4.8) we find a periodic solution

$$\varphi_1 = \alpha_* + b \sin \alpha \cos \tau \quad (4.10)$$

where α_* is an arbitrary constant. Substituting this expression into equation (4.9) we come to the equation

$$\varphi_2'' = h_0 b \sin\alpha \sin\tau - \sin\alpha - b(\alpha_* + b\sin\alpha\cos\tau) \qquad (4.11)$$

For this equation to allow a periodic solution it is necessary and sufficient that the constant coefficient in the expansion of the right part of equation (4.11) should become zero. This requirement leads to the relationship

$$P(\alpha) = \sin\alpha + \frac{1}{4}b^2 \sin 2\alpha = 0 \qquad (4.12)$$

From equation (4.12) we determine two essentially different values of the constant $\alpha = \varphi_0$ which can be answered by the periodic motions of the pendulum

$$\alpha_1 = 0, \quad \alpha_2 = \pi \qquad (4.13)$$

and in case the conditions $b^2 > 2$ are satisfied, two more values are found

$$\alpha_{3-4} = \pm\arccos(-2/b^2) \qquad (4.14)$$

The investigation which consists in considering the equation in variations (which we will not dwell on here), leads to the conclusion that the stable motions correspond to those values of α for which the following condition is satisfied

$$P'(\alpha) = \cos\alpha + \frac{1}{2}b^2 \cos\alpha > 0 \qquad (4.15)$$

Hence the classical result follows directly: the solution $\varphi_0 = \alpha_1 = 0$, corresponding to the lower position of the pendulum, is always stable; along with that, if the condition

$$b^2 > 2 \qquad (4.16)$$

is satisfied, or, if according to the designation (4.3), the conditions

$$G^2\Omega^2 > 2gl_0 \qquad (4.17)$$

are satisfied, then the upper (overturned) position of the pendulum $\varphi_0 = \pi$ is stable to; the motions, corresponding to $\varphi_0 = \alpha_{3-4}$ are unstable.

A strictly-mathematical approach requires to supplement the above investigation by determining the radius of convergence of the series (4.5). (This investigation guarantees convergence if the values of the parameter $\varepsilon = p_0/\Omega$ are small enough). In the majority of cases however this is not done, partly because it is too complicated and also because the assumptions, made in its process, lead to rather pessimistic estimations.

4.4 The use of Floquet-Lyapunov's Theory and of Ince-Strutt's Diagram

Equation (4.4) allows solutions $\varphi = 0$ and $\varphi = \pi$, corresponding to the "usual" lower and "upside-down" upper equilibrium positions of the pendulum. The investigation of the stability of these positions is reduced to studying the equation in variations, i.e. to equation (4.4), linearized near those positions

$$\varphi_2'' = -\varepsilon h_0 \varphi' - \varepsilon \sigma (\varepsilon + b \cos \tau) = 0 \qquad (4.18)$$

where $\sigma = 1$ for the lower position and $\sigma = -1$ for the upper position. Equation (4.18) is a linear differential equation of a second order with a periodic coefficient, well knows as Mathieu's equation.

Floquet-Lyapunov's theory makes it possible to investigate stability of the solutions of systems of linear differential equations with periodic coefficients of any order. This investigation, however presuppose obtaining a fundamental system of solutions of the equations, which, as a rule, cannot be done analytically. If such a system $-x_{sj}(\tau)$, $s,j = 1, \ldots, n$ (n being the order of the system), satisfying a unit matrix of the initial conditions $x_{sj}(0) = \delta_{sj}$, (δ_{sj} being the Kronecker symbol) is known, then the characteristic equation of the n-th power can be built

$$\left| x_{sj}(T) - \delta_{s,j} \rho \right| = 0 \quad (s, j = 1, \ldots n) \qquad (4.19)$$

where T is the period of the solution.

Solution of the question of stability depends on the roots of that equation ρ : if the absolute values of all the roots are less than unit, the solution is stable, and if there is at least one root whose absolute value is more than unit, the solution is unstable. The matrix $\left\| x_{sj}(T) \right\|$ is called the monodromy matrix.

The fundamental system can be built analytically, in particular in those cases when the system of differential equations contains the small parameter ε, then with $\varepsilon = 0$ we get the system of equations with constant coefficients. However, in these cases too the investigation is rather bulky and tedious, even in the case of the equation of the second order (4.18). Therefore simpler methods of investigating stability have been elaborated: those based on using the method of small parameter (see, say, [20, 26]). As for the equation of Mathieu, the so-called *diagram of Ince-Strutt* was built for it. This diagram makes it possible to judge about the stability of solutions, depending on the parameters of the equation.

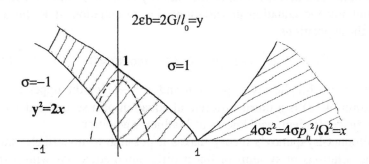

Figure 4.2. Stability region of the upper and lower positions of the pendulum with a vertically vibrating axis of suspension (fragment of the diagram of Ince-Strutt)

A fragment of this diagram, referring to equation (4.18) at $h_0 = 0$ (which is of no importance here) is shown in Fig. 4.2. The smallness of the parameter ε means that part of the diagram enclosed inside a certain semi-oval with the center at the origin of the coordinates refers to equation (4.18). The regions of the stability of solutions are shown on the diagram by hatching. The right part of the oval corresponds to the lower position of the pendulum $\varphi = 0$, and the left part corresponds to the upper "overturned" position of the pendulum $\varphi = \pi$.

As is seen, the lower position, on the assumption of the smallness of $\varepsilon = p_0/\Omega$, is always stable, while the upper position is stable if the condition $y^2 > -2x$ i.e. the inequality $\left(2\dfrac{G}{l_0}\right)^2 > -2\sigma \dfrac{4 p_0^2}{\Omega^2}$ are satisfied.

Taking into account designations (4.3) and the fact that for the upper position $\sigma = -1$, the latter inequality coincides with the classic condition (4.17).

4.5 Asymptotic Method

To illustrate the asymptotic method we will take, with an accuracy to the designations, the solution of the problem, given by N.N.Bogolyubov [9]. An important element of this solution is the change of the variables, found by N. N. Bogolyubov,

$$\varphi = \alpha + \varepsilon \cos\tau \sin\alpha, \quad \varphi' = \varepsilon q + \varepsilon \sin\tau \sin\alpha \qquad (4.20)$$

to transfer from the variable φ to two variables α and q. As a result, instead of equation (4.4) we obtain a system with two equations of the first order, in the so-called standard form

$$\alpha' = \varepsilon q + \varepsilon^2 ...,$$
$$q' = \varepsilon(-b^2 \cos^2\tau \sin\alpha \cos\alpha - \sin\alpha + b\omega \sin\tau \cos\alpha \qquad (4.21)$$
$$- h_0\omega + h_0 b \sin\tau \sin\alpha) + \varepsilon^2 ...$$

Equation averaged over the variable τ for the period of 2π (equations of the first approximation) have in this case the form

$$\alpha' = \varepsilon q, \quad q' = -\varepsilon[h_0\omega + \sin\alpha(1 + \dfrac{1}{2}b^2 \cos\alpha)] \qquad (4.22)$$

and are reduced to the equations of the second order with respect to α

$$\alpha'' + \varepsilon h_0 \alpha' + \varepsilon^2(1 + \dfrac{1}{2}b^2 \cos\alpha)\sin\alpha = 0 \qquad (4.23)$$

Like in section 4.3, this equation when $b^2 > 2$ allows four stationary solutions

$$\alpha_1 = 0, \quad \alpha_2 = \pi, \quad \alpha_{3-4} = \pm\arccos(-2/b^2) \qquad (4.24)$$

Forming equations in variations for these solutions, it is easy to come to the conclusion that the solution $\alpha_1 = 0$, corresponding to the lower position of the pendulum, will always be stable, while the upper ("overturned") position is stable provided condition (4.16) is satisfied, and the intermediate positions are unstable. In other words, we have come to the conclusions which agrees with the results of sections 4.3 and 4.4.

4.6 Method of Multiple Scales

Using the method of multiple scales [22] we will consider the differential equation

$$\frac{d^2\varphi}{dT_0^2} + \varepsilon(\varepsilon + b\cos T_0)\sin\varphi = 0 \qquad (4.25)$$

which differs from the initial one (4.4) by the fact that the variable $\tau = \Omega t$ is designated by T_0 and to simplify the calculations the term $\varepsilon h_0\dot{\varphi}$, characterizing the viscous resistance, is omitted. In accordance with the procedure of the method, and confining ourselves to a two-scale decomposition, we will seek a periodic solution of equation (4.25) in the form of the series

$$\varphi = \varphi_0(T_0,T_1) + \varepsilon\varphi_1(T_0,T_1) + \varepsilon^2\varphi_2(T_0,T_1) + \ldots \qquad (4.26)$$

where

$$T_1 = \varepsilon T_0 = \varepsilon\tau \qquad (4.27)$$

and φ_0, φ_1, φ_2,... are 2π-periodic functions of T_0

In this case in the process of solving the variables T_0 and T_1 are considered to be independent and equation (4.25) when series (4.26) is substituted, takes the form of the equation in partial derivatives.

4.6. Method of Multiple Scales

Taking into consideration the relationships

$$\frac{d}{d\tau} = \frac{d}{dT_0} = \frac{\partial}{\partial T_0} + \varepsilon \frac{\partial}{\partial T_1},$$

$$\frac{d^2}{d\tau^2} = \frac{d^2}{dT_0^2} = \frac{\partial^2}{\partial T_0^2} + 2\varepsilon \frac{\partial^2}{\partial T_0 \partial T_1} + \varepsilon^2 \frac{\partial^2}{\partial T_1^2},$$
(4.28)

we substitute series (4.26) into equation (4.25). Then we come to the relationship (recorded with an accuracy to the terms with ε^2 inclusively)

$$\frac{\partial^2 \varphi_0}{\partial T_0^2} + 2\varepsilon \frac{\partial^2 \varphi_0}{\partial T_0 \partial T_1} + \varepsilon \frac{\partial^2 \varphi_1}{\partial T_0^2} + 2\varepsilon^2 \frac{\partial^2 \varphi_1}{\partial T_0 \partial T_1} + \varepsilon^2 \frac{\partial^2 \varphi_0}{dT_1^2} + \varepsilon^2 \frac{\partial^2 \varphi_2}{dT_0^2} \quad (4.29)$$
$$+ \varepsilon b \sin \varphi_0 \cos T_0 + \varepsilon^2 \sin \varphi_0 + \varepsilon^2 b \varphi_1 \cos T_0 \cos \varphi_0 = 0$$

Equating to zero the terms to certain powers of ε, we arrive at equations in partial derivatives

$$\frac{\partial^2 \varphi_0}{\partial T_0^2} = 0 \qquad (4.30)$$

$$\frac{\partial^2 \varphi_1}{\partial T_0^2} + 2\frac{\partial^2 \varphi_0}{\partial T_0 \partial T_1} + b \sin \varphi_0 \cos T_0 = 0 \qquad (4.31)$$

$$\frac{\partial^2 \varphi_2}{\partial T_0^2} + \frac{\partial^2 \varphi_0}{\partial T_1^2} + 2\frac{\partial^2 \varphi_1}{\partial T_0 \partial T_1} + \sin \varphi_0 + b \varphi_1 \cos \varphi_0 \cos T_0 = 0 \quad (4.32)$$

According to equation (4.30)

$$\varphi_0 = A(T_1) + B(T_1)T_0 \qquad (4.33)$$

For φ_0 to be periodic it is necessary to assume that $B(T_1) = 0$, so that

$$\varphi_0 = A(T_1) \qquad (4.34)$$

Then equation (4.31) takes the form

$$\frac{\partial^2 \varphi_1}{\partial T_0^2} + b \sin A(T_1) \cos T_0 = 0 \qquad (4.35)$$

and the periodic solution of this equation will be

$$\varphi_1 = b \sin A(T_1) \cos T_0 \qquad (4.36)$$

since when determining the functions $\varphi_1, \varphi_2, \ldots$ we can confine ourselves to a partial solution of the corresponding equations.

Substituting expressions (4.34) and (4.36) into equation (4.32), we obtain

$$\frac{\partial^2 \varphi_2}{\partial T_0^2} - 2 \frac{dA(T_1)}{dT_1} \cos A(T_1) \cos T_0 + \frac{d^2 A(T_1)}{dT_1^2} \qquad (4.37)$$
$$+ \sin A(T_1) + b^2 \cos^2 T_0 \sin A(T_1) \cos A(T_1) = 0$$

The condition of the periodicity of the function φ_2 leads to the equation

$$\frac{d^2 A(T_1)}{dT_1^2} + \sin A(T_1) + \frac{1}{2} b^2 \sin A(T_1) \cos A(T_1) = 0 \qquad (4.38)$$

or, after the change of $A(T_1)$ for φ_0 and of T_1 for $\varepsilon T_0 = \varepsilon \tau = p_0 t$ to the equation

$$\frac{d^2 \varphi_0}{dt^2} + p_0^2 \sin \varphi_0 (1 + \frac{b^2}{2} \cos \varphi_0) = 0 \qquad (4.39)$$

This equation coincides (with an accuracy to the designations and the term with the first derivative, omitted by us) with equation (4.23). Therefore the corresponding conclusions about the stability of positions of the pendulum also coincide.

4.7 Methods of Harmonic Balance and of Bubnov—Galerkin

It seems interesting to make sure that the regularities of the behavior of the pendulum which are under consideration are in agreement with

4.7. Methods of Harmonic Balance and of Bubnov-Galerkin

those obtained by the method of harmonic balance, or with the method of Bubnov—Galerkin, furnishing the same procedure.

We will consider, like in section 4.4, the behavior of the pendulum near the positions $\varphi = 0$ and $\varphi = \pi$, described by equation (4.18), ignoring for the sake of simplicity the moment of viscous resistance. We will seek the 2π - periodical over τ solution of equation (4.18) in the form of a series

$$\varphi = a_0(t) + a_1 \cos\tau + ... \qquad (4.40)$$

where $a_0(t)$ is a certain slowly changing (as compared to $\cos\tau$) function of time, and a_1 is a constant. We will limit ourselves to just written terms of the decomposition.

Substituting expression (4.40) into equation (4.18) (at $h_0 = 0$), we obtain the relationship (mind that the prime marks the differentiation with respect to $\tau = \Omega t$):

$$\frac{1}{\Omega^2} \frac{d^2 a_0(t)}{dt^2} - a_1 \cos\tau + \sigma\varepsilon(\varepsilon + b\cos\tau)[a_0(t) + a_1 \cos\tau] = 0 \quad (4.41)$$

Now assuming that the coefficient $a_0(t)$ plays the role of a constant coefficient in the decomposition in the 2π - periodic over τ function , we multiply equality (4.41) successively by the constant (or by $a_0(t)$) and by $\cos\tau$ and average it over τ for the 2π - period. Then we get the relationships

$$\frac{1}{\Omega^2} \frac{d^2 a_0(t)}{dt^2} + \sigma\varepsilon^2 a_0(t) + \frac{1}{2}\sigma\varepsilon a_1 b = 0 \qquad (4.42)$$

$$-a_1 + \sigma\varepsilon^2 a_1 + \sigma\varepsilon b a_0(t) = 0$$

Having found the coefficient a_1 from the second equality and substituting it into the first one, we get (in order to find $a_0(t)$ with an accuracy to the terms with $\varepsilon^2 = p_0^2/\Omega^2$ inclusively) the following equation

$$\frac{d^2 a_0(t)}{dt^2} + p_0^2 (\sigma + \frac{1}{2}b^2) a_0(t) = 0 \qquad (4.43)$$

Hence follows the same conclusion that when we use other methods: the lower position of the pendulum ($\sigma = 1$) is also stable while the upper position ($\sigma = -1$) is stable provided the condition $b^2 > 2$ is satisfied, or, in the initial designations, when the inequality $G^2\Omega^2 > 2gl_0$ is satisfied.

The authors realize that the methods, named in the headline of the section are used here somewhat wider, and also that this expansion (4.40) has been made in association with the results, obtained by methods of many scales and of direct separation of motions (see below). However, the simple calculations, given here, and their results still seem to be of interest: they show that if there is certain information about the type of the solution sought, the methods of Bubnov—Galerkin and of harmonic balance easily bring us to the correct result. But such information is also beneficial when using other methods!

4.8 Method of Direct Separation of Motions

In accordance with the method of direct separation of motions [14, 15, 1, 2, 23] (see also chapter 2 of this book) we seek the solution of equation (4.2) in the form

$$\varphi = \alpha(t) + \psi(t,\tau) \qquad (4.44)$$

where t is the slow time, and $\tau = \Omega t$ is the fast time, α is a slowly changing and ψ is a fast changing 2π - periodic over τ function with

$$\langle \psi(t,\tau) \rangle = 0 \qquad (4.45)$$

i.e. the mean over the period 2π with respect to τ value of the function ψ for definiteness sake is believed to be equal to zero. To determine the functions α and ψ, instead of equation (4.2) two equations are obtained

$$I\ddot{\alpha} = -h\dot{\alpha} - mgl\sin\alpha + V \qquad (4.46)$$

$$I\ddot{\psi} = \Psi \qquad (4.47)$$

where

4.8. Method of Direct Separation of Motions

$$\Psi = -h\dot\psi - mgl[\sin(\alpha+\psi) - \langle\sin(\alpha+\psi)\rangle]$$
$$- ml\Omega^2 G \quad \alpha+\psi \quad \tau - \langle G \quad \alpha+\psi \quad \tau\rangle [\text{ si} \quad (4.48)$$
$$V = mgl[\sin\alpha - \langle\sin(\alpha+\psi)\rangle] - ml\Omega^2\langle G\sin(\alpha+\psi)\cos\tau\rangle$$

Equation (4.46) is obtained by substituting expression (4.44) into equation (4.2) and averaging over the period 2π with respect to the fast time τ, and equation (4.47) is obtained from the condition satisfying equation (4.2) via expression (4.44).

In accordance with the method for composing (with a satisfactory accuracy) equation (4.48) (equation of slow motions, the "main equation of vibrational mechanics") the equation (4.47) can be solved approximately. Besides, in the process of solving the value α can be regarded as constant ("frozen"). In view of the suppositions (4.3) we can assume that $\Psi = \varepsilon\Psi_1$ and present equation (4.47) in the form

$$I\ddot\psi = \varepsilon\Psi_1 \qquad (4.49)$$

and seek its periodic solution in the form of decomposition $\psi = \psi_0 + \varepsilon\psi_1 + \ldots$. For the first approximation in view of the condition (4.45) we obtain $\psi_0 = 0$ and the first approximation ψ_1 is determined by the equation

$$I\varepsilon\ddot\psi_1 = -ml\Omega^2 G\sin\alpha\cos\tau \qquad (4.50)$$

on which in view of (4.45) we find

$$\psi \approx \varepsilon\psi_1 = \frac{ml}{I}G\sin\alpha\cos\tau \qquad (4.51)$$

As a result, according to (4.48) we find with an accuracy to ε

$$V = mgl[\sin\alpha - \langle\sin(\alpha+\varepsilon\psi_1)\rangle] - ml\Omega^2\langle G\sin(\alpha+\varepsilon\psi_1)\cos\tau\rangle$$
$$= -ml\Omega^2\varepsilon\langle\psi_1 \quad \tau\rangle \quad \alpha \qquad \cos \quad \cos \qquad (4.52)$$

and having performed the operation of averaging, we get

$$V = -\frac{(mlG\Omega)^2}{4I}\sin 2\alpha \qquad (4.53)$$

In view of (4.53) the equation of slow motions (the equation of vibrational mechanics) takes the form

$$I\ddot\alpha + h\dot\alpha + mgl\sin\alpha + \frac{(mlG\Omega)^2}{4I}\sin 2\alpha = 0 \qquad (4.54)$$

Expression (4.53), called "vibrational moment" (see chapter 2 of this book) allows a distinct physical interpretation: it is an additional "slow" moment which must be added to all the ordinary moments, acting on the pendulum, so that the action of vibration might be taken into consideration. Equation (4.54) is interpreted adequately. Linearizing this equation near the equilibrium positions of the pendulum $\alpha = 0$ and $\alpha = \pi$, we obtain the equation

$$I\ddot\alpha + h\dot\alpha + [\sigma mgl + \frac{(mlG\Omega)^2}{2I}]\alpha = 0 \qquad (4.55)$$

where like before we have $\sigma = 1$ for the lower position and $\sigma = -1$ for the upper one. The presence of the second term in the square brackets can be interpreted as the appearance of invisible springs, caused by vibration, with the angular rigidity

$$c_V = \frac{1}{2}(mlG\Omega)^2 / I \qquad (4.56)$$

"attracting" the pendulum to both upper and lower equilibrium positions (Fig. 4.3).

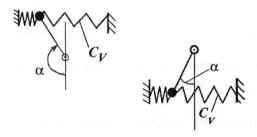

Figure 4.3. Physical interpretation of the stability of the upper position of the pendulum and accelaration of the tick of the clocks caused by vibration – the seeming appearance of the strings drawing the pendulum to the equilibrium positions.

As a result, the lower position of equilibrium remains (within the limits of the suppositions made) always stable, with the period of free slow oscillations of the pendulum increasing, (the clock on the vibrating base is always fast). The upper position, unstable in the absence of vibration, acquires stability if the condition

$$\frac{ml(G\Omega)^2}{2gl} > 1 \tag{4.57}$$

is satisfied. This condition with an accuracy to the designations coincides with the classic condition (4.17) and also with the corresponding conditions, obtained by other methods.

In the conclusion of this item we will mark the importance of the physical clearness of this method, allowing to solve a number of important applied problems. In particular, it is apparently due to this circumstance it was possible to obtain a number of interesting results, developing the investigation [14, 15], made by P.Kapitsa, quoted in general outline here. Among such results we may mention establishing the possibility of raising the stability of elastic systems by means of vibration [10, 11, 6], the discovery of the effect of conjugacy of resonances and bifurcations under the action of the two-frequency vibration on systems with the positional non-linearity [7] (see also chapter 9) and also establishing the possibility of parrying of the natural instability of the conducting body in the electrostatic fields by means of imposing high-frequency oscillations [21, 24] (see also chapter 10).

4.9 Discussion

Summing up we may conclude that the method of direct separation of motions, at least when solving the classic problem that has been considered, has certain advantages over any other methods :

1. As compared to all the above-considered methods, both the process of solving and the result are characterized by physical clearness, laconic brevity, and simplicity of application.

2. Unlike the Poincare-Lyapunov method or the Floke-Lyapunov theory, this method makes it possible to get the equation, describing not only the periodic motions and their stability, but also the slow non-stationary processes.

3. As compared to the asymptotic methods, the result is easier to get, in particular, since it is not necessary to transform the equation to the new variables and to perform the inverse transformation.

The use of the method of harmonic balance and the method of Bubnov—Galerkin is not more complicated than that of the direct separation of motions, but it requires a somewhat wider interpretation and the use of the information about the type of solutions.

It should be emphasized that what has been said refers only to those advantages of the method of direct separation of motions which have been displayed by the example of a comparatively simple problem, considered here. The consideration of more complicated systems [2] makes it possible to display a number of other advantages, for instance a wider region of their applicability, in particular – to solving equations with non-smooth and discontinuous right-hand sides (see also chapter 5).

To the essential advantages of the method of multiple scales we should refer the distinct formalization of the principle of separating motions according to the tempo of their change.

Perhaps the opinions, stated here, reflect the personal partiality and experience of the authors. However, the authors hope this work will help the reader to form his own opinion on this question.

Chapter 5

Conclusion: On the Main Peculiarities and Advantages of the Approaches of Vibrational Mechanics and of the Method of Direct Separation of Motions as Compared to Other Methods of Nonlinear Mechanics

I. I. Blekhman

*Institute for Problems of Mechanical Engineering,
Russian Academy of Sciences
and
Mekhanobr-Tekhnika Corp.
22 Liniya 3, V.O., 199106, St. Petersburg, Russia*

5.1 Peculiarities and Limitations

1. The approach, called vibrational mechanics and the method of direct separation of motions have been adapted to investigate processes presentable in the form of the sum of the main "slow" motion $X(t)$ and the fast (not necessarily small) oscillations $\psi(t,\omega t)$ i.e. in the form

$$x = X(t) + \psi(t,\omega t) \tag{5.1}$$

Here lies its distinction from, e.g., the classical version of Van der Pol's method when the solution sought has the form

$$x = A(t)\sin[\omega(t)t + \beta(t)] \tag{5.2}$$

where $A(t)$, $\omega(t)$ and $\beta(t)$ are the slowly changing functions of time.

As can be seen from the content of the book [1, 2] and of this book, cases when solutions of the type of (5.1) that are of interest are quite numerous and highly diversified.

2. When using in practice the method of direct separation of motions, the way of obtaining an approximate solution of the equation of fast motions (2.7) is of special importance. The main limitation in using this method lies in the fact that the component X changes slowly enough as compared to the component ψ, i.e. that the frequency ω is sufficiently large. That fact imposes certain constraints on the relative orders of the right sides of equations (2.7) and (2.8). Usually it is sufficient that the typical period of changing the component X should be at least three times larger than that of the component ψ.

Another limitation lies in the requirement that the equations of fast motions (2.7) should have solutions asymptotically stable by the corresponding variables. Though for the separate equation for the variable X (not necessarily slow!) to be valid, it is sufficient to satisfy only the second of the above mentioned limitations.

5.2 Advantages

Many of the advantages of the method and approach listed below are connected with their peculiarity that only the slow component X of the motion is of interest. These advantages are as follows:

1. The main advantage of this method and this approach lies in the fact that the main equations (the equations of slow motions) are obtained in the form of equations of dynamics. The result of the action of vibration on the nonlinear system is expressed in these equations by means of adding to the "ordinary" slow forces the so-called vibrational forces V. The appearance of these forces, as well as the process of obtaining them, have in this case a distinct physical meaning.

The indicated advantage of this method is illustrated by the content of the books [1, 2] and of this book. In particular it plays an essential part when considering the following problems:

1) When revealing the so-called potential on the average dynamic systems – systems for which vibrational forces have a potential. Slow motions are described in this case by the potential system, while the initial system is essentially non-potential (a more exact definition is given in chapter 4 of book [2] where several classes of the potential on the average dynamic systems are considered).

2) When solving complicated problems in case it is difficult even to compose the initial differential equations of motion of the system. In some cases expressions for vibrational forces, found as a result of solving simple "model" problems, can be used with certain modifications to solve much more complicated problems. Problems on slow flows of the granular material, caused by vibration (see chapter 15 of book [2]) are the examples.

3) When solving problems on the action of vibration on systems with dry friction and impacts. In these cases, instead of the initial non-smooth or discontinuous system, for the description of slow motions one gets a system with "smooth" vibrational forces. Such systems are considered in chapters 9 and 13 – 15 of book [2]).

4) When solving problems of controlling the properties of non-linear mechanical systems, including vibro-rheological properties of non-linear media by means of vibration (see chapter 18 of book [2]).

5) When solving problems of the inventors. For the creators of new machines and technologies, it is very convenient and expedient to take the position of the observer V and add to all the ordinary slow forces the additional force V. It gives an opportunity "to neglect" the laws of mechanics which prevent achieving the aim. Thus the inventor can obtain quite unexpected positive effects. This approach – the temporary ignoring in the process of inventing the physical laws – is known under the name of "*fantastic analogy*" suggested by Gordon and proved rather effective [13].

2. Unlike asymptotic methods the method of direct separation of motions does not require a transformation of the initial equations to the so-called canonic form. Such transformation is not always easy and sometimes is even impossible.

3. The order of the system of slow motion can be much less than the order of the initial system, e.g. in the problem of section 6.3 of book [2] the order of the initial system is $2(n+1)$, $(n=1,2,...)$ and the order of the equation of slow motion is 1.

4. The method of direct separation of motions is "not very sensitive" to the accuracy of finding the fast component of motion ψ when obtaining the equation for the slow component X (the main equation of vibrational mechanics). This is connected with the fact that the fast component enters into the expression for the vibrational force only under the sign of integration. Due to that to obtain an acceptable result it is often sufficient to find a crudely approximate solution of the equation of

fast motions. In particular, we can confine ourselves to the so-called purely inertional approach (see e.g. the solution of certain problems in this book and in sections 9.4 and 10.1 of book [2]).

5. As a result, instead of the difficulties of solving the initial differential equations, we face the difficulties, connected with solving separately the equations of fast motions and those of slow motions. As follows from what has been said, these latter difficulties are far lesser than the first.

6. As has been shown in chapter 2 of the book [2], vibrational mechanics can be considered within the frames of a much more general conception – the conception of systems with the ignored (hidden) motions. Hence vibrational mechanics can be regarded as quite well grounded. The method of direct separation of motions is also well grounded, at least with regard to the majority of problems, met in applications.

5.3 Final Remarks

The contents of this book and of book [2] show that the enumerated peculiarities and limitations of the method and approach, discussed above, do not prevent one from solving numerous applied problems on the action of vibration on nonlinear systems. At the same time their merits make it possible to get considerable advantages when solving such problems. Though, a question arises: whether it is possible to get the same results by other methods. For most of the problems the answer will be definitely positive, see, e. g. chapter 4. In particular some results, stated in the books [1, 2], were first obtained by the author by means of other methods. For instance the solution of the problems on self-synchronization of mechanical vibro-exciters (chapter 7 of book [2] and part III of this book) was obtained by Poincare-Lyapunov's method of small parameter, and the solution of the problems of the theory of vibrational displacement (chapters 9 and 10 of book [2]) was got by exact methods. The reader, just like the author, can also make sure that by method of direct separation of motions the solution can be obtained in a much simpler and often in a much more general form; the reasons, causing this circumstance, are discussed in i.i. 3 and 4 of section 5.2. As an example we refer to the solution of the problem concerning the transportation of a body up a vertical vibrating pipe (see section 10.1 of book [2]). At present there are a number of complicated problems of the theory of vibrational displacement and of the theory of synchronization of vibro-exciters whose solution was obtained by the method of direct separation of motions and which has not yet been obtained (and apparently cannot be obtained) by any other method. Such problems were mentioned above in section 5.2.

References to Part 1

1. Blekhman I. I., *Vibrational Mechanics* (Fizmatlit, Moscow, 1994), p. 394 (in Russian).
2. Blekhman I. I., *Vibrational Mechanics*. (World Scientific, Singapore, 2000), p. 509.
3. Blekhman I. I., *Synchronization of Dynamic Systems*, (Nauka, Moscow, 1971), p. 894 (in Russian).
4. Blekhman I. I. and G. Yu. Dzhanelidze, On stability of vibro-linearized systems, *App. Math. and Mech.*, **25**, 1 (1961) 173—176 (in Russian).
5. Blekhman I. I., A. D. Myshkis, and Ya. G. Panovko, *Mechanics and applied mathematics: Logic and peculiarities of applications of mathematics* (Nauka, Moscow, 1990, 2nd ed.), p. 356 (in Russian). German translation: Veb Deutscher Verlag der Wissenschaftn, 1989, Berlin, p. 350.
6. Blekhman I. I., Dresig H, Shishkina E.V. About the Indian magic rope, *Proc. of the XXIX Summer School "Actual Problems in Mechanics"* (St. Petersburg, Russia, 2001)
7. Blekhman I. I., and Landa P.S., Conjugate resonances in nonlinear systems under biharmonic excitation. Vibro-induced bifurcations, *Izvestiya VUZ, Applied Nonlinear Dynamics*, **10**, 1-2 (2002) 44-51 (in Russian).
8. Bogolyubov N. N., *Selected works in 3 volumes*, **1** (Naukova Dumka, Kiev, 1969), p. 447 (in Russian).
9. Bogolyubov N. N. and Yu. A. Mitropolsky, *Asymptotic methods in the theory of nonlinear oscillations* (Nauka, Moscow, 1974), p. 503 (in Russian).
10. Chelomey V. N., On the possibility of the increase of elastic systems stability by means of vibrations, *DAN SSSR*, **110**, 3 (1956) 341–344 (in Russian).
11. Chelomey S. V., On dynamic stability of elastic systems, *DAN SSSR*, **252**, 2 (1980) 307–310 (in Russian).
12. Chelomey S. V., Dynamic stability at high-frequency parametric excitation, *DAN SSSR*, **257**, 4 (1981) 853–857 (in Russian).
13. Jones J. C., *Design methods. Seeds of human future* (A Wiley-Interscience Publication, London, 1980), p. 326.
14. Kapitsa P. L, Pendulum with vibrating axis of suspension, *Uspekhi fizicheskich nauk*, **44**, 1 (1954) 7—20 (in Russian).
15. Kapitsa P. L, Dynamic stability of the pendulum when the point of suspension is oscillating, *Jour. of Exper. and Theor. Physics*, **21**, 5 (1951) 588–597 (in Russian).
16. Kirgetov A. V., On stability of the quasiequilibrium positions of the Chelomey pendulum, *Izv. AN SSSR. MTT*, 6 (1986) 57–62 (in Russian).

17. Kolovsky M. Z., Investigation of the dynamics of the stationary motion of the machine aggregate with an elastic transmitting mechanism, *Izv. AN SSSR. Mashinovedenie*, 2 (1985) 40–47 (in Russian).
18. Loitsyansky L. G., *Mechanics of Liquid and Gases*, 6th ed. (Nauka, Moscow, 1987), p. 840 (in Russian). English translation: Begell-Haus Inc., NY, Walligford UK, ed. Robert H. Nann, 1995.
19. Malakhova O. Z., *Extremal signs of the stability of motion and their use in the creating vibrational devices*, PhD thesis (Leningrad, Mekhanobr, 1990), p. 154 (in Russian).
20. Malkin I. G., *Certain problems of the theory of nonlinear oscillations* (Gostekhizdat, Moscow, 1956), p. 491 (in Russian).
21. Martinenko Yu. G., *Motion of a solid body in electric and magnetic fields* (Nauka, Fizmatlit, Moscow, 1988), p.368 (in Russian).
22. Nayfeh A. H. and B. Balachandran, *Applied Nonlinear Dynamics: Analytical, Computational, and Experimental Methods* (Wiley, New York, 1995).
23. Ragulskis K. M, Mechanisms on a vibrating foundation, *Problems of dynamics and stability* (In-te of Energetics and Electrical Engineering of AN Lietuva SSR, Kaunas, 1963), p. 232 (in Russian).
24. Skubov D. Yu., Electro-mechanical systems in the static and alternating magnetic fields (Textbook, St. Petersburg State Technical University, St. Petersburg, 1999), p. 44 (in Russian).
25. Volosov V. M. and B. I. Morgunov, *Method of averaging in the theory of nonlinear oscillatory systems* (Moscow State Univ., Moscow, 1971), p. 507 (in Russian).
26. Yakubovich V. A. and V. M. Starzhinsky, *Linear differential equations with periodic coefficients and their applications* (Nauka, Moscow, 1972), p. 718 (in Russian).
27. Yudovich V. I., Vibrodynamic of systems with constraints, *DAN*, **354**, 5 (1997), 622–624.
28. Yudovich V. I., Dynamics of a material particle on a smooth vibrating surface, *Applied Mathematics and Mechanics*, **62**, 6 (1998) 968–976.

Part II

Pendulum and Pendulum Systems under High-Frequency Excitation — Non-Trivial Effects

Chapter 6

Quasi-equilibrium Positions and Stationary Rotations of the Pendulums with a Periodically Vibrating Axis

I. I. Blekhman[1], H. Dresig[2], P. Rodionov[3]

[1] *Institute for Problems of Mechanical Engineering,*
Russian Academy of Sciences and
Mekhanobr-Tekhnika Corp
22 Liniya 3, V.O., 199106, St. Petersburg, Russia

[2, 3] *TU Chemnitz, Institute of Mechanics,*
Str.der Nationen 62, D-09107, Chemnitz, Germany

6.1 Preliminary Remarks, Equation of Motion

The classical problem about a pendulum with a vibrating axis of suspension (see [1, 2, 3, 4, 5] and chapter 4 of this book) allows a generalization for the case when the axis performs arbitrary periodic oscillations in two mutually perpendicular directions. Such generalization can be obtained quite easily by using the method of direct separation of motions. Then both oscillatory and rotary motions can be studied.

The consideration of this generalization is of interest, particularly in the connection with the idea of balancing the mechanisms by placing pendulums on their links. This idea was proposed by the authors in their work [6], some results of which are stated below.

Let the axis of the pendulum perform periodic oscillations $x(\omega t)$, $y(\omega t)$ in two mutually perpendicular directions (Fig. 6.1) with a certain period $T = 2\pi/\omega$, circumscribing in this case a certain closed curve. The

6. Quasi-Equilibria and Rotations of the Pendulums with Vibrating Axis

equation of motion of such a pendulum looks as

$$I\ddot{\varphi} + h\dot{\varphi} + mgl\sin\varphi = ml(\ddot{x}\cos\varphi - \ddot{y}\sin\varphi) \qquad (6.1)$$

Here φ is the angle of deviation of the pendulum from the normal, l, m and I are respectively the length, the mass and the moment of inertia with respect to its axis; g is the free fall acceleration, h is the viscous damping coefficient.

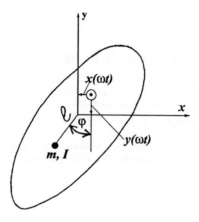

Figure 6.1. Pendulum with a periodically vibrating axis.

6.2 Regimes of Quasi-Equilibrium

We find the solution of this equation in the form

$$\varphi = \alpha(t) + \psi(t, \omega t) \qquad (6.2)$$

where α is the slow component and ψ is the fast component. We use the method of direct separation of motion in the way similar to that used in section 4.8 of this book and under the same assumption. Then we will obtain after calculation the equation for the main slow component (the main equation of vibrational mechanics).

$$I\ddot{\alpha} + h\dot{\alpha} + mgl\sin\alpha = V(\alpha) \qquad (6.3)$$

where
$$V(\alpha) = \frac{(ml)^2}{2I}\left[\left\langle \dot{x}^2 - \dot{y}^2 \right\rangle \sin 2\alpha + 2\langle \dot{x}\dot{y}\rangle \cos 2\alpha \right] \quad (6.4)$$

is the vibrational moment.

If x and y are given as series

$$x = \sum_{n=1}^{\infty} H_n \sin(n\omega t + \beta_n), \quad y = \sum_{n=1}^{\infty} G_n \cos(n\omega t + \gamma_n), \quad (6.5)$$

then

$$\langle \dot{x}^2 \rangle = \frac{1}{2}\omega^2 \sum_{n=1}^{\infty} n^2 H_n^2, \quad \langle \dot{y}^2 \rangle = \frac{1}{2}\omega^2 \sum_{n=1}^{\infty} n^2 G_n^2,$$
$$\langle \dot{x}\dot{y}\rangle = -\frac{1}{2}\omega^2 \sum_{n=1}^{\infty} n^2 H_n G_n \sin(\gamma_n - \beta_n) \quad (6.6)$$

Introducing the designation

$$-\frac{(ml)^2}{2I}\langle \dot{x}^2 - \dot{y}^2 \rangle = B\cos 2q, \quad 2\frac{(ml)^2}{2I}\langle \dot{x}\dot{y}\rangle = B\sin 2q, \quad (6.7)$$

we will present (6.4) in the form

$$V(\alpha) = -B\sin 2(\alpha - q) \quad (6.8)$$

and then equation (6.3) will take the following simple form

$$I\ddot{\alpha} + h\dot{\alpha} + mgl\sin \alpha + B\sin 2(\alpha - q) = 0 \quad (6.9)$$

The stationary solutions of this equation, corresponding to the positions of quasi-equilibrium of the pendulum, $\alpha = \alpha^*$, will be found from the equation

$$P(\alpha^*) = mgl\sin \alpha^* + B\sin 2(\alpha^* - q) = 0 \quad (6.10)$$

A certain solution α^* of this equation will be answered by the asymptotically stable position of the quasi-equilibrium of the pendulum, provided the following inequality is fulfilled

$$P(\alpha^*) = mgl\cos \alpha^* + B\cos 2(\alpha^* - q) > 0 \quad (6.11)$$

Introducing the dimensionless parameter

$$\beta = B/mgl \qquad (6.12)$$

we can write the relationships (6.10) and (6.11) respectively in form:

$$\sin\alpha^* + \beta\sin 2(\alpha^* - q) = 0 \qquad (6.13)$$

$$\cos\alpha^* + \beta\cos 2(\alpha^* - q) > 0 \qquad (6.14)$$

The solution of equation (6.13) depends on two parameters β and q. Using equation (6.13) we can write the stability condition (6.14) in form:

$$p \equiv \cos\alpha^* - 2\sin\alpha^*\cot 2(\alpha^* - q) > 0 \qquad (6.15)$$

Fig. 6.2 shows the dependencies $\beta(\alpha^*)$ (solid lines) and $\eta'(\alpha^*)$ (dashed lines) defined respectively by formulas (6.13) and (6.15) with $q = -0.286$.

We can see that with $0 < \beta < 0.831$, i.e. for relatively non-intensive vibrations of the axis of pendulum, there are two solutions, one of them is stable and another is unstable (line I; sign "+" marks stable solution and sign "-" marks unstable solution). With $\beta > 0.831$ we have solutions, two of them are stable and two are unstable (line II).

The calculations show a good accordance of the obtained approximate solution and the direct obtained solution of the initial equation (6.1) with the same values of parameters. In particular, the mean component of the solution of the initial equation in the domain of stability coincide practically with the finding slow component.

If the gravity force mgl is small relative to the vibrational moment B, or of theoscillations occur in a horizontal plane, then the equation (6.13) has four essentially different solutions

$$\alpha_1^* = q, \quad \alpha_2^* = q + \pi, \quad \alpha_3^* = q + \frac{\pi}{2}, \quad \alpha_4^* = q + \frac{3}{2}\pi \qquad (6.16)$$

(other solutions are different by $2\pi n$). The first two solutions are in this case, according to (6.14), answered by the asymptotically stable positions of the quasi-equilibrium of the pendulum, while the third and the fourth solutions – by the unstable positions.

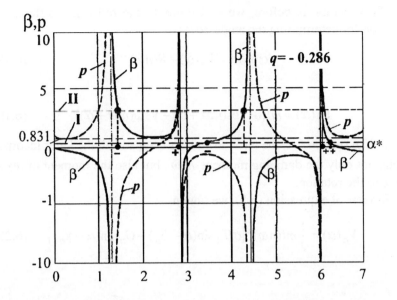

Figure 6.2. To the determination of the quasi-equilibrium positions of the pendulum and its stability.

It is easy to make sure that in the case of harmonic oscillations of the axis of the pendulum, when

$$H_1 = H, \quad H_2 = H_3 = ... = 0, \quad G_1 = G, \quad G_2 = G_3 = ... = 0, \quad \gamma_1 = \theta,$$

the obtained results coincide with those found in the book [5].

The pendulum is "attracted" to the major semiaxes of the elliptic (in this case!) trajectory of oscillations of the axis.

6.3 Regimes of Rotation

When investigating such regimes we will assume that

$$\varphi = \sigma[n\omega t + \alpha(t) + \psi(t, \omega t)] \qquad (6.17)$$

where $\alpha(t)$ is the main slow component as before, and ψ is the fast component, σ is equal to 1 or -1, depending on the direction of the rotation under investigation, and n is a positive integer.

Then, acting as before, we will come to the following differential equation for determining the angle $\alpha(t)$:

$$I\ddot{\alpha} + h\dot{\alpha} = V_n(\alpha) - R(n\omega) \tag{6.18}$$

where

$$V_n(\alpha) = ml\langle \sigma \ddot{x} \cos(n\omega t + \alpha) - \ddot{y} \sin(n\omega t + \alpha) \rangle \tag{6.19}$$

and unlike section 6.2, the term $R(n\omega)$ has been added, the latter incorporating not only the damping moment $hn\omega$, but also the moment of resistance to the rotation.

In view of expression (6.5), we obtain

$$V_n(\alpha) = \frac{1}{2} ml(n\omega)^2 [\sigma H_n \sin(\alpha - \beta_n) - G_n \sin(\alpha - \gamma_n)] \tag{6.20}$$

At

$$\beta_n = 0, \quad \gamma_n = \theta, \quad H_n = H, \quad G_n = G \text{ and } \alpha = \alpha_1 - \sigma \frac{\pi}{2}, \quad n = 1, \quad l = \varepsilon_1$$

this formula coincides with the one received in [5] for the case of simple harmonic oscillations of the rotor axis when the trajectory of oscillations has the form of an ellipse. As we see, in the general case of periodic oscillations, every harmonic component whose frequency is $n\omega$ can be answered by a stationary rotation of the rotor with the mean angular velocity $\langle \dot{\varphi} \rangle = \pm n\omega$.

The value of the angle $\alpha = \alpha^*$ in every such regime is determined from the equation

$$V_n(\alpha^*) = R(n\omega) \tag{6.21}$$

According to (6.20) there may be two such solutions, but only that motion will be stable for which

$$V'_n(\alpha^*) < 0 \tag{6.22}$$

Thus, for the existence o this regime equation (6.21) must allow real solutions.

The same refers to the condition

$$[V_n(\alpha)]_{\max} > R(n\omega) \tag{6.23}$$

6.3. Regimes of Rotation

According to what was said in [2], the latter condition can be brought to the following simple form

$$F_n A_n > R(n\omega) \tag{6.24}$$

where

$$F_n = ml(n\omega)^2, \quad A_n = \frac{1}{2}(a_n + \sigma^* b_n), \tag{6.25}$$

a_n and b_n being respectively the major and the minor semiaxis of the elliptical trajectory of oscillations, and the number $\sigma^* = 1$ if one considers the rotation of the rotor in the direction of motion of its axis along the same trajectory, and in the opposite case $\sigma^* = -1$.

So, conditions of the existence of rotation of the rotor with the frequency $n\omega$ depend in the above approximation only on the presence of the n-th harmonic in the oscillations of its axis. (In other words, in the above approximation the rotation of the rotor with the frequency $n\omega$ can take place only in case there is the n-th harmonic in the oscillations of its axis.) In principle, regimes of rotation with the mean frequencies $\langle \dot\varphi \rangle = (p/q)\omega$, where p and q are integers, are also possible [5]; the interaction of the harmonic components of oscillations also takes place; and the chaotic motions can also be possible. Calculations show, however, that the practical meaning of these circumstances is inessential.

Like result can be obtained in the case when instead of a pendulum we have a roller or a ball, rolled in a cavity [5].

References

1. Stephenson A., On a new type of dynamic stability, *Mem. Proc. Manch. Lit. Phil. Soc.* **52** (1908) 1–10.
2. Stephenson A., On induced stability, *Phylosophical Magazine* **15** (1908) 233–236.
3. Kapitsa P. L, Pendulum with vibrating axis of suspension, *Uspekhi fizicheskich nauk*, **44**, 1 (1954) 7—20 (in Russian).
4. Blekhman I. I, The action of vibration on mechanical systems, *Vibrotekhnika*, Issue 3(20) (Mintis, Vilnus, 1973) 369–373 (in Russian).
5. Blekhman I. I., *Vibrational Mechanics*. (World Scientific, Singapore, 2000), p. 509.
6. Dresig, H., Blekhman, I., Rodionov, P., Use of an additional pendulum for balancing the mechanisms, *Actual Problems in Mechanics. Proc. of the XXVIII Summer School*, St. Petersburg, Russia, June 1–10, (2000) 187–197.

Chapter 7

Non-Trivial Effects of High-Frequency Excitation for Pendulum Systems

J. S. Jensen[1], J. J. Thomsen[1], D. M. Tcherniak[2]

[1]*Department of Mechanical Engineering, Solid Mechanics, Technical University of Denmark, Building 404, DK-2800 Lyngby, Denmark*

[2]*Brüel & Kjær Sound & Vibration A/S, Skodsborgvej 307, DK-2850 Nærum, Denmark*

7.1 Preliminary Remarks

Recently there has been renewed interest in using high-frequency vibrations for controlling the low-frequency properties of structures, e.g. their equilibrium states, stability, effective natural frequencies, and vibration amplitudes [5, 6, 7, 12, 15, 16, 17, 20, 25, 26, 27, 28, 29, 30, 34, 40, 51, 52, 53, 55, 58, 59, 60, 61].

Several of these works consider mechanisms of the pendulum type. Pendulum type systems are important as simplified idealizations of many kinds of real systems, but are also relevant in their own right due to the relative ease by which they can be realized experimentally.

In the following three main sections we present examples of the work that we have done on pendulum systems subjected to high-frequency vibrations:

Chelomei's Pendulum, an extension of the well-known upright pendulum on a vibrating support, serves as a simple device for introducing some important effects of high-frequency excitation of pendulum type systems. Below we describe its seemingly paradoxical behaviour that has only recently been fully explained.

Then we turn to double pendulums subjected to follower-type forces. The elastic double pendulum with a partial follower-force is considered with its base of support subjected to high-frequency vibration in two perpendicular directions. The effect of high-frequency excitation is analysed of with respect to stability of equilibrium positions and nonlinear dynamics including chaos. Lastly, we present a study of a double pendulum consisting of two rigid pipes conveying fluid pulsating with high frequency. This study further sheds light on the effect of high-frequency vibration here stemming from a completely different source.

Throughout a common method of analysis is applied: the Method of Direct Partition of Motion, MDPM (see chapter 2 of this book and [7])*, which in a number of studies has proven a convenient tool for obtaining useful approximate solutions for systems subjected to high-frequency, non-resonant excitation. The following examples show how the MDPM facilitates the analyses of the dynamical behaviour of systems with high-frequency excitation.

7.2 Chelomei's Pendulum – Resolving a Paradox

7.2.1 Introduction

In 1982 former Russian chief rocket designer V.N. Chelomei described an interesting experiment. It involved a pendulum with a vertically

vibrating support and a sliding disk at its rod, similar to the one in Figure 7.1(a). With the support vibrating at high frequency and small amplitude, the pendulum was observed pointing upwards, against gravity – while the disk floated almost at rest somewhere along the rod [13]. Chelomei's

---

* Note of the editor: MDPM (Method of Direct Partition of Motions) is the same as MDSM (Method of Direct Separation of Motions). Unlike the other chapters of the book, this chapter retains the terminology adopted in the works of the Danish group of researches "Fast Vibrational Club".

own explanation of this paradoxical phenomenon has been subjected to much controversy. Below we summarise in some detail a recent study [62], which provides a coherent theory that explains the phenomenon, and gives quantitative predictions that are supported by numerical simulation and laboratory experiments.

7.2.2 History of the Problem and the Controversy

Chelomei, in an interesting popular description [13] of several paradoxical phenomena, presents two equations of motions for the pendulum (one for the pendulum swings and one for the disk position), and states that "...it is easy to find the quasi-static solutions of these [averaged] equations that give the values of the equilibrium points ...", and "A determination of stability ... presents no special difficulties", and "These ... results were checked ... on a computer ... results are in close agreement with the experiment". No details were given of the experimental set-up, or of the theoretical or numerical analysis. In the following years other groups attempted to provide the details to support the theoretical part of Chelomei's conclusion, though without success. Using different techniques, Blekhman and Malakhova [8], Menyailov and Movchan [39], and Kirgetov [33] all reached identical conclusions: The disk cannot stabilize on the inverted rod of a uniform rigid pendulum with purely vertical support excitation of small amplitude (see also [7]), and thus Chelomei's rigid pendulum model cannot explain the experimental observations.

One study [35] claims to confirm Chelomei's conclusions, by using numerical integration of Chelomei's model equations. However, using identical parameters and two different well-proven commercial numerical differential equation solvers, we have been unable to reproduce these numerical results. Even if this were possible, the parameter values used certainly do not describe Chelomei's experimental pendulum: In [35] the sliding disk has a mass of 0.4 mg and the pendulum has a length of 100 mm and a mass of 40 mg, i.e. if in the form of a circular steel rod, the diameter would be only 0.25 mm. The base of this very thin and supposedly rigid pendulum is vibrated at amplitude 10 mm and frequency 2364 Hz; this corresponds to a peak support acceleration of about 225,000 times the acceleration of gravity.

Thus, there is no convincing evidence that Chelomei's 2-dof rigid pendulum model can explain the phenomenon observed; in fact there is substantial evidence that it cannot. The problem then is to suggest the simplest possible model that will explain the phenomenon, that is, to pinpoint those physical mechanisms that are responsible for balancing both the pendulum rod and the sliding disk against gravity. Here Blekhman and Malakhova [8] have shown that the presence of flexural vibrations of the rod might be responsible for stabilization of the disk. By a nonlinear interaction mechanism, such vibrations may create a "wave pressure" pushing the sliding mass towards the antinodes of vibrations, i.e. towards the point of maximum deflection. (It was later proposed to utilise this nonlinear effect for vibration damping, see [1] and [55, 56, 57]). However, Blekhman and Malakhova do not suggest which mechanism should cause such vibrations in the first place, and consequently no predictions are given for the stable states of the system in dependence of the physical parameters of the pendulum and the excitation.

Then in [62] we showed how a small symmetry breaking imperfection, such as a slight deviation from perfect vertical excitation, can provide a mechanism for generating stationary flexural rod vibrations. We also confirmed experimentally that such vibrations exist and are pronounced under conditions where the disk floats. Also, using perturbation analysis, analytical expressions for the prediction of stable states of the system were set up, providing frequency responses that agreed closely with numerical simulation, and agreed qualitatively with experimental observations. Next we describe the experiments, the theoretical model, the perturbation analysis of the nonlinear system, and the comparison of experimental observations and theoretical predictions.

7.2.3 Experimental Observations

No data of Chelomei's original experiment or similar experiments were available, other than the photograph in [13]. Further, there was no theory for making predictions of what to expect. On that basis we initiated experiments using *Experimental Pendulum 1*, to be described below, which has more or less arbitrary data. The results obtained triggered theoretical developments, which in turn called for another set-up, *Experimental Pendulum 2*. Different types of results were obtained using these pendulums, so we report on both of them.

Experimental pendulum 1A

Experimental Pendulum 1A (Figure 7.1(a)) consists of a steel rod (Ø3×150 mm; "Ø" denoting diameter) with a plastic disk (Ø12.2/Ø3×5.3 mm, 0.8 grams) that can slide freely with little slip. The pendulum hinge is attached to a driven crank mechanism, oscillating vertically or almost vertically with fixed amplitude (2 mm) and adjustable frequency (0–230 Hz). The natural frequency of free pendulum oscillations was measured to 1.5 Hz, the fundamental frequency for flexural vibrations in the pendulum plane to 400 Hz, and the frequency of out-of-plane vibrations to 96 Hz.

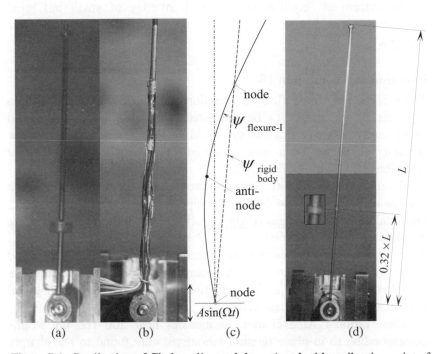

Figure 7.1. Realisation of Chelomei's pendulum: A rod with a vibrating point of support and a disk that can slide along the rod. (a,b) Experimental pendulums 1A and 1B; (c) Two fundamental mode shapes for in-plane motions; (d) Experimental pendulum 2, having a very small and light disk; the support is oscillating at frequency ≈ 158 Hz, tilted $\phi = 0.034 \approx 2°$ with respect to vertical, and the disk floats about one third up the rod. From Thomsen & Tcherniak (2001)[62].

The following observations were made: Increasing the frequency of support oscillations from zero, near 200 Hz the pendulum could be stabilized in a nearly inverted position with the disk floating 30–40 mm above the support (Figure 7.1(a)). This disk position is about 10 mm below the vibration node of the fundamental hinged-free flexure mode (*cf.* Figure 7.1(c)). As observed also by Chelomei, the disk was not absolutely at rest, but performed small but rapid oscillations about a "fixed" (or at least slowly changing) position. It required precise tuning and frequent re-tuning of the excitation frequency to make the disk float, and the results were hard to reproduce consistently under seemingly similar conditions. The response of the disk and the pendulum was further observed to be chaotically modulated, that is, the orderly picture of the system at "equilibrium" with an overlay of small but rapid oscillations, was further overlaid by minor seemingly unpredictable changes.

Experimental pendulum 1B

The presence of strong flexural vibrations of the pendulum rod plays a key role in the theory to be presented. One observation that could immediately disprove this theory would be the presence of only weak flexural vibrations, or a lack of coincidence between such flexural vibrations and floating of the disk.

To measure possible flexural vibrations, an exact copy of Pendulum 1A was made, but with no disk, and equipped instead with four pairs of 1 mm miniature strain gauges (Figure 7.1 (b)). Two of these pairs measure dynamic strains corresponding to in-plane flexural vibrations, whereas the other two measure out-of-plane vibrations. Each pair is mounted and wired so that only strains corresponding to flexural deformations are recorded, whereas there is negligible sensitivity to longitudinal deformations and temperature changes.

Observations: At excitation frequencies near 200 Hz, the strains corresponding to in-plane flexural vibrations were found to rise sharply to about 500 µstrain, with a main component at double the value of the excitation frequency, $\approx$ 400 Hz, which happens to be also the resonance frequency for fundamental in-plane flexural vibrations of the rod. Strains of this magnitude must be considered quite large, with 500 µstrain corresponding to maximum surface stresses of 105 MPa, or about one third of the yield stress of the rod material. However, the corresponding deformations are still small, about 0.1 mm, and thus almost invisible by the naked eye. Based on a series of such measurements, and comparing with the results obtained with pendulum 1A, it was concluded that the disk floats on the nearly inverted rod for excitation frequencies that causes strong resonant flexural vibrations in the rod. For other frequencies the disk just bounces repeatedly against a rod end.

Experimental pendulum 2

Experimental Pendulum 2 was build after the significance of flexural vibrations had become clear, and after simple analytical expressions had been derived for predicting stationary responses. This pendulum has a steel rod ∅2×200 mm of mass 5 grams (totally 17.6 grams including a bob and 2 nuts), and a very small plastic disk ∅2/∅3×2 mm of mass 0.01 grams. The natural frequency of free pendulum oscillations is 1.37 Hz, the two lowest fundamental frequencies for flexural vibrations in the pendulum plane are 158 Hz and 512 Hz, and those for out-of-plane vibrations are 36 Hz and 225 Hz. The rest of the set-up was unchanged.

This pendulum differs from Pendulum 1 in two key points: First the natural frequency for the fundamental in-plane flexural mode is lower, so that it can be excited directly by the available equipment. This contrasts to pendulum 1, which can only have this mode excited by secondary mechanisms because the relevant natural frequency is larger than the maximum excitation frequency of the equipment used. Second, the mass of the disk was made so small that its influence on motions of the pendulum rod could be considered a second order effect. With this small mass the theoretical predictions become so simple that it is possible to gain insight into the mechanisms at work (and after all we consider this the whole purpose of Chelomei's pendulum: to illustrate an interesting phenomenon in the simplest possible setting). Finally, the direction of excitation was tilted a slight angle from vertical, by sloping the base plane of the apparatus and measuring the inclination with a digital level meter.

Observations: The mass stabilized at the nearly inverted pendulum when the excitation frequency is about 160 Hz, i.e. near the natural frequency for fundamental in-plane flexural vibrations of the rod. Figure 7.1(d), shows the system in a stabilized position. The frame lines of this picture are horizontal / vertical, the direction of excitation is tilted 2° with respect to vertical, and the pendulum stabilizes in a position 5° from vertical with the disk about one third up the rod. However, typically this state would be slowly changing, so that the rod and the disk could be observed in other positions. Careful frequency tuning was required, chaotic modulation of the orderly response was observed, and consistent results were hard to reproduce. Increasing the excitation frequency slightly, the disk ceased to float. *Decreasing* the frequency slightly, the rod started to vibrate violently in the fundamental flexural mode (clearly visible by eye), the disk approached the antinode of that mode, the pendulum started to oscillate wildly in the rigid body mode, and the experiment had to be stopped to prevent damage.

7.2.4 A Theoretical Model

Figure 7.2 (a) illustrates Chelomei's original pendulum model, and Figure 7.2 (b) the model to be considered in this study. Our model differs from Chelomei's by the capability of the rod to flex, and by a slight offset from vertical of the direction of excitation. It will appear that including both factors make the model capable of explaining and predicting the phenomenon under consideration.

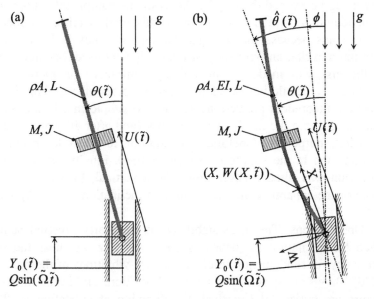

Figure 7.2. (a) Chelomei's pendulum: A rigid rod with a vertically oscillating support and a sliding disk; (b) The system considered in this work: Chelomei's pendulum with a flexible rod, and a slight deviation from perfect vertical support oscillation. From Thomsen & Tcherniak [62].

7.2. Chelomei's Pendulum – Resolving a Paradox

In Figure 7.2 (b) $U(\tilde{t})$ denotes the position at time $\tilde{t}$ of a solid disk having mass M and rotary inertia J, sliding without slip along the rod. The rod is uniform with length L and mass per unit length ρA. It is assumed to be flexible with bending stiffness EI, where E is Young's modulus and I is the area moment of inertia. The instantaneous state of the rod is described by a rigid body rotation $\theta(\tilde{t})$ with respect to a vertical gravity field g, and a plane flexural deformation $W(X,\tilde{t})$, where X is the length-wise coordinate along the rigid body axis. The supporting hinge of the rod performs prescribed oscillations, $Y_0(\tilde{t}) = Q\sin(\tilde{\Omega}\tilde{t})$, along a line that is tilted a small angle ϕ from vertical; here the amplitude of the excitation is very small, $Q \ll L$, whereas its frequency is very large, $\tilde{\Omega} \gg \sqrt{g/L}$.

We assume here that the mass M is sufficiently small for the rod to vibrate in a single dominant mode $\psi(X)$ with instantaneous amplitude $V(\tilde{t})$, that is

$$W(X,\tilde{t}) = V(\tilde{t})\psi(X) \tag{7.1}$$

This assumption seems justified when the disk mass is small and stationary conditions have been reached, as with Experimental Pendulum 2 whose rod was clearly vibrating in a stationary spatial shape $\psi(X)$.

Setting up expressions for the kinetic and potential energies and using Lagrange's equations with a Rayleigh dissipation function, one obtains three equations governing, respectively, rigid body rod motions $\theta(\tilde{t})$, flexural rod deformations $V(\tilde{t})$, and disk motions $U(\tilde{t})$. In nondimensional form they become, for the rigid body motions θ:

$$\left[1 + \gamma + \alpha u^2 + \left(3 + \alpha\psi_u^2\right)v^2\right]\ddot{\theta} + \left[2\beta_0 + 2\alpha\left(u + \psi_u\psi_u'v^2\right)\dot{u} + \left(6 + 2\alpha\psi_u^2\right)v\dot{v}\right]\dot{\theta}$$
$$- (1 + \tfrac{2}{3}\alpha u)(\sin\theta + \tfrac{3}{2}\ddot{y}_0\sin(\theta - \phi)) - (2\psi_{001} + \tfrac{2}{3}\alpha\psi_u)v(\cos\theta + \tfrac{3}{2}\ddot{y}_0\cos(\theta - \phi))$$
$$+ 2\beta_0\psi_0'\dot{v} + \alpha\left[\psi_u\left(u\ddot{v} - \ddot{u}v\right) + \psi_u'\left(2u\dot{u}\dot{v} + u\ddot{u}v\right) + \psi_u''u\dot{u}^2v\right]$$
$$+ \gamma\left[\psi_u'\ddot{v} + \psi_u''\left(2\dot{u}\dot{v} + \ddot{u}v\right) + \psi_u'''\dot{u}^2v\right] = 0$$

$$\tag{7.2}$$

for the nondimensional modal coefficient describing rod flexure v:

$$\left[1+\tfrac{1}{3}\alpha\psi_u'^2 +\tfrac{1}{3}\gamma\psi_u''^2\right]\ddot{v}+\left[2\beta_r\omega+\tfrac{2}{3}(\alpha\psi_u+\gamma\psi_u'')\psi_u'\dot{u}\right]\dot{v}$$
$$+\left[\omega^2-(1+\tfrac{1}{3}\alpha\psi_u^2)\dot\theta^2+\tfrac{1}{3}(\alpha\psi_u+\gamma\psi_u'')\psi_u'\ddot{u}+\tfrac{1}{3}(\alpha\psi_u\psi_u''+\gamma\psi_u'\psi_u''')\dot{u}^2\right]v$$
$$+\tfrac{1}{3}(\alpha\psi_u u+\gamma\psi_u')\ddot\theta+\tfrac{2}{3}(\alpha\psi_u\dot u+\beta_0\psi_0')\dot\theta$$
$$-\tfrac{2}{3}(\psi_{001}+\tfrac{1}{3}\alpha\psi_u)(\sin\theta+\tfrac{3}{2}\ddot y_0 \sin(\theta-\phi))=0 \tag{7.3}$$

and for nondimensional disk motions u:

$$\left[\alpha(1+\psi_u'^2 v^2)+\gamma\psi_u''^2 v^2\right]\ddot u+\left[2\alpha\beta_d+2(\alpha\psi_u'^2+\gamma\psi_u''^2)v\dot v+(\alpha\psi_u'+\gamma\psi_u''')\psi_u'v^2\dot u\right]\dot u$$
$$+\alpha(\psi_u'\ddot\theta v-\dot\theta^2)u+\tfrac{2}{3}\alpha\left[(\cos\theta+\tfrac{3}{2}\ddot y_0\cos(\theta-\phi))-\psi_u'v(\sin\theta+\tfrac{3}{2}\ddot y_0\sin(\theta-\phi))\right]$$
$$-(\alpha\psi_u-\gamma\psi_u'')\ddot\theta v+(\alpha\psi_u+\gamma\psi_u'')\psi_u'v\ddot v-\alpha\psi_u\dot\theta(2\dot v+\psi_u'\dot\theta v^2)=0 \tag{7.4}$$

with the external kinematic excitation given by

$$y_0(t)=q\sin(\Omega t) \tag{7.5}$$

where $(\cdot)\equiv d/dt$, and the nondimensional variables and parameters are:

$$x=X/L,\quad u=U/L,\quad v=V/L,\quad y_0=Y_0/L,\quad t=\omega_0\tilde t,$$
$$\omega_0=\sqrt{\tfrac{3}{2}g/L},\quad \Omega=\tilde\Omega/\omega_0,\quad \omega=\sqrt{\psi_{022}EI/\rho AL^4}/\omega_0,$$
$$q=Q/L,\quad \alpha=M/(\tfrac{1}{3}\rho AL),\quad \gamma=J/(\tfrac{1}{3}\rho AL^3), \tag{7.6}$$
$$2\beta_0=3c_0/\omega_0,\quad 2\beta_r=\left(\psi_{042}c_r+\psi_0'^2 c_0\right)/\sqrt{\psi_{022}EI/\rho AL^4},\quad 2\beta_d=c_d/\omega_0,$$
$$\psi_u=\psi(u),\quad \psi_0=\psi(0),\quad \psi_{ijk}=\int_0^1 x^i\left(d^j\psi/dx^j\right)^k dx,\quad \psi_{002}=1$$

where c_d is the coefficient of viscous damping for the disk sliding on the rod, c_0 is the coefficient of viscous damping of the hinge, and c_r is the coefficient of internal damping accompanying flexural vibrations of the rod. Note that ω_0 is the linear natural frequency for small rigid body oscillations of the rod when there is no disk, Ω describes the frequency of the excitation and ω the natural frequency for small flexural vibrations of the rod, α and γ represent, respectively, disk mass and rotary inertia, q denotes the amplitude of support motions, and ψ_u, ψ_0, and ψ_{ijk} are determined by the mode shape $\psi(x)$, which is normalised so that $\psi_{002}=1$.

For the nondimensional function $\psi(x)$ we use a single mode shape for a hinged-free slender beam of unit length (e.g. [18], p. 61-9):

7.2. Chelomei's Pendulum – Resolving a Paradox

$$\psi = G_1\left(\kappa(1-x)\right) - \frac{G_1(\kappa)}{G_2(\kappa)} G_2\left(\kappa(1-x)\right) \qquad (7.7)$$

where $G_1 = \cosh u + \cos u$ and $G_2 = \sinh u + \sin u$, and κ is a solution to the frequency equation $\tan \kappa = \tanh \kappa$. Table 7.1 lists values of κ and the relevant mode shape integrals as given by (7.7) and the expression for ψ_{ijk} in (7.6) for the lowest three modes. Here $\psi_{002} = 1$ by normalisation, and one can show that ψ_{101} vanishes identically for all modes ($\psi_{002} = 1$ and $\psi_{101} = 0$ has been inserted in (7.2)–(7.4).

Table 7.1. Mode shape coefficients for a hinged-free slender beam. From Thomsen & Tcherniak [62].

mode	κ	ψ_{001}	ψ_{101}	ψ_{002}	ψ_{022}	ψ_{042}	ψ_0'
1	3.9266	−0.3703	0	1	0.2377×10^3	56.51×10^3	−5.400
2	7.0686	0.1998	0	1	2.496×10^3	6.233×10^6	10.01
3	10.210	−0.1385	0	1	10.87×10^3	118.1×10^6	−14.44

Equations (7.2)–(7.4) with (7.5) are the ordinary differential equations governing motions of the system under the assumptions stated. They are nonlinear, and further coupled by linear and nonlinear terms; in the following they will be termed the "full" equations of motions.

7.2.5 Predicting Stationary Solutions

Deriving equations governing the slow components of motion

We here predict stationary solutions for the full equations of motion (7.2)–(7.5). Closed form solutions are unavailable, since the equations are highly nonlinear. Approximate perturbation methods can then be applied, defining the reciprocal of the (supposedly high) excitation frequency as a small parameter of a perturbation problem. Rather than using standard perturbation schemes such as Averaging or Multiple Scales (e.g. see chapter 4 of this book and [41, 57, 37]) we use the so-called Method of Direct Partition of Motions (see chapter 4 of this book and [7, 25, 27]). With this method the solutions are assumed to consist of slowly changing components overlaid (as opposed to multiplied / modulated) by rapidly oscillating motions. It is especially well suited for problems as the present, characterised by high-frequency excitation, and is then typically easier to apply than more traditional perturbation approaches. To use it we split up motions of the system in slow and fast components:

$$\theta(t) = \theta_0(t) + \varepsilon\theta_1(t,\tau)$$
$$v(t) = \varepsilon v_1(t,\tau) \qquad (7.8)$$
$$u(t) = u_0(t) + \varepsilon u_1(t,\tau)$$

where $\tau = \Omega t$ is the fast time, $\varepsilon = \Omega^{-1} \ll 1$ is a small parameter, θ_0 and u_0 hold the slow components of θ and u, respectively, and $\varepsilon\theta_1$, εv_1, and εu_1 are rapidly oscillating overlays of small amplitude, which are 2π-periodic in the fast time τ and have zero fast-time average, i.e.:

$$\langle\theta_1(t,\tau)\rangle = \langle v_1(t,\tau)\rangle = \langle u_1(t,\tau)\rangle = 0; \quad \langle f(t,\tau)\rangle \equiv \frac{1}{2\pi}\int_0^{2\pi} f(t,\tau)d\tau \quad (7.9)$$

where $\langle\ \rangle$ defines the "short time" averaging operator (it is understood that the integration is carried out considering the slow time t as a constant. The assumptions on the relative orders of magnitude of the terms in (7.8) can (and was) be checked a posteriori. By these definitions one has:

$$\langle\theta\rangle = \theta_0(t); \quad \langle v\rangle = 0; \quad \langle u\rangle = u_0(t); \qquad (7.10)$$

so that θ_0 and u_0 describe the short-time averages of θ and u. A slow component v_0 could be included, however, it turns out to be negligible. Inserting this and (7.8) into (7.2)–(7.5) one obtains, on Taylor-expanding the trigonometric terms and the ψ_u-terms for small ε, for the rigid body rotations θ of the pendulum rod:

$$\varepsilon^{-1}\{\theta_1'' + \tfrac{3}{2}q\Omega\sin(\theta_0 - \phi)\sin\tau\}$$
$$+ \varepsilon^0\{\ddot{\theta}_0 + 2\dot{\theta}_1' + \overline{\alpha}(u_0^2 + \overline{\gamma})\theta_1'' + 2\beta_0(\dot{\theta}_0 + \theta_1') - \sin\theta_0$$
$$+ \left[\tfrac{3}{2}(\theta_1 + 2\psi_{001}v_1)\cos(\theta_0 - \phi) + \overline{\alpha}u_0\sin(\theta_0 - \phi)\right]q\Omega\sin\tau \qquad (7.11)$$
$$+ 2\beta_0\psi_0'v_1' + (\overline{\alpha}\psi_{u_0}u_0 + \overline{\gamma}\psi_{u_0}')v_1''\} + O(\varepsilon) = 0$$

for the flexural vibrations v of the pendulum rod:

7.2. Chelomei's Pendulum – Resolving a Paradox

$$\varepsilon^{-1}\left\{v_1'' + 2\beta_r\bar{\omega}v_1' + \bar{\omega}^2 v_1 + \psi_{001} q\Omega \sin(\theta_0 - \phi)\sin\tau\right\}$$

$$+ \varepsilon^0 \left\{2\dot{v}_1' + \tfrac{1}{3}\bar{\alpha}(\psi_{u_0}^{\prime\,2} + \hat{\gamma}\psi_{u_0}^{\prime\prime\,2})v_1''\right.$$

$$+ 2\beta_r\bar{\omega}\dot{v}_1 + \tfrac{2}{3}\beta_0\psi_0'(\dot\theta_0 + \theta_1') + \tfrac{1}{3}\bar\alpha(\psi_{u_0}u_0 + \hat\gamma\psi_{u_0}')\theta_1'' - \tfrac{2}{3}\psi_{001}\sin\theta_0 \qquad (7.12)$$

$$\left. + \left[\psi_{001}\theta_1\cos(\theta_0-\phi) + \tfrac{1}{3}\bar\alpha\psi_{u_0}\sin(\theta_0-\phi)\right]q\Omega\sin\tau\right\} + O(\varepsilon) = 0$$

and for the instantaneous position u of the disk:

$$\varepsilon^0\left\{u_1'' - q\Omega\cos(\theta_0-\phi)\sin\tau\right\} + \varepsilon^1\left\{\ddot u_0 + 2\dot u_1' + 2\beta_d(\dot u_0 + u_1') + \psi_{u_0}'\theta_1'v_1 u_0\right.$$

$$- \left(\dot\theta_0^2 + \theta_1'^2 + 2\dot\theta_0\theta_1'\right)u_0 + \tfrac{2}{3}\cos\theta_0 + \left(\theta_1 + \psi_{u_0}'v_1\right)q\Omega\sin(\theta_0-\phi)\sin\tau$$

$$\left. - \left(\psi_{u_0} - \hat\gamma\psi_{u_0}''\right)\theta_1'v_1 + \left(\psi_{u_0} + \hat\gamma\psi_{u_0}''\right)\psi_{u_0}'v_1 v_1'' - 2\psi_{u_0}(\dot\theta_0 + \theta_1')v_1'\right\} + O(\varepsilon^2) = 0$$

$$(7.13)$$

where now $(\dot{}) \equiv \partial/\partial t$ and $()' \equiv \partial/\partial\tau$, and where $\bar\omega \equiv \varepsilon\omega = \omega/\Omega = O(1)$, $\bar\alpha = \alpha/\varepsilon = O(1)$, and $\bar\gamma = \gamma/\varepsilon = O(1)$ to indicate that ω is large (of the same order as Ω), α and γ are small, and $\hat\gamma = \bar\gamma/\bar\alpha$. It is further assumed that $q\Omega = O(1)$, which since $\Omega = \varepsilon^{-1} \gg 1$ means that $q = O(\varepsilon)$, i.e. the amplitude of kinematic excitation is small.

We first determine approximations to the fast components of motion (θ_1, v_1, u_1) by considering only the dominating terms of (7.11)–(7.13), i.e. we consider:

$$\theta_1'' = -\tfrac{3}{2}q\Omega\sin(\theta_0-\phi)\sin\tau + O(\varepsilon)$$

$$v_1'' + 2\beta_r\bar\omega v_1' + \bar\omega^2 v_1 = -\psi_{001}q\Omega\sin(\theta_0-\phi)\sin\tau + O(\varepsilon) \qquad (7.14)$$

$$u_1'' = q\Omega\cos(\theta_0-\phi)\sin\tau + O(\varepsilon); \quad \bar\alpha \neq 0$$

This is a system of linear partial differential equations in the two time scales t and τ, however, with all time derivatives being with respect to the fast time τ. Hence the stationary solution can readily be obtained (employing the conditions in (7.9) to eliminate the constants of integration):

$$\theta_1 = \tfrac{3}{2}q\Omega\sin(\theta_0-\phi)\sin\tau + O(\varepsilon)$$

$$v_1 = -\psi_{001}q\Omega R(\bar\omega,\beta_r)\sin(\theta_0-\phi)\sin(\tau-\chi) + O(\varepsilon) \qquad (7.15)$$

$$u_1 = -q\Omega\cos(\theta_0-\phi)\sin\tau + O(\varepsilon)$$

where R is a *resonance function* for the pendulum rod and χ the phase shift:

$$R(\bar{\omega}, \beta_r) = \left((\bar{\omega}^2 - 1)^2 + (2\beta_r \bar{\omega})^2\right)^{-1/2}; \quad \tan\chi = 2\beta_r \bar{\omega}/(\bar{\omega}^2 - 1) \quad (7.16)$$

To find the corresponding solutions for the slow components (θ_0, u_0), we start by averaging (7.11) and (7.13) in fast time, i.e. the $\langle\ \rangle$-operator is applied to all terms, and (7.9)–(7.10) are utilised to obtain:

$$\ddot{\theta}_0 + 2\beta_0 \dot{\theta}_0 - \sin\theta_0 + q\Omega\left(\tfrac{3}{2}\langle\theta_1 \sin\tau\rangle + 3\psi_{001}\langle v_1 \sin\tau\rangle\right)\cos(\theta_0 - \phi) = O(\varepsilon)$$

$$\ddot{u}_0 + 2\beta_d \dot{u}_0 + \left(\psi'_{u_0}\langle\theta'_1 v_1\rangle - \dot{\theta}_0^2 - \langle\theta_1'^2\rangle\right)u_0$$
$$+ q\Omega\left(\langle\theta_1 \sin\tau\rangle + \psi'_{u_0}\langle v_1 \sin\tau\rangle\right)\sin(\theta_0 - \phi) + O(\varepsilon)$$
$$= -\tfrac{2}{3}\cos\theta_0 + \left(\psi_{u_0} - \hat{\gamma}\psi''_{u_0}\right)\langle\theta'_1 v_1\rangle - \left(\psi_{u_0} + \hat{\gamma}\psi''_{u_0}\right)\psi'_{u_0}\langle v_1 v'_1\rangle + 2\psi_{u_0}\langle\theta'_1 v'_1\rangle$$
(7.17)

Using the definition for $\langle\ \rangle$ in (7.7) to calculate averages, inserting the approximate results (7.15) and using (7.16), one arrives at the following equations governing the slow motions θ_0 and u_0:

$$\ddot{\theta}_0 + 2\beta_0 \dot{\theta}_0 - \sin\theta_0 + (1 + \omega_\theta^2)\sin(\theta_0 - \phi)\cos(\theta_0 - \phi) = O(\varepsilon)$$
$$\ddot{u}_0 + 2\beta_d \dot{u}_0 - \dot{\theta}_0^2 u_0 = f_1(\theta_0, u_0) + O(\varepsilon) \quad (7.18)$$

where

$$\omega_\theta^2(\Omega) \equiv \tfrac{9}{8}(q\Omega)^2\left(1 - \frac{\tfrac{4}{3}\psi_{001}^2(\bar{\omega}^2 - 1)}{(\bar{\omega}^2 - 1)^2 + (2\beta_r \bar{\omega})^2}\right) - 1$$

$$f_1(\theta_0, u_0) \equiv -\tfrac{2}{3}\cos\theta_0 - \tfrac{3}{4}(q\Omega)^2 \sin^2(\theta_0 - \phi)\left\{1 - \tfrac{3}{2}u_0\right.$$
$$\left. + \psi_{001}R^2\left[(\bar{\omega}^2 - 1)\left(\psi_{u_0} + (u_0 - \tfrac{2}{3})\psi'_{u_0} + \hat{\gamma}\psi''_{u_0}\right) - \tfrac{2}{3}\psi_{001}(\psi_{u_0} + \hat{\gamma}\psi''_{u_0})\psi'_{u_0}\right]\right\}$$
(7.19)

Here the function f_1 contains all the "static" forces (depending only on the state (θ_0, u_0)) that affect motions of the disk.

Specializing to the case of interest: $\theta_0 \approx \phi \approx 0$

For the phenomenon of concern the pendulum rod is observed close to vertical, and the direction of excitation is close to vertical, i.e. $\theta_0 \ll 1$ and

7.2. Chelomei's Pendulum – Resolving a Paradox

$\phi \ll 1$. It is reasonable then to include only the first term of the Taylor expansion of the trigonometric terms in (7.18). This gives, assuming $\theta_0^3 = O(\varepsilon)$ and $(\theta_0 - \phi)^3 = O(\varepsilon)$ and retaining only first order terms:

$$\ddot{\theta}_0 + 2\beta_\theta \dot{\theta}_0 - \theta_0 + (1+\omega_\theta^2)(\theta_0 - \phi) = 0$$
$$\ddot{u}_0 + 2\beta_d \dot{u}_0 - \dot{\theta}_0^2 u_0 = f_2(\theta_0, u_0) \tag{7.20}$$

where

$$f_2(\theta_0, u_0) \equiv -\tfrac{2}{3}(1 - \tfrac{1}{2}\theta_0^2) - \tfrac{3}{4}(q\Omega)^2(\theta_0 - \phi)^2\{1 - \tfrac{3}{2}u_0$$
$$+\psi_{001}R^2\left[(\bar{\omega}^2 - 1)\left(\psi_{u_0} + (u_0 - \tfrac{2}{3})\psi'_{u_0} + \hat{\gamma}\psi''_{u_0}\right) - \tfrac{2}{3}\psi_{001}(\psi_{u_0} + \hat{\gamma}\psi''_{u_0})\psi'_{u_0}\right]\}$$
$$\tag{7.21}$$

where the first term in f_2 originates from the action of gravity.

Stationary solutions: quasi-equilibria

Stationary solutions to the equations of slow motion (7.20) are found by equating to zero all velocities and accelerations and solving for θ_0 and u_0. It is found that the stationary rigid body angle of the pendulum rod is:

$$\theta_0(t) = \theta_{0s} = (1 + \omega_\theta^{-2})\phi = \left[1 - \tfrac{8}{9}(q\Omega)^{-2}\left(1 - \frac{\tfrac{4}{3}\psi_{001}^2(\bar{\omega}^2 - 1)}{(\bar{\omega}^2 - 1)^2 + (2\beta_r\bar{\omega})^2}\right)^{-1}\right]^{-1}\phi$$
$$\tag{7.22}$$

while the corresponding stationary position u_{0s} of the disk is governed by the following algebraic equation:

$$0 = (1 - \tfrac{1}{2}\theta_{0s}^2) + \tfrac{8}{9}(q\Omega)^2(\theta_{0s} - \phi)^2\{1 - \tfrac{3}{2}u_0$$
$$+\psi_{001}R^2\left[(\bar{\omega}^2 - 1)\left(\psi_{u_{0s}} + (u_{0s} - \tfrac{2}{3})\psi'_{u_{0s}} + \hat{\gamma}\psi''_{u_{0s}}\right) - \tfrac{2}{3}\psi_{001}(\psi_{u_{0s}} + \hat{\gamma}\psi''_{u_{0s}})\psi'_{u_{0s}}\right]\}$$
$$\tag{7.23}$$

where the first term describes the action of gravity. This equation is transcendental due to the mode shape related functions ψ. However, inserting the solution for θ_{0s} from (7.22), it can readily be solved numerically to find solutions for u_{0s} in the interval $[0;1]$.

Substituting these results back into (7.8), together with (7.15) and $\varepsilon = \Omega^{-1}$ and $\tau = \Omega t$, it is found that the rigid body rod angle and the disk position at stationarity is

$$\theta(t) = \theta_{0s} + O(\varepsilon) + \tfrac{3}{2} q \sin(\theta_{0s} - \phi)\sin(\Omega t) + O(\varepsilon^2; \Omega t)$$
$$u(t) = u_{0s} + O(\varepsilon) - q\cos(\theta_{0s} - \phi)\sin(\Omega t) + O(\varepsilon^2; \Omega t) \quad (7.24)$$

where $O(\varepsilon^2; \Omega t)$ denote very small terms that are rapidly oscillating. We term this state a *quasi-equilibrium*, since the system will appear "at rest", except for a rapidly oscillating overlay of very small amplitude (recall that $q \ll 1$ and $\Omega \gg 1$)[**].

A straight-forward analysis (*cf.* [62]) shows that the equilibrium $(\theta_0, u_0) = (\theta_{0s}, u_{0s})$ is asymptotically stable when $\omega_\theta^2 > 0$ and $\partial f_2/\partial u_0|_{(\theta_{0s}, u_{0s})} < 0$, i.e. when the following conditions are both fulfilled:

$$\tfrac{9}{8}(q\Omega)^2 \left(1 - \frac{\tfrac{4}{3}\psi_{001}^2(\bar{\omega}^2 - 1)}{(\bar{\omega}^2 - 1)^2 + (2\beta_r\bar{\omega})^2}\right) > 1 \quad (7.25)$$

$$\psi_{001} R^2 \left[(\bar{\omega}^2 - 1)\left(2\psi'_{u_{0s}} + (u_{0s} - \tfrac{2}{3})\psi''_{u_{0s}} + \hat{\gamma}\psi'''_{u_{0s}}\right)\right.$$
$$\left. - \tfrac{2}{3}\psi_{001}\left((\psi'_{u_{0s}} + \hat{\gamma}\psi'''_{u_{0s}})\psi'_{u_{0s}} + (\psi_{u_{0s}} + \hat{\gamma}\psi''_{u_{0s}})\psi''_{u_{0s}}\right)\right] > \tfrac{3}{2} \quad (7.26)$$

7.2.6 Analysing the Results

Before analysing and discussing the above results for a realistic case, we consider first some simple limit cases that will provide insight into the mechanisms at work.

Limit case I: rigid rod without disk

With a rigid rod and no disk one has $v(t) = \alpha = \gamma = 0$, $\bar{\omega} \to \infty$, $\psi = 0$, and (7.2)–(7.5) describe motions of a pendulum with distributed mass and near-vertical harmonic support excitation:

$$\ddot{\theta} + 2\beta_0\dot{\theta} - \sin\theta + \tfrac{3}{2}q\Omega^2 \sin(\Omega t)\sin(\theta - \phi) = 0 \quad (7.27)$$

The quasi-equilibrium near $\theta = 0$ for this case becomes, by (7.22)

$$\theta_{0s} = \left[1 - \tfrac{8}{9}(q\Omega)^{-2}\right]^{-1} \phi \quad (7.28)$$

[**] See the footnote at the beginning of chapter 10

7.2. Chelomei's Pendulum – Resolving a Paradox

which according to (7.25) is stable when

$$(q\Omega)^2 > \tfrac{8}{9} \tag{7.29}$$

With this condition fulfilled the pendulum rod performs small oscillations (*cf.* (7.24)) about an equilibrium position θ_{0s} that is set off from zero, i.e. from vertical. It further appears from (7.28)–(7.29) that stable angles θ_{0s} of the rod always exceed ϕ, the angle of direction of excitation. However, as $\Omega \to \infty$ then $\theta_{0s} \to \phi$, i.e. when the excitation frequency is increased the rod tends to line up with the direction of excitation.

According to Kapitza [32], and Panovko and Gubanova [47], the stability condition for the upright equilibrium of a pendulum with a vertically ($\phi = 0$) oscillating support is $(q\Omega)^2 > 2$ (see also chapter 4 of this book). This holds for a pendulum with a mass concentrated at the end, whereas the result with uniformly distributed mass is obtained by changing $q \to \tfrac{3}{2} q$, which makes the condition agree with (7.29).

Kapitza's results hold only for a perfectly vertically excited pendulum ($\phi = 0$), i.e. with zero imperfection the pendulum will be able to point straight up as soon as the magnitude of $q\Omega$ exceeds the critical threshold. However, in any real experiment there will be imperfections breaking the perfect symmetry, as modelled by the parameter ϕ in this study. Equation (7.28) then shows that for non-zero ϕ the pendulum will stabilize in a position that differs from vertical, and when the magnitude of the excitation just exceeds the critical threshold, then this deviation from vertical can be quite large even if ϕ is very small. A similar observation was made by Sudor and Bishop [50] in their numerical study of a tilted pendulum, which concludes that ".. the stable 'inverted' position will never be truly inverted and may only approach this state asymptotically".

Limit case II: rigid rod with disk
With a rigid rod with disk one has $v(t) = 0$, and the equations of motion (7.2)–(7.4) reduce to

$$\left(1 + \gamma + \alpha u^2\right)\ddot{\theta} + \left(2\beta_0 + 2\alpha u \dot{u}\right)\dot{\theta} - \left(1 + \tfrac{2}{3}\alpha u\right)\left(\sin\theta + \tfrac{3}{2}\ddot{y}_0 \sin(\theta - \phi)\right) = 0$$

$$\ddot{u} + 2\beta_d \dot{u} - \dot{\theta}^2 u + \tfrac{2}{3}\left(\cos\theta + \tfrac{3}{2}\ddot{y}_0 \cos(\theta - \phi)\right) = 0$$

$$\tag{7.30}$$

The case $\phi = 0$ corresponds to Chelomei's original model, and then the equations reduce to what can be found in equivalent form elsewhere, e.g., in [13] (with a typo in the term corresponding to $2\alpha u \dot{u} \dot{\theta}$), and in [7, 8, 35, 57]. The quasi-equilibrium for the rod and the disk becomes, by (7.22)–(7.23):

$$\theta_{0s} = \left[1 - \tfrac{8}{9}(q\Omega)^{-2}\right]^{-1}\phi; \quad u_{0s} = \tfrac{2}{3} + \frac{\tfrac{2}{3}(1 - \tfrac{1}{2}\theta_{0s}^{2})}{\tfrac{8}{9}(q\Omega)^{2}(\theta_{0s} - \phi)^{2}} \quad (7.31)$$

which according to (7.25) and (7.26) is stable when

$$\tfrac{9}{8}(q\Omega)^{2} > 1 \quad \text{and} \quad 0 > \tfrac{3}{2} \quad (7.32)$$

Since the second condition cannot be satisfied, the quasi-equilibrium is never stable. Further, without imperfection ($\phi = 0$) one finds $\theta_{0s} = 0$ and $u_{0s} \to \infty$, i.e. there are no equilibria for the disk on the rod, neither stable nor unstable.

This agrees with results obtained independently by Blekhman and Malakhova [8], Menyailov & Movchan [39], and Kirgetov [33], who all conclude that the disk cannot stabilize on the inverted rod of a uniform rigid pendulum with purely vertical support excitation.

Limit case III: isolated effect of flexural vibrations

We here consider what happens to the disk when the rod performs strong flexural vibrations, without considering how such vibrations should occur. To this end we assume the rod to vibrate in flexural resonance and to have negligible rotary inertia, that is: $\Omega \approx \omega$, which implies $\bar{\omega} \approx 1$ so that $R \approx 1/(2\beta_r)$ (cf.(7.16)), and $\hat{\gamma} = 0$. Then Equation (7.23) for determining the stationary disk position u_{0s} becomes

$$(1 - \tfrac{1}{2}\theta_{0s}^{2}) + \tfrac{8}{9}(q\Omega)^{2}(\theta_{0s} - \phi)^{2}\left\{1 - \tfrac{3}{2}u_{0s} - \tfrac{1}{6}\psi_{001}^{2}\beta_{r}^{-2}\psi_{u_{0s}}\psi'_{u_{0s}}\right\} = 0 \quad (7.33)$$

Ignoring for this particular purpose the first term describing gravity (e.g. the pendulum could swing in a horizontal plane), then the second term must vanish. It will do so if $\theta_{0s} = \phi$, i.e. if the pendulum rod lines up with the direction of excitation. However, this will never happen for an imperfect pendulum, as shown above for limit case I, and hence the content of the curly brackets must vanish. With small damping, $\beta_r \ll 1$, the term describing the effect of flexural vibrations will dominate inside the curly brackets. Hence under these conditions u_{0s} is the solution of $\psi_{u_{0s}}\psi'_{u_{0s}} = 0$, that is: The disk is at (quasi-)equilibrium at the vibration nodes of the rod (where $\psi = 0$), and at the antinodes (where $\psi' = 0$).

The stability condition (7.26) becomes in this case, using the same assumptions as above (this includes neglecting the effects of rigid body pendulum motions, i.e. the right-hand side of (7.26) is put to zero):

$$\left(\psi_{u_{0s}}\psi'_{u_{0s}}\right)' < 0 \qquad (7.34)$$

This condition should be evaluated at $u = u_{0s}$, i.e. at the vibration nodes and antinodes of the rod; it can be shown that it is only fulfilled at the antinodes.

Thus we conclude that the isolated effect of flexural vibrations of the rod is to force the sliding disk towards the antinodes of vibration, which are stable positions, and to repel it from the nodes of vibration, which are unstable positions.

This effect originates from nonlinear terms of the original equations, most importantly the term $\alpha\psi_u\psi'_u v\ddot{v}$ in (7.4), which describes a small longitudinal projection of transverse inertia forces. This projection is non-zero only at rotating rod segments, and hence the disk comes to rest where there are no rotations, i.e. at the antinodes. In general this is a very weak nonlinear effect, so quite strong vibrations are required to make it compete in magnitude with linear terms arising e.g. from gravity. Similar conclusions have been obtained for rigid bodies sliding on vibrating strings and beams [7, 8, 1, 56, 57]. Also, it can be shown that for sufficiently large values of rotary inertia $\hat{\gamma}$, the stability of nodes and antinodes may reverse, so that the nodes become the stable positions for the disk [7, 55].

Case: realistic

Limit case I above shows why the rod with imperfect excitation may stabilize in a position deviating from the vertical. Case II then shows why the disk cannot float on a rigid rod, and case III shows why it might float on a vibrating flexible rod. Having thus qualitatively explained the mechanisms involved, we proceed here to make quantitative predictions for a realistic case, and to compare these with results of numerical simulation and laboratory experiments.

Time histories

Figure 7.3 illustrates typical numerical solutions (DIVPAG from the IMSL Math/Library, using Gear's BDF method for stiff systems) of the full equations of motion (7.2)–(7.5), with parameters as given in the Figure caption. These non-dimensional parameters correspond to the physical parameters for Experimental Pendulum 2. The Figure also shows the theoretical predictions for stationary values of the slow components of motion, as given by expressions (7.22)–(7.23) for θ_0s and u_{0s}, and for the amplitude of the flexural rod vibrations amp(εv_1) (*cf.* (7.8)), equal to the coefficient of $\sin(\tau-\chi)$ in (7.15).

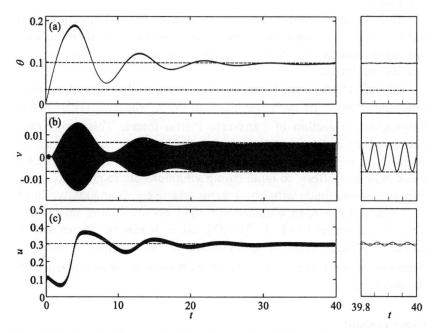

Figure 7.3. Results of numerical simulation of the full equations of motion (7.2)–(7.5). (a) Rod inclination $\theta(t)$ (with $\theta = \phi$ indicated in dash-dotted line), (b) flexural rod vibrations $v(t)$, and (c) position $u(t)$ of the disk. Dashed lines indicate the corresponding theoretical predictions for the quasi-stationary states, respectively, θ_{0s}, amp(εv_1), and u_{0s}. Parameters: $q = 0.01125$, $\Omega = 115.5$, $\omega = 115.6$, $\alpha = 0.003$, $\gamma = 0$, $\beta_0 = 0.15$, $\beta_r = 0.02$, $\beta_d = 10$, $\phi = 0.034$ ($\approx 2°$), $\theta(0) = 0.001$, $u(0) = 0.05$, $v(0) = \dot\theta(0) = \dot v(0) = \dot u(0) = 0$. From Thomsen & Tcherniak (2001) [62].

As appears from Figure 7.3 (a), when the pendulum is released from the inverted vertical position, its rigid body rotation θ performs damped oscillations and then stabilizes at the theoretically predicted value (in dashed line). This value of $\theta_{0s} = 0.10$ corresponds to about $5.7°$, which is somewhat larger than the tilt $\phi = 0.0034 \approx 2°$ (shown in dash-dotted line), as predicted. The thick appearance of the time history curve reflects that the large and slow oscillations are overlaid by small and rapid oscillations, as appears from the close-up of a short time interval at the right of the figure. This is also in close agreement with the theoretical predictions given by (7.8). Figure 7.3 (b) shows how – in response to the off line excitation of the pendulum rod – flexural rod vibrations grow up and stabilize at the predicted amplitude (in dashed line). Figure 7.3 (c) then shows how – in response to the average effect of flexural vibrations from the rod, gravity, etc. – the disk climbs from the rod hinge to the position predicted theoretically (in dashed line). This position, according to (7.23), is $u_{0s} = 0.30$, which is just below the antinode at $u = 0.38$ (obtained as a zero of ψ' with ψ given by (7.7)) for first-mode vibrations of the hinged-free rod.

Comparing the stationary state with experimental observations
The above theoretical predictions of $\theta_{0s} \approx 5.7°$ and $u_{0s} = 0.30$ agrees quite closely with the values $\theta_{0s,exp} \approx 5°$ and $u_{0s,exp} \approx 0.32$ that can readily be measured on the photo of the experimental pendulum (Figure 7.1(d)). However, one should not put too much into this agreement of numbers: As mentioned the experimental pendulum was hard to keep in the same position for more than a few seconds, and results were hard to reproduce due to difficulties with maintaining a sufficiently fine frequency tuning. Other photos (not shown) show stable disk positions in a range from about $u = 0.19$ to about 0.32, and accompanying deviations in θ.

Frequency responses
Figure 7.4 shows a frequency response for the system. The parameters are as for Figure 7.3, except that the frequency of support excitation Ω is varied near the flexural resonance ω of the rod. The stationary values of θ, u and v are then depicted versus Ω. Curves represent theoretical predictions according to (7.22)–(7.23), and solid and dashed curves represent stable and unstable solutions, respectively. Circled points mark the results of numerical solutions of the full equations of motion (7.2)–

(7.5), showing good agreement with the stable branches of the theoretical predictions.

Figure 7.4 (a) shows the frequency response for the stationary value θ_{0s} of rigid body rotations of the rod. It appears that the near-upright position of the rod becomes stable when Ω reaches a sufficiently high frequency, very close to the flexural resonance frequency of the rod (about 0.998ω for the actual parameters). Increasing Ω beyond that value, the rod tends to line up closer and closer to the direction of excitation ϕ (shown in dash-dotted line). It is important to note that just above the critical value of Ω, the offset from vertical and from ϕ is quite pronounced. Figure 7.4 (b) shows the accompanying growth in the amplitude v of flexural vibrations. As already explained, this is possible only because the pendulum is not completely vertical. Increasing Ω beyond ω, the excitation tunes out of resonance with the rod, and the rod comes closer to the direction of excitation ϕ. Thus two effects combine to result in a significant drop in amplitude above ω, so that pronounced vibrations exist only for a narrow range of excitation frequencies. Figure 7.4 (c) shows the associated values of stationary positions u_{0s} of the sliding disk. As appears, stable positions for the disk exist only for a narrow range of excitation frequencies near resonance. This is quite natural, since the disk is pushed along the rod and held against gravity by strong flexural vibrations of the rod, which in turn exist only in this narrow range (*cf.* Figure 7.4(b)). It should be noted, that as the rod vibrations v become strongest (i.e. at $\Omega \approx 0.998\omega$ in Figure 7.4(b)), then the other forces affecting the disk, such as gravity, play only a minor role, so that the disk behaves as in limit case III above. This is apparent from the results in Figure 7.4(c), showing one stable position for the disk near the antinode of the rod at $u = 0.38$, and two unstable positions near the nodes at $u = 0$ and $u = 0.74$.

Comparing frequency responses with experimental observations

The changes in response depicted by Figure 7.4 occur in a range of frequency that is too narrow to be resolved in any detail with the experimental set-up used. For example, to cover the theoretical resonant range $\Omega/\omega \in [0.996; 1.000]$ using just three experimental points, the frequency of excitation would need to be adjustable and maintainable in steps of $0.002\times\omega$, i.e. within 0.2%.

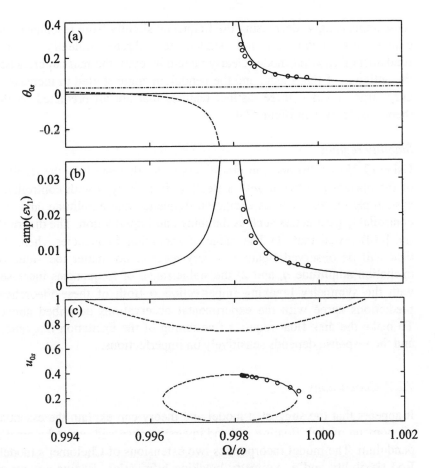

Figure 7.4. Analytical predictions of the frequency response for Chelomei's pendulum. (a) quasi-stationary rod inclination θ_{0s} (with $\theta = \phi$ indicated in dash-dot line), (b) amplitude of stationary pendulum rod vibrations εv_1, and (c) position u_{0s} of the disk on the pendulum rod. (———)/(- - - -) stable/unstable analytical predictions; (O) Numerical integration of the full equations of motion (7.2)–(7.5). Parameters as for Figure 7.3. From Thomsen & Tcherniak (2001) [62].

However, the experimental observations reported above for Experimental Pendulum 2 all confirm the tendency predicted by the curves in Figure 7.4: Floating of the disk occurs only at excitation frequencies near flexural resonance. It occurs more easily, and closer to the fundamental antinode, the more the rod is tilted from vertical. And most impressingly, decreasing the frequency slightly from resonance, the experimental pendulum rod started to vibrate violently in the fundamental flexural mode (clearly visible by eye), the mass approached the antinode of that mode, and the pendulum angle started to increase to very large values – three distinct features that are all predicted by the theoretical curves in Figure 7.4.

Stability regions

Using (7.25)–(7.26) one can plot diagrams to illustrate how the stability of the upright pendulum with a floating disk changes with controllable system parameters such as excitation frequency and amplitude, and less controllable parameters such as damping and imperfection. The diagrams (*cf.* [60]) reveal that: 1) The range of excitation frequencies where the disk will be observed floating is very narrow, no matter the value of excitation amplitude q, and 2) the stable range of frequencies increases with the symmetry breaking imperfection ϕ. Both of these theoretical predictions agree with the experimental observations described above: To make the disk float requires fine tuning of the excitation frequency, and the response depends sensitively on imperfections.

7.2.7 Conclusions

It appears that the suggested model and theory can explain the essential phenomena of disk floating and rod inversion observed with Chelomei's pendulum. The model incorporates two extensions of Chelomei's model: Rod flexibility and a symmetry breaking bifurcation. Excluding any of these, the model fails to reproduce the phenomena in question. Thus it is believed that the model is as simple as possible and as complicated as necessary.

The main hypothesis is that the disk floats due to resonant flexural vibrations of the pendulum, which are excited by the vertical support excitation through a symmetry breaking imperfection. Support is given by the two main results: I) Experimental observations, showing strong flexural vibrations to co-exist with floating of the disk near vibration antinodes; and II) A mathematical 3-dof model of the system, which is capable of reproducing the experimental observations only when the rod is flexible and there is some imperfection. Also, it appears from the theory, the disk will stabilize on the inverted pendulum rod only for a narrow range of excitation frequencies, which further depends critically on highly uncertain parameters describing the rod damping and imperfection. Thus, the theory also explains why consistent experimental results are hard to reproduce.

7.3 Nonlinear Dynamics of the Follower-Loaded Double Pendulum with Added Support-Excitation

Structures subjected to follower-type forces are seen in the industry as well as in civil applications. Some examples are turbo machinery and compressors subjected to fluid loading, pipes/tubes conveying fluid and bridges, antennas and panels experiencing wind loading. These structures may lose the stability of the original configuration by dynamic instability (flutter) or by static instability (divergence), due to the action of the follower forces (e.g. [10, 21, 36]).

High-frequency excitation acting upon follower-loaded structures has been shown to change the stability properties and also the dynamic behaviour in case of instability. Quantification of the effect of high-frequency excitation is essential for predicting stability and dynamic behaviour of a given structure and also for possible utilization of high-frequency excitation for tailoring the dynamic behaviour.

This section is devoted to the analysis of a simple physical model of a follower-loaded structure with high-frequency excitation. Despite being a simple model, the elastic double-pendulum with a partial follower-load displays the two basic instability mechanisms encountered in real engineering systems. The material presented here is a compact and edited version of the material published in [25, 28].

7.3.1 The Model and Model Equations

Figure 7.5 shows the model. The system considered consists of two rigid massless rods of equal length l. The rods are connected to each other and to the support by hinges with equal linear torsional stiffness coefficient k and equal linear viscous damping coefficient $\tilde{c}$ The rods carry two masses $2m$ and m positioned at the end of the first and second rod respectively. The rotation angles of the rods are given as θ_1 and θ_2. The system is subjected to a partially follower-load $\tilde{p}$ acting at the free rod end. The load arrangement is characterized by the parameter α, where

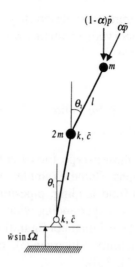

Figure 7.5. The elastic partially follower-loaded double pendulum with added support-excitation. From Jensen (1998) [23].

$\alpha=1$ corresponds to pure follower loading and $\alpha=0$ corresponds to pure conservative loading. The support is subjected to a harmonic displacement parallel to the upright position given as $\tilde{w}\sin\tilde{\Omega}t$, where $\tilde{w}$ is the amplitude and $\tilde{\Omega}$ the frequency of displacement. Small-amplitude, off-resonant (high-frequency) excitation is considered, implying that $\tilde{w}<<l$ and $\tilde{\Omega}>>\omega_2$, where ω_2 is the highest natural frequency of the pendulum.

The equations of motion are set up using Lagrange's equations [25]:

$$3\ddot{\theta}_1 + \left(\cos(\theta_1-\theta_2)\ddot{\theta}_2 + \sin(\theta_1-\theta_2)\dot{\theta}_2^2\right) + 2c\dot{\theta}_1 - c\dot{\theta}_2 + 2\theta_1 - \theta_2$$
$$= p((1-\alpha)\sin\theta_1 + \alpha\sin(\theta_1-\theta_2)) - 3w\Omega^2 \sin\theta_1 \sin\Omega\tau, \quad (7.35)$$

$$\ddot{\theta}_2 + \left(\cos(\theta_1-\theta_2)\ddot{\theta}_1 - \sin(\theta_1-\theta_2)\dot{\theta}_1^2\right) + c\dot{\theta}_2 - c\dot{\theta}_1 - \theta_1 + \theta_2$$
$$= p(1-\alpha)\sin\theta_2 - w\Omega^2 \sin\theta_2 \sin\Omega\tau, \quad (7.36)$$

where the following non-dimensional quantities have been introduced:

$$\tau \equiv \sqrt{\frac{k}{ml^2}}t, \quad \Omega \equiv \sqrt{\frac{ml^2}{k}}\tilde{\Omega}, \quad p \equiv \frac{\tilde{p}l}{k}, \quad w \equiv \frac{\tilde{w}}{l}, \quad c \equiv \frac{\tilde{c}}{\sqrt{kml^2}}. \quad (7.37)$$

7.3.2 Direct Partition of Motion

The model equations (7.35) and (7.36) contain parametric excitation terms. This makes stability analysis and nonlinear analysis using theoretical tools cumbersome. Also, numerical integration of the equations is time-consuming due to small time-steps needed with high-frequency excitation.

The method of direct partition of motion (MDPM) (see chapter 2 of this book and [7]) is conveniently applied to eliminate the time-dependent terms in Equations (7.35) and (7.36). MDPM is an averaging technique where the fast motion (comparable with the excitation frequency Ω) and the slow motion (comparable with the natural frequencies ω_i, $i=1, 2$) are de-coupled. For this purpose the equations of motion are rewritten in the form, with $\theta = \{\theta_1 \; \theta_2\}^T$:

$$\ddot{\theta} = \mathbf{f}(\dot{\theta}, \theta) + \Omega \mathbf{q}(\theta), \qquad (7.38)$$

In Equation (7.38), terms with explicit dependence of $\Omega\tau$ have been collected in the vector $\mathbf{q}=\{q_1 \; q_2\}^T$, and the remaining terms have been collected in the vector $\mathbf{f}=\{f_1 \; f_2\}^T$ [25].

Two independent time-scales are now introduced: a slow time scale $T_0 \equiv \tau$ and a fast time scale $T_1 \equiv \Omega\tau = \Omega T_0$. It is assumed that the solution to Equation (7.38) can be written as a sum of a 'slow' function $\mathbf{x}(T_0) = \{x_1(T_0) \; x_2(T_0)\}^T$ superposed with a 'fast' function $\psi(T_1) = \{\psi_1(T_1) \; \psi_2(T_1)\}^T$. The assumed solution is written on the form:

$$\theta(T_0, T_1) = \mathbf{x}(T_0) + \mu\psi(T_1), \qquad (7.39)$$

where $\mu \equiv 1/\Omega$ ($\mu \ll 1$) has been introduced as a small parameter. It is assumed that the fast function $\psi(T_1)$ is a 2π-periodic function of T_1 with $\langle\psi\rangle = 0$, where $\langle \; \rangle \equiv \frac{1}{2\pi}\int_0^{2\pi}(\;)d(T_1)$ is a linear averaging operator, i.e. averaging with respect to T_1 over 2π (a forcing period).

The solution assumption in Equation (7.39) is inserted into Equation (7.38):

$$D_0^2\mathbf{x} + \mu^{-1}D_1^2\psi = \mathbf{f}(D_0\mathbf{x} + D_1\psi, \mathbf{x} + \mu\psi) + \mu^{-1}\mathbf{q}(\mathbf{x} + \mu\psi), \quad (7.40)$$

where the notation $D_i^j \equiv \partial^j/\partial T_i^j$ denoting partial differentiation with respect to the individual time-scales has been introduced.

7. Non-Trivial Effects of High-Frequency Excitation

The averaging operator < > is applied to both sides of Equation (7.40). Using that $<D_0^2x> = D_0^2x$ and $<D_1^2\psi>=0$ and applying the first-order Taylor expansion $q(x+\mu\psi) \approx q(x)+\mu\psi \cdot \partial q(x)/\partial x$ yields:

$$D_0^2 x = \left\langle f(D_0 x + D_1 \psi, x + \mu\psi) + \mu^{-1} q(x) + \psi \cdot \frac{\partial q(x)}{\partial x} \right\rangle, \quad (7.41)$$

$$D_1^2 \psi = \mu f(D_0 x + D_1 \psi, x + \mu\psi) + q(x) + \mu\psi \cdot \frac{\partial q(x)}{\partial x}$$
$$- \left\langle \mu f(D_0 x + D_1 \psi, x + \mu\psi) + q(x) + \mu\psi \cdot \frac{\partial q(x)}{\partial x} \right\rangle. \quad (7.42)$$

The zero-order approximation $f(D_0x+D_1\psi, x+\mu\psi) \approx f(D_0x+D_1\psi, x)$ is now applied. Using also that $<q(x)>=0$, yields the following two equations:

$$D_0^2 x = \left\langle f(D_0 x + D_1 \psi, x) + \psi \cdot \frac{\partial q(x)}{\partial x} \right\rangle, \quad (7.43)$$

$$D_1^2 \psi = q(x) + O(\mu). \quad (7.44)$$

The set of differential equations (7.43) and (7.44) represent a zero-order approximation of Equation (7.38) with the assumption of the solution form in Equation (7.39).

Solving Equation (7.44) for ψ yields:

$$\psi = \frac{-w\Omega}{3-\cos^2(x_1-x_2)} \left\{ \begin{array}{l} \sin x_2 \cos(x_1-x_2) - 3\sin x_1 \\ 3\sin x_1 \cos(x_1-x_2) - 3\sin x_2 \end{array} \right\} \sin T_1. \quad (7.45)$$

Solving Equation (7.43) using (7.45), and writing the obtained solution in the same form as (7.35) and (7.36), yields the two equations governing the slow motion of the system:

$$3\ddot{x}_1 + \left(\cos(x_2-x_1)\ddot{x}_2 - \sin(x_2-x_1)\dot{x}_2^2\right) + c(2\dot{x}_1 - \dot{x}_2) + 2x_1 - x_2$$
$$= p((1-\alpha)\sin x_1 - \alpha\sin(x_2-x_1)) + V_1, \quad (7.46)$$

$$\ddot{x}_2 + \left(\cos(x_2-x_1)\ddot{x}_1 + \sin(x_2-x_1)\dot{x}_1^2\right) + c(\dot{x}_2 - \dot{x}_1) - x_1 + x_2$$
$$= p(1-\alpha)\sin x_2 + V_2, \quad (7.47)$$

where

$$\begin{Bmatrix} V_1 \\ V_2 \end{Bmatrix} = \frac{v}{(\cos 2(x_1-x_2)-5)^3} \begin{Bmatrix} 102\sin 2x_1 - 2\sin(2x_1-4x_2) - \\ \sin(2x_1-4x_2) + \sin(6x_1-4x_2) - \\ 10\sin(4x_1-2x_2) + 20\sin(2x_1-2x_2) - 2\sin(4x_1-4x_2) - 30\sin 2x_2 \\ 10\sin(4x_1-2x_2) - 20\sin(2x_1-2x_2) + 2\sin(4x_1-4x_2) + 10\sin 2x_2 \end{Bmatrix},$$

(7.48)

with $v \equiv 3/4(w\Omega)^2$ introduced as a measure of the vibrational force.

The two functions V_i, $i=1, 2$, represent the vibrational forcing acting on the slow motion of the system, i.e. motion comparable to the natural frequencies of the pendulum, as a result of the high-frequency support-excitation. Note that by means of MDPM the explicit time-dependence has been eliminated in the new model equations (7.46) and (7.47). The effect of the high-frequency excitation is instead approximated by equivalent static forces.

7.3.3 Linear Stability of the Upright Pendulum Position

From Equations (7.46)–(7.48) it is noted that $\{x_1\ x_2\}^T = \{0\ 0\}^T$ is an equilibrium position, also with added support-excitation. In this section, the effect of the support-excitation on the linear stability of the straight upright pendulum position is investigated.

Linearizing Equations (7.46) and (7.47) around $\{x_1\ x_2\}^T = \{0\ 0\}^T$ and writing the equations as a system of four first order differential equations, using the notation $\mathbf{y} = \{x_1\ x_2\ \dot{x}_1\ \dot{x}_2\}^T$, yields:

$$\dot{\mathbf{y}} = \mathbf{A}\mathbf{y},$$

(7.49)

Exponential time-dependence of the form $\mathbf{y} = \mathbf{u}\cdot\exp(\lambda\tau)$ is inserted into Equation (7.49). Solving the determinant equation $\det(\mathbf{A}-\lambda\mathbf{I})=0$, yields the following characteristic equation for the eigenvalue λ:

$$a_0\lambda^4 + a_1\lambda^3 + a_2\lambda^2 + a_3\lambda + a_4 = 0,$$

(7.50)

where

$$a_0 = 2,\ a_1 = 7c,\ a_2 = 2p(\alpha-2) + c^2 + 7 + 8v,$$
$$a_3 = c(3p(\alpha-1) + 2 + 3v),\ a_4 = (3-p+4v)p(\alpha-1) + (2v+1)(v+1).$$

(7.51)

The zero solution of Equation (7.49) is stable only if all roots of λ in Equation (7.50) have negative real parts. The zero solution loses stability by a static instability (divergence) if a single real eigenvalue passes the origin. The instability is dynamic (flutter) if a pair of complex conjugate eigenvalues passes the imaginary axis.

Figure 7.6 shows the regions of stability of the upright pendulum position in an (α, p) parameter-space for four different values of the vibrational forcing parameter v. It is indicated whether stability of the upright position is lost by divergence (dashed lines) or by flutter (solid lines). More detailed stability diagrams can be found in [54] and [21] for the system without added support-excitation.

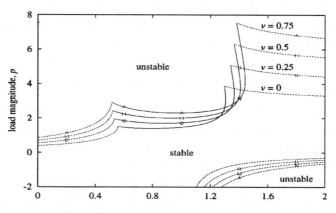

Figure 7.6. Stability borders for the straight upright pendulum position for four different amounts of support-excitation, solid lines: Hopf bifurcations, dashed lines: pitchfork bifurcations, markers: stability borders calculated by numerical integration. Parameter-values: c=0.1, Ω=25. From Jensen (1998)[25].

With vibrational forcing added ($v\neq0$) the region in parameter-space with a stable upright position is broadened: Stability borders are moved up (down) for p>0 (p<0) respectively. Especially for α≈1.4, p>0 the stability of the upright position is improved significantly with support-excitation. Near α=1.3 the presence of vibrational forcing may destabilize the pendulum implying that loads which previously did not affect the upright position, may now cause the pendulum to flutter (see Figure 7.7).

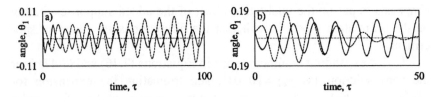

Figure 7.7. Lower rod angle θ_1 versus time τ, based on numerical integration of Equations (3.35) and (3.36), solid lines: $v=0.5$, dashed lines: $v=0$, a) $p=2$, $\alpha=1.2$, b) $p=3$, $\alpha=1.32$. Parameter-values: $c=0.1$, $\Omega=25$. Initial conditions: (0.05, -0.07, 0, 0). From Jensen (1998) [25].

The markers in Figure 7.6 represent stability borders computed by numerical integration of the full model equations (7.35) and (7.36) by using a standard Runge-Kutta algorithm. Very good agreement is noted, indicating that MDPM accurately captures the full effect of the support-excitation when it comes to predicting linear stability.

Figure 7.7 exemplifies how added support-excitation can stabilize or destabilize the upright pendulum position. Transient behaviour of the lower rod angle θ_1 is shown for $v=0$ shown as dashed lines, and for $v=0.5$ shown as solid lines. For $p=2$ and $\alpha=1.2$ (Figure 7.7a), the initially applied disturbance is amplified for $v=0$, whereas with sufficient support-excitation present ($v=0.5$) the vibrations are damped out to approach the stable zero solution. For another set of load parameters, $(\alpha, p)=(1.32, 3)$ (Figure 7.7b), the upright position is stable for $v=0$, whereas support-excitation destabilized the zero solution. This causes the initial disturbance of the pendulum to be amplified, until the motion is finally limited by stabilizing nonlinear terms. In Figure 7.7 it is noted that the frequency of vibrations is increased when the vibrational forcing is added. This feature is common for systems subjected to high-frequency excitation, and reflects an apparent stiffening effect (e.g. [7, 51, 25, 58, 59]).

Local periodic and non-zero static solutions

Local bifurcations of the zero solution are now investigated using the methods of center manifold reduction and normal forms. Only behaviour associated with simple bifurcations (codimension one), i.e. either pure flutter or pure divergence, is analyzed. Codimension two bifurcations, i.e. coupled flutter and divergence, is beyond the scope of this paper.

Emphasis is put on the qualitative effect of the added support-excitation on the bifurcation types, i.e. whether the bifurcations are super- or sub-critical.

Taylor expanding nonlinear terms in Equations (7.46) and (7.47) near the zero solution $\{y_1\ y_2\}^T = \{0\ 0\}^T$, and truncating the expansion for $y_i^p y_j^q y_k^r$, $i, j, k=1,\ldots, 4$, $p+q+r>3$, yields a system of four first order differential equations:

$$\dot{\mathbf{y}} = \mathbf{A}\mathbf{y} + \mathbf{f}(\mathbf{y}), \qquad (7.52)$$

where $\mathbf{f}$ is a vector containing cubical nonlinearities [25].

Equation (3.52) is now written in Jordan canonical form using the coordinate transformation $\mathbf{y} = \mathbf{P}\mathbf{z}$ with $\mathbf{z} = \{z_1\ z_2\ z_3\ z_4\}^T$:

$$\dot{\mathbf{z}} = \Lambda \mathbf{z} + \mathbf{g}(\mathbf{z}), \qquad (7.53)$$

where $\Lambda = \mathbf{P}^{-1}\mathbf{A}\mathbf{P}$ and $\mathbf{g}(\mathbf{z}) = \mathbf{P}^{-1}\mathbf{f}(\mathbf{P}\mathbf{z})$, and $\mathbf{P}$ is the so-called modal matrix composed of imaginary and real parts of the eigenvectors, see e.g. [48].

The matrix Λ is in block form which allows for a de-coupling of the linear part of Equation (7.53) into an essential part Λ_{cri} associated with the critical eigenvalues (the ones with zero real part) and a stable part Λ_{stable} associated with eigenvalues with negative real parts. With a subsequent de-coupling of the nonlinear function $\mathbf{g}$ (center manifold reduction), the essential system behaviour can be traced on a lower dimensional sub-system [11, 19]. Further simplification of the system can then be made using normal forms [42]. In the following, Hopf- and pitchfork-bifurcations will be studied separately in a non-detailed manner. For further details on the application of the methods of center manifold reduction and normal forms in similar systems see e.g. [38] and [46].

Hopf bifurcations

The nonlinear dynamics associated with a Hopf bifurcation can be analyzed by examining the two-dimensional sub-system given as, with $\mathbf{z}_{cri} = \{z_1\ z_2\}^T$, $\mathbf{g}_{cri}(\mathbf{z}) = \{g_1(\mathbf{z})\ g_2(\mathbf{z})\}^T$:

$$\dot{\mathbf{z}}_{cri} = \begin{bmatrix} \kappa & -\omega \\ \omega & \kappa \end{bmatrix} \mathbf{z}_{cri} + \mathbf{g}_{cri}(\mathbf{z}), \qquad (7.54)$$

where ω and κ are the imaginary and real parts of the critical eigenvalue. At the critical point ω represents the flutter frequency whereas κ vanishes.

The two scalar nonlinear functions g_1 and g_2 are still coupled to the remaining (stable) system. The tangent-space approximation: $g_{cri}(z) \approx g_{cri}(z_{cri}, 0)$ can be applied with no additional approximations introduced. This leads to the following set of equations:

$$\dot{z}_1 = \kappa z_1 - \omega z_2 + g_1(z_1, z_2), \qquad (7.55)$$

$$\dot{z}_2 = \omega z_1 + \kappa z_2 + g_2(z_1, z_2). \qquad (7.56)$$

Equations (7.55) and (7.56) are now to be simplified using the method of normal forms. Since the behaviour near the bifurcation point is of interest, the chosen outfolding parameter, e.g. the load magnitude p, is perturbed so that $p=p_{cri}+\delta$. From the perturbation parameter δ, the changes in the real and imaginary parts of the eigenvalue δ_1 and δ_2 can be computed.

Equations (7.55) and (7.56) become:

$$\dot{z}_1 = \delta_1 z_1 - (\omega + \delta_2)z_2 + a_1 z_1^3 + a_2 z_1^2 z_2 + a_3 z_1 z_2^2 + a_4 z_2^3, \qquad (7.57)$$

$$\dot{z}_2 = (\omega + \delta_2)z_1 + \delta_1 z_2 + a_5 z_1^3 + a_6 z_1^2 z_2 + a_7 z_1 z_2^2 + a_8 z_2^3. \qquad (7.58)$$

The constants a_i, $i=1,\ldots, 8$, depend on the system parameters.

By using the polar coordinate transformation $z_1 = r \cdot \cos\phi$ and $z_2 = r \cdot \sin\phi$, the normal form of Equations (7.57) and (7.58) becomes [42]:

$$\dot{r} = r(\delta_1 + ar^2), \qquad (7.59)$$

$$\dot{\phi} = \omega + \delta_2 + br^2, \qquad (7.60)$$

where $a = \frac{1}{8}(3a_1 + a_3 + a_6 + 3a_8)$ and $b = -\frac{1}{8}(a_2 + 3a_4 - 3a_5 - a_7)$. In Equation (7.59) and (7.60), r governs the amplitude of oscillations, whereas ϕ is the phase angle.

Steady state values of r are found by letting $\dot{r} = 0$ in Equation (7.59). Two solutions for r ($r \geq 0$) are found:

$$r = 0, \quad r = \sqrt{-\frac{\delta_1}{a}} \quad \text{for} \quad \frac{\delta_1}{a} < 0, \qquad (7.61)$$

that is, the zero solution and a nonlinear limit cycle solution. With known oscillation amplitude r, Equation (7.60) can be solved for the steady state phase angle velocity $\dot{\phi}$. The obtained solution can then, if desired, be transformed back to be expressed in terms of the original variable y.

Here the type of bifurcation is of main interest. For $a<0$, Equation (7.61) predicts a supercritical Hopf bifurcation whereas $a>0$ corresponds to a subcritical bifurcation. The value of a turns out to depend strongly upon the amount of added support-excitation. If v is increased, a increases also, implying that the added support-excitation might turn supercritical Hopf bifurcations into subcritical ones.

For $\alpha=0.8$, i.e. rather close to a perfectly follower-loaded system, the bifurcation is supercritical without added support-excitation. However the value of a is computed to turn from negative to positive for $v \approx 0.318$, and thus subcritical bifurcations are predicted for higher values of v. Figure 7.8a shows limit cycle solutions obtained with the algorithm PATH [31], based on numerical integration of the autonomous model Equations (7.48) and (7.49) . PATH is a path-following algorithm capable of following stable and unstable solution branches. Hopf bifurcations of the zero solution are shown for three different values of v. Figure 7.8a shows supercritical bifurcations for the two lower values of v, and a subcritical bifurcation for $v=0.5$, as predicted by the theoretical

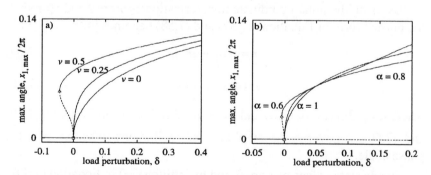

Figure 7.8. Max. lower rod angle $x_{1,\,max}/2\pi$ versus load perturbation δ for flutter instabilities, based on numerical integration of Equations (3.46) and (3.47) using PATH, a) $\alpha=0.8$ for three different values of v and b) $v=0.25$ for three different values of α. Solid and dashed lines, stable and unstable solutions respectively. Hopf bifurcation: O, saddle-node bifurcation: Δ. Parameter-values: $c=0.1$, $\Omega=25$. From Jensen (1998)[25].

7.3. Nonlinear Dynamics of the Follower-Loaded Double Pendulum

analysis. Figure 7.8b shows three Hopf bifurcations of the zero solution for $v=0.25$, for three different values of α. The theoretical value of a in this case increases for decreasing values of α, and is computed turn negative for $\alpha \approx 0.686$. It is seen that for $\alpha=1$ and $\alpha=0.8$, the corresponding Hopf bifurcations are supercritical, whereas for $\alpha=0.6$ the bifurcation has turned subcritical, as predicted by theory.

Thus the results presented show that sufficiently strong support-excitation may be destabilize the upright pendulum position in favour of flutter oscillations when a strong disturbance is applied, even though the upright position is linearly stable. It should be noted also, that secondary saddle-node bifurcations occur in the subcritical cases. These secondary bifurcations are not predicted by Equation (7.61), which is based on a third order nonlinear model, but is captured by PATH. To capture these bifurcations theoretically, higher order approximations are called for.

Pitchfork bifurcations

The nonlinear behaviour associated with pitchfork bifurcations is now analyzed. The essential behaviour is governed by the critical one-dimensional sub-system given as:

$$\dot{\mathbf{z}}_{cri} = \kappa \mathbf{z}_{cri} + \mathbf{g}_{cri}(\mathbf{z}), \qquad (7.62)$$

with $\mathbf{z}_{cri}=\{z\}$, $\mathbf{g}_{cri}=\{g(\mathbf{z})\}$, and where the real part of the eigenvalue κ, vanishes at the critical point. The decoupling of the nonlinear function $\mathbf{g}_{cri}(\mathbf{z})$ is performed as $\mathbf{g}_{cri}(\mathbf{z}) \approx \mathbf{g}_{cri}(\mathbf{z}_{cri}, \mathbf{0})$, without any additional approximations introduced. The nonlinear behaviour is thus governed by the single scalar equation:

$$\dot{z} = \delta_1 z + a z^3, \qquad (7.63)$$

where the change in the real part of the eigenvalue δ_1 is computed from a chosen perturbation parameter δ, and a is a constant that depends on the system parameters.

The steady state solution for z is found directly from Equation (7.63) by letting $\dot{z}=0$. This yields:

$$z = 0, \quad z = \pm\sqrt{-\frac{\delta_1}{a}} \quad \text{for} \quad \frac{\delta_1}{a} < 0, \qquad (7.64)$$

that is, the zero solution and two static non-zero solution branches. As for Hopf bifurcations $a<0$ implies a supercritical bifurcation, whereas $a>0$ corresponds to a subcritical bifurcation. Also in this case does the value of a depend strongly upon the amount of support-excitation, i.e. the value of v.

Bifurcation types corresponding to the lower divergence region of Figure 7.6, i.e. for $\alpha \approx 1.1$ and $p<0$, are illustrated by use of PATH, for different amounts of support-excitation and for different values of α. Figure 7.9a shows pitchfork bifurcations for $\alpha=1.7$, for three different values of the v. Only positive values of $x_1/2\pi$ are shown since the curves are symmetrical around the abscissa. The transition from supercritical to subcritical, i.e. a turning positive, is computed to occur for $v \approx 0.255$. This agrees with what is shown in Figure 7.9a. For $v=0$ and $v=0.25$ the bifurcations are supercritical and for $v=0.5$, PATH shows a subcritical bifurcation. For $v=0.25$, the transition from super- to sub-critical is predicted to occur for $\alpha \approx 1.600$ for decreasing α. The bifurcation curves in Figure 7.9b, computed by use of PATH, shows a supercritical bifurcation for $\alpha=1.7$, and for $\alpha=1.5$ (and $\alpha=1.3$), as predicted, subcritical ones.

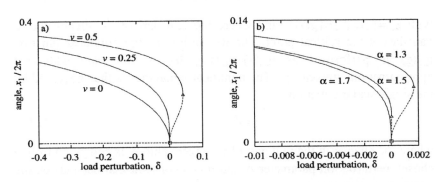

Figure 7.9. Lower rod angle $x_1/2\pi$ versus load perturbation δ for divergence instabilities, based on numerical integration of Equations (3.46) and (3.47) using PATH, a) $\alpha=1.7$ for three different values of v and b) $v=0.25$ for three different values of α. Solid and dashed lines, stable and unstable solutions respectively. Pitchfork bifurcation: □, saddle-node bifurcation: △. Parameter-values: $c=0.1$, $\Omega=25$. From Jensen (1998)[25].

Thus, also for divergence instabilities, the added support-excitation may turn bifurcations subcritical, i.e. even with the pendulum being linearly stable, a sufficiently strong disturbance may cause the pendulum to occupy another static point of equilibrium.

Local bifurcation diagram

Means have been provided for determining the type of bifurcation, i.e. either supercritical or subcritical, of the upright pendulum position in case of linear instability. Both the cases of Hopf bifurcations (flutter) and pitchfork bifurcations (divergence) were treated.

The stability diagram (Figure 7.6) is now redrawn with special attention given to the type of bifurcations. Figure 7.10 shows this bifurcation diagram. Supercritical bifurcations are represented by dotted lines, whereas solid lines represent bifurcations of the subcritical kind. As in Figure 7.6, the bifurcation curves are shown for the case of no support-excitation and for three non-zero values of v.

For $v=0$, supercritical bifurcations are seen to exist for most load arrangements. However, for the upper right divergence region, as well as for a small section of the flutter curve near the right codimension two point, the corresponding bifurcations are subcritical. When support-

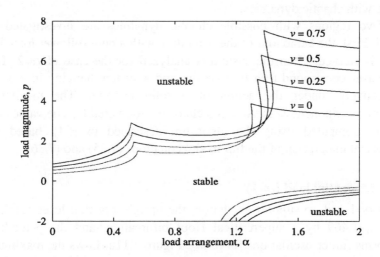

Figure 7.10. Local bifurcation diagram, solid lines: subcritical bifurcations, dotted lines: supercritical bifurcations. Parameter-values: $c=0.1$, $\Omega=25$. From Jensen (1998) [25].

excitation is added, parts of the supercritical bifurcations turn into subcritical. This happens both for the flutter and for the divergence curves. With $v=0.75$ almost all bifurcations shown in Figure 7.10 are subcritical, except for a small part of the upper left divergence region near the left codimension two points. Also for higher values of v this small supercritical region prevails, with decreasing size however. It should be remembered though, that near codimension two points even more complicated dynamics may exist. The transitions between bifurcation types, as shown in Figure 7.10, have been verified by PATH as well as by numerical integration of the original model equations.

Thus the local bifurcation investigation has revealed that added support-excitation tend to turn supercritical bifurcations into subcritical ones. This has shown to occur both for divergence and for flutter instabilities.

7.3.4 Global Dynamic Behavior

Local bifurcations were studied, and consequently the system behaviour near the straight upright pendulum position. In this section, the effect of the added support-excitation on the global dynamics of the pendulum, is studied. Special attention is given to regions in the (α, p) parameter-space with chaotic dynamics.

Two regions with possible chaotic dynamics are investigated in detail. First, the behaviour of the pendulum with a pure follower load, i.e. $\alpha=1$, is studied. Then the system is analyzed for the case of $\alpha=2$. The two cases correspond to a free system and a system hanging in gravity respectively, under the action of a follower-force. The maximum Lyapunov exponent λ_1 is used as a characteristic variable. The values for λ_1 are computed using the algorithm described in [63], based on numerical integration of the full model equations (7.35) and (7.36) .

Pure tangential load ($\alpha=1$)

With $\alpha=1$ and no support-excitation, the upright position loses stability for $p_{cri} \approx 1.469$ by a supercritical Hopf bifurcation, and the pendulum performs flutter oscillations for $p > p_{cri}$. Figure 7.11a shows the maximum Lyapunov exponent λ_1 versus the load magnitude p with the small initial disturbance: (0.05, -0.07, 0, 0). Negative values of λ_1 are seen for $p < p_{cri}$ indicating a static equilibrium, in this case the upright position. For

$p>p_{cri}$, λ_1 approximately vanishes which indicates periodic motion. No signs of chaotic dynamics ($\lambda_1>0$) appears in Figure 7.11a, but for very high load magnitudes ($p>\approx 15$), λ_1 turns positive and the motion will be chaotic.

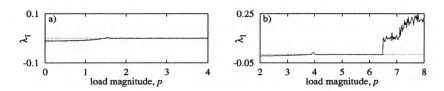

Figure 7.11. Largest Lyapunov exponent λ_1, plotted versus the load magnitude p, a) $\alpha=1$, $\nu=0$, b) $\alpha=1$, $\nu=2$, total sampling time: 2000 s., transient cut-off: 1000 s., sampling frequency: 20 Hz. Initial conditions: (0.05, -0.07, 0, 0). Parameter-values: $c=0.1$, $\Omega=50$. From Jensen (1998) [25].

If strong support-excitation ($\nu=2$) is added (Figure 7.11b), the stability of the upright position is increased, as seen also in Figures 7.6 and 7.10. The zero solution loses stability for $p_{cri}\approx 3.800$ by a subcritical bifurcation, and is replaced by periodic oscillations of the pendulum. For $p>\approx 6.5$ positive values of λ_1 are seen, indicating chaotic motion. Figure 7.11b was obtained with the same initial conditions as in Figure 7.11a. With a larger initial disturbance, periodic motion is seen also for values of p slightly less than critical due to the subcritical bifurcation.

More detailed information about the system behaviour for $\nu=2$ ($\alpha=1$) is obtained using the path-following algorithm PATH. For values of p near the critical value, the bifurcational behaviour can be shown to resemble that of Figure 7.8a for $\nu=0.5$, i.e. a stable periodic motion exists after a secondary saddle-node bifurcation of the unstable periodic solution.

The case of $\alpha=2$

With $\alpha=2$, added support-excitation may also significantly change the global nonlinear behaviour of the pendulum. For $\nu=0$ the bifurcation associated with the divergence instability, occurring for $p_{cri}\approx 3.304$, is subcritical. Figure 7.12a and Figure 7.12c shows the largest Lyapunov coefficient plotted versus the load magnitude p, for a weak and a strong initial disturbance, respectively. The upright pendulum position is in both cases replaced by a non-zero static equilibrium when the load is increased beyond the critical level. This is seen as a jumps in λ_1 (λ_1 is still negative). Due to the subcritical bifurcation, the stronger disturbance (Figure 7.12c) pushes the system to the new static equilibrium point also for values of p slightly less than critical. With a further increase in p, the new equilibrium position loses stability by a Hopf bifurcation, leading to chaotic motion ($\lambda_1 > 0$). Numerical simulation reveals that the motion may eventually settle down on a static equilibrium point after a unpredictably long transient period of chaos. Also, a very strong initial disturbance has been shown to may cause the system to settle on large amplitude periodic motion for $p < p_{cri}$ [54].

Figure 7.12b and Figure 7.12d show λ_1 versus p for added support-excitation, corresponding to $v=0.75$. The Figures correspond like Figures 7.8a,c to a weak and a strong initial disturbance. From Figure 7.12b it is seen that the upright pendulum position is now replaced directly by chaotic pendulum motion when p is increased beyond the critical level.

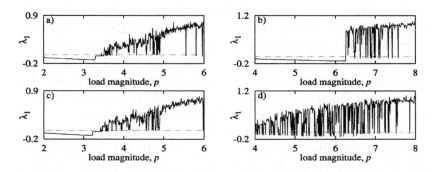

Figure 7.12. Largest Lyapunov exponent λ_1, plotted versus the load magnitude p, a) $\alpha=2$, $v=0$, b) $\alpha=2$, $v=0.75$, c) $\alpha=2$, $v=0$, d) $\alpha=2$, $v=0.75$, total sampling time: 2000 s., transient cut-off: 1000 s., sampling frequency: 20 Hz. Initial conditions, a-b): (0.05, -0.07, 0, 0), c-d): (1, 1.5, 0.5, 0.75). Parameter-values: $c=0.1$, $\Omega=25$. From Jensen (1998) [23].

Actually, chaos appears for load levels less than critical, which shows that even the weak disturbance is sufficient to destabilize the linearly stable upright position. With a stronger initial disturbance, as shown in Figure 7.8d, chaos appears far into the pre-critical range, however with several narrow 'windows' where the pendulum settles on a static equilibrium point. As was the case for $v=0$, the chaotic motion may also for added support-excitation finally settle down on a static equilibrium. This happens however, if ever, after a long and unpredictable time-range.

For the case of the left and the lower right divergence regions (see e.g. Figure 7.2), no signs of chaotic behaviour can be traced, with or without excitation of the support. But for the regions studied by examples in this and in the previous section, the results have pointed out that chaotic behaviour of the pendulum is more likely to occur with added support-excitation, compared to the case of a fixed support. Importantly, chaotic dynamics was shown to be possible for load levels less than critical, i.e. even with the upright pendulum position linearly stable.

7.3.5 Effects of Bi-Directional Support-Excitation

Figure 7.13 shows the modified model. Excitation is now added to the system in form of the prescribed displacements of the support: $w_y\sin\Omega t$ in the longitudinal direction and $w_x\cos\Omega t$ in the transverse direction. It is analyzed how the presence of transverse excitation affects the stability and dynamics of the system. The material presented is a compact and edited version of [28].

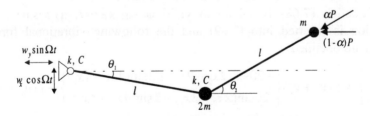

Figure 7.13. System configuration. From Jensen (2000a) [28].

Lagrange's equations are used to set up the model equations ($i=1, 2$),

$$3\ddot{\theta}_1 + \left(\cos(\theta_2 - \theta_1)\ddot{\theta}_2 - \sin(\theta_2 - \theta_1)\dot{\theta}_2\right) + c(2\dot{\theta}_1 - \dot{\theta}_2) + 2\theta_1 - \theta_2 =$$
$$p((1-\alpha)\sin\theta_1 - \alpha\sin(\theta_2 - \theta_1)) + 3\Omega^2\left(w_x\cos\theta_1\cos\Omega t - w_y\sin\theta_1\sin\Omega t\right),$$
(7.65)

$$\ddot{\theta}_2 + (\cos(\theta_2-\theta_1)\ddot{\theta}_1 + \sin(\theta_2-\theta_1)\dot{\theta}_1) + c(\dot{\theta}_2-\dot{\theta}_1) + \theta_2 - \theta_1 =$$
$$p(1-\alpha)\sin\theta_2 + \Omega^2(w_x\cos\theta_2\cos\Omega t - w_y\sin\theta_2\sin\Omega t),$$
(7.66)

Separation of fast and slow motion

The fast and the slow motion in the system are now separated in the same manner as previously. The model equations are rewritten as,

$$\ddot{\theta} = \mathbf{f}(\dot{\theta},\theta) + \mu\mathbf{q}_1(\theta)\cos T_0 + \mu\mathbf{q}_2(\theta)\sin T_0, \qquad (7.67)$$

where $\mathbf{q}_1 = \{q_{11}\ q_{12}\}^T$ and $\mathbf{q}_2 = \{q_{21}\ q_{22}\}^T$ contain the coefficients to $\cos T_0$ and $\sin T_0$ respectively, and $\mathbf{f} = \{f_1\ f_2\}^T$ contains the remaining terms [28]. The μ^0-order approximation of the obtained equations becomes:

$$D_0^2\psi = \mathbf{q}_1(\mathbf{x})\cos T_0 + \mathbf{q}_2(\mathbf{x})\sin T_0. \qquad (7.68)$$

$$D_1^2 = \mathbf{f}(D_1\mathbf{x},\mathbf{x}) +$$
$$\left\langle \mathbf{f}(D_1\mathbf{x}+D_0\psi,\mathbf{x}) - \mathbf{f}(D_1\mathbf{x},\mathbf{x}) + \psi\frac{\partial \mathbf{q}_1(\mathbf{x})}{\partial \mathbf{x}}\cos T_0 + \psi\frac{\partial \mathbf{q}_2(\mathbf{x})}{\partial \mathbf{x}}\sin T_0\right\rangle, \qquad (7.69)$$

where $\mathbf{f}(D_1\mathbf{x},\mathbf{x})$ has been added outside and subtracted inside the averaging brackets in (7.69). The averaging bracket contains the additional static terms due to the added excitation.

Equation (7.68) is solved to yield $\psi = -\mathbf{q}_1(\mathbf{x})\cos T_0 - \mathbf{q}_2(\mathbf{x})\sin T_0$. The solution is inserted into (7.69) and the following vibrational forcing terms are obtained:

$$\begin{cases}V_1\\V_2\end{cases} = f(v_y+v_x)\begin{cases}-20\sin(2x_2-2x_1)+2\sin(4x_2-4x_1)\\20\sin(2x_2-2x_1)-2\sin(4x_2-4x_1)\end{cases} + f(v_y-v_x)$$
$$\begin{cases}102\sin 2x_1 + 2\sin(4x_2-2x_1)+10\sin(2x_2-4x_1)-30\sin 2x_2\\-\sin(4x_2-6x_1)-\sin(4x_2-2x_1)+10\sin(2x_2-4x_1)+10\sin 2x_2\end{cases},$$

(7.70)

with $f \equiv (\cos(2x_1-2x_2)-5)^{-3}$, and where $v_x \equiv \sqrt[3]{4}w_x^2\Omega^2$ and $v_y \equiv \sqrt[3]{4}w_y^2\Omega^2$ denote the intensities of excitation. The expressions in Equation (7.70) reduce to those in Equation (7.48) if no transverse excitation is present ($v_x=0$).

7.3. Nonlinear Dynamics of the Follower-Loaded Double Pendulum

Stability of the straight pendulum

The eigenvalues of **A** determine the stability of the straight position and are found from the determinant equation det($\mathbf{A}-\lambda\mathbf{I}$)=0, where **I** is the identity matrix and λ is the eigenvalue. Solving the determinant equation yields the characteristic polynomial $a_0\lambda^4+a_1\lambda^3+a_2\lambda^2+a_3\lambda+a_4=0$ where:

$$a_0 = 2, \quad a_1 = 7c, \quad a_2 = c^2 + 7 + 2p(\alpha - 2) + 2(4v_y - v_x),$$

$$a_3 = c(2 + 3p(\alpha - 1) + (3v_y - 2v_x)), \quad (7.71)$$

$$a_4 = p(3 - p + 4v_y - 2v_x)(\alpha - 1) + (2v_y - 2v_x + 1)(v_y + 1),$$

from which the eigenvalues can be computed. When all eigenvalues have negative real parts, the straight position is stable and otherwise it is unstable.

The stability of the straight position is lost by divergence is an eigenvalue passes the origin from the left to the right half-plane. The condition for a zero eigenvalue is $a_4=0$. From Equation (7.71) it can be shown that the straight pendulum position is unstable without load applied if the following condition is fulfilled, $v_x > v_y + \frac{1}{2}$, i.e. when $w_x > (w_y^2 + 2/(3\Omega^2))^{\frac{1}{2}}$.

Stability is lost by flutter if a pair of complex conjugate eigenvalues crosses the imaginary axis from the left into the right half-plane. The condition for onset of flutter can be found as $a_1 a_2 a_3 - a_0 a_3^2 - a_4 a_1^2 = 0$, with the corresponding frequency of oscillation given as $\omega = (a_3/a_1)^{\frac{1}{2}}$.

Figure 7.14 shows the effect of added excitation on the divergence and flutter stability boundaries. In Figures 7.14a, 14b and 14c the middle curve-segment corresponds to flutter whereas the two other segments correspond to divergence. In the three figures, the stability boundaries for the system without excitation added have been included for comparison (shown with dashed lines).

With transverse excitation added (Figure 7.14b), the stability limits for divergence instabilities are moved to lower loads whereas the stability with respect to flutter is increased. The α-range with flutter is seen to be significantly decreased in favour of a larger range where divergence can occur.

Figure 7.14c shows an example of a specific choice of excitation in order to obtain a desired behaviour. The intensity of longitudinal and transverse excitation is chosen so that the right divergence stability boundary is kept almost unchanged, whereas the stability with respect to flutter is increased.

If the intensity of the transverse excitation component exceeds the critical value $v_x > v_y + \frac{1}{2}$, the stability diagram changes qualitatively (Figure 7.14d). If no load is applied, the straight position is unstable. For $\alpha > 1$ the straight position can be stabilized by adding a small load, but is destabilized again for $p > \approx 3.3$. There exists also a small stable region for $\alpha \approx 0.8$-0.95 and $p \approx 4$.

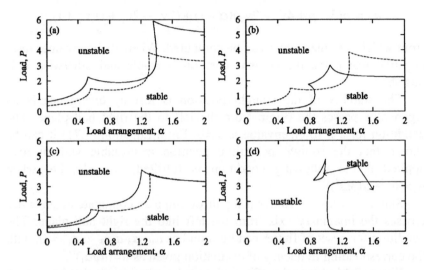

Figure 7.14. Stability boundaries for the straight pendulum position. Dashed lines: boundaries without added excitation. Parameter values: (a) $w_x=0$, $w_y=0.03$, (b) $w_y=0$, $w_x=0.03$, (c) $w_x=0.027$, $w_y=0.02$, (d) $w_x=0.05$, $w_y=0.0375$. In all figures, $\Omega=25$, $c=0.1$. From Jensen (2000a)[28].

Periodic motion

Periodic pendulum motion in case of flutter instability is now examined. The method of multiple scales using a fifth-order model expansion is applied, see e.g. also [57] or [26]. An approximate solution for the amplitude of periodic motion is derived and an estimate for the lower load-boundary for existence of periodic motion is obtained.

7.3. Nonlinear Dynamics of the Follower-Loaded Double Pendulum

The fifth-order expansion is written as,

$$\dot{\mathbf{y}} = \mathbf{A}(p,\alpha,c)\mathbf{y} + \varepsilon\bigl(\mathbf{f}(\mathbf{y},p,\alpha,c) + \mathbf{g}(\mathbf{y},p,\alpha,c)\bigr), \qquad (7.72)$$

where a bookkeeping parameter ε is introduced to indicate that nonlinear terms are assumed to be small compared to linear terms, and where $\mathbf{f}(\mathbf{y},p,\alpha,c)$ and $\mathbf{g}(\mathbf{y},p,\alpha,c)$ contain third- and fifth-order nonlinear terms respectively [28].

An approximate solution is looked for in form of the uniformly valid expansion $\mathbf{y}=\mathbf{y}_0(T_1,T_2)+\varepsilon\mathbf{y}_1(T_1,T_2)$, where two different time scales $T_1=t$ and $T_2=\varepsilon t$ are introduced. The load p is given a small perturbation σ from the critical value $p=p_{cri}+\varepsilon\sigma$, where ε indicates that σ is small. Expanding also $\mathbf{A}$, $\mathbf{f}$ and $\mathbf{g}$ in powers of ε and inserting into Equation (7.72) yields to order ε^0 and ε^1,

$$\varepsilon^0: \qquad \partial\mathbf{y}_0/\partial T_1 - \mathbf{A}_0\mathbf{y}_0 = \mathbf{0}, \qquad (7.73)$$

$$\varepsilon^1: \qquad \partial\mathbf{y}_1/\partial T_1 - \mathbf{A}_0\mathbf{y}_1 = -\partial\mathbf{y}_0/\partial T_2 + \sigma\mathbf{A}_1\mathbf{y}_0 + \mathbf{f}_0(\mathbf{y}_0) + \mathbf{g}_0(\mathbf{y}_0), \qquad (7.74)$$

where $\mathbf{f}_0$ and $\mathbf{g}_0$ denote $\mathbf{f}$ and $\mathbf{g}$ evaluated for $p = p_{cri}$.

Equation (7.73) is solved to yield $\mathbf{y}_0=A(T_2)\mathbf{u}\exp(i\omega T_1)+cc$, where $\mathbf{u}$ is the right eigenvector of the corresponding eigenvalue problem $(\mathbf{A}_0-i\omega\mathbf{I})\mathbf{u}=\mathbf{0}$, and cc denotes complex conjugates of the preceding terms.

The solution to Equation (7.74) will grow unbounded in time unless the product between $\mathbf{v}^T$, where $\mathbf{v}$ is the left eigenvector, and the parentheses on the r.h.s. in Equation (7.74), vanishes. The following condition for solvability is then obtained,

$$\partial A/\partial T_2 - \sigma\kappa A - \gamma A^2\overline{A} - \chi A^3\overline{A}^2 = 0, \qquad (7.75)$$

where κ, γ and χ are complex quantities.

The polar notation $A=a\cdot\exp(i\phi)$ is introduced. Stationary solutions can be found by setting $\partial a/\partial T_2=0$, which yields the following non-zero positive solution for a,

$$a = \sqrt{\frac{-\gamma_R \pm \sqrt{\gamma_R^2 - 4\sigma\kappa_R\chi_R}}{2\chi_R}}, \qquad (7.76)$$

where the subscript R denotes the real part of the complex number. An approximate solution for the amplitude of oscillation for the free end can now be found as $r \approx 2a|u_2|$, where u_2 is the second component of the right eigenvector **u**.

The type of the corresponding bifurcation is determined from the sign of γ_R. For $\gamma_R<0$, the bifurcation is supercritical and stable periodic solutions exist for post-critical loads, i.e. for $\sigma>0$. If $\gamma_R>0$, the bifurcation is subcritical and unstable periodic solutions exist for $\sigma<0$. If $\gamma_R>0$ and $\chi_R<0$, a saddle-node bifurcation of the unstable pre-critical solutions occur, and pre-critical stable periodic solutions co-exist with the stable zero-solution. The location of a saddle-node bifurcation is given as,

$$\sigma = p - p_{cri} = \frac{\gamma_R^2}{4\chi_R \kappa_R}. \tag{7.77}$$

Equation (7.77) defines an approximate global stability load for the subcritical bifurcation case, i.e. the lowest load for the existence of stable finite-amplitude periodic solutions.

The effects of the high-frequency excitation on the nonlinear dynamics is exemplified by load-response curves based on Equation (7.76), showing the amplitude of oscillation r of the free end.

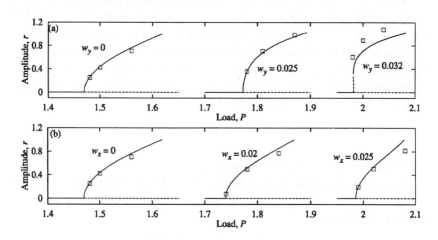

Figure 7.15. Nonlinear responses. Solid lines: stable solutions, dashed lines: unstable solutions. Discrete markers: numerical integration of the full model equations. Parameter values, (a) $w_x=0$, (b) $w_y=0$. In both figures: $\alpha=1$, $\Omega=25$, $c=0.1$. From Jensen [28].

In Figure 7.15a longitudinal excitation is added. The critical load is increased with increased w_y and the shape of the response curve is changed. For $w_y=0.032$, the corresponding bifurcation has turned subcritical and if w_y is further increased stable solutions exist for loads well below the critical load. When transverse excitation is added (Figure 7.15b) the critical load is increased but subcritical bifurcations are not created. It is noted that the approximate solution fairly accurately captures the behaviour of the full system. This accuracy is diminished, however, when the bifurcations turn subcritical.

7.3.6 Conclusions

We have analyzed linear stability and nonlinear behaviour of the partially follower-loaded elastic double pendulum with small-amplitude high-frequency (off-resonant) excitation of the support.

The method of direct partition of motion was used to turn the governing model equations into autonomous form, by approximating the added support-excitation by equivalent static forces. Linear stability was investigated, and a local nonlinear analysis was performed using the techniques of center manifold reduction/normal forms and the multiple scales perturbation method, as well as by numerical integration of the original equations and by the use of PATH (a path-following algorithm) applied to the autonomous set of equations. Lyapunov exponents were computed also, in order to study the global behaviour.

The presence of longitudinal support-excitation has been shown to strongly affect the linear stability of the follower-loaded pendulum, i.e. for most values of α, excitation of the support stabilized the system. The nonlinear behaviour was shown to change, both quantitatively and qualitatively, when support-excitation was added, e.g. supercritical bifurcations was turned into subcritical ones when a sufficient amount of support-excitation was added. Additionally, the parameter domain of chaotic dynamics increases.

By adding two perpendicular high-frequency excitation components to the partially follower-loaded double pendulum, is was possible to change and adjust the stability of the straight position with respect to flutter and divergence instabilities. The stability could be increased for flutter and increased for divergence, or it could be increased for flutter and left unchanged for divergence, etc. It was shown also how the nonlinear properties could be modified in order to create or avoid precritical periodic pendulum motion.

7.4 Articulated Pipes Conveying Fluid Pulsating with High Frequency

A two degree-of-freedom articulated model of a cantilever pipe conveying unsteady flowing fluid is considered next. The effect of a high-frequency pulsating component on the stability of the downward hanging pipe position and on the nonlinear dynamic behaviour is studied.

Fluid-conveying pipes have in the last four decades been the subjected extensive analysis. Previous works on continuous and articulated models include those of Benjamin [4], Chen [14], Bohn & Herrmann [9], Païdoussis & Issid [43, 45], Rousselet & Herrmann [49], Bajaj [2, 3], Païdoussis & Li [44].

The origin and nature of the fast excitation is inherently different for this system of fluid-conveying pipes, as compared to the pendulum system with support excitation, and thus the effect on the system behaviour is different. The motivation for this study was, apart from studying specifically the system of fluid conveying pipes, to gain further insight into the linear and nonlinear effects of high-frequency excitation. The work presented is a compact and edited version of the article by Jensen [26].

7.4.1 The Model and Model Equations

Figure 7.16 shows the model. The system considered consists of two articulated rigid pipes hanging downwards in gravity and conveying fluid. The pipes are connected to each other and to the support by identical connecting joints modelled by the linear stiffness coefficient k and the linear viscous damping coefficient $\tilde{c}$. The two pipes are assumed to be of equal length l and with equal mass per unit length m. Fluid is pumped through the pipes with the prescribed time-dependent flow speed $U(t)=U_0(1+p\cos\tilde{\Omega} t)$, where U_0 is the mean flow speed, and p and $\tilde{\Omega}$ the amplitude and frequency of the pulsating flow component, respectively. The frequency $\tilde{\Omega}$ is considered to be much larger than the characteristic frequencies for natural oscillations of the system. The fluid mass per unit pipe-length is M.

7.4. Articulated Pipes conveying Fluid Pulsating with High Frequency

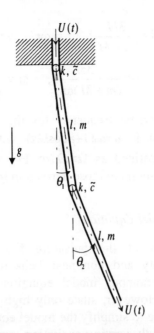

Figure 7.16. System configuration. Two rigid pipes hanging in gravity are connected to each other and the support by linear springs and dampers and are conveying fluid. The fluid speed has a high-frequency pulsating component. From Jensen (1999a)[26].

It can be shown that the following set of coupled nonlinear differential equations govern the motion of the upper and lower pipe [26]:

$$4\ddot{\theta}_1 + \tfrac{3}{2}\left(\ddot{\theta}_2 \cos(\theta_2 - \theta_1) - \dot{\theta}_2^2 \sin(\theta_2 - \theta_1)\right) + 2c\dot{\theta}_1 - c\dot{\theta}_2 + 2\theta_1 - \theta_2 \\ + 3g\sin\theta + \beta(\dot{u} + u^2)\sin(\theta_2 - \theta_1) + 2\beta u \dot{\theta}_2 \cos(\theta_2 - \theta_1) + \beta u \dot{\theta}_1 = 0, \quad (7.78)$$

$$\ddot{\theta}_2 + \tfrac{3}{2}\left(\ddot{\theta}_1 \cos(\theta_2 - \theta_1) + \dot{\theta}_1^2 \sin(\theta_2 - \theta_1)\right) + c\dot{\theta}_2 - c\dot{\theta}_1 + \theta_2 - \theta_1 \\ + g\sin\theta_2 + \beta u \dot{\theta}_2 = 0. \quad (7.79)$$

Equations (7.78) and (7.79) have been put in nondimensional form using the following set of variables:

$$\beta \equiv \frac{3M}{m+M}, \quad \tau \equiv \omega t, \quad u \equiv \frac{U}{\omega l},$$

$$g \equiv \frac{3\tilde{g}}{2l\omega^2}, \quad c \equiv \frac{3\tilde{c}}{(m+M)\omega l^3}, \quad \omega \equiv \sqrt{\frac{3k}{(m+M)l^3}}, \quad (7.80)$$

where ω is a characteristic frequency for the system and the time-dependent flow speed is $u=u_0(1+p\cos\Omega\tau)$ with the nondimensional excitation frequency defined as $\Omega \equiv \tilde{\Omega}/\omega$. For steady flow, Equations (7.78) and (7.79) correspond to those derived in [4].

7.4.2 Autonomous Model Equations

To study the effects of the pulsating fluid, analytical solutions concerning the stability and nonlinear behaviour of the system are desired. The non-autonomous model equations call for complicated methods of analysis. However, since only high-frequency excitation is considered it is possible to simplify the model equations considerably by approximating the rapidly-varying excitation terms with slowly-varying terms. This is done using an averaging method, the Method of Direct Partition of Motion (MDPM), see chapter 2 of this book and [7] or [27], which is based on separation of the slow and fast varying terms in the model equations.

The equations of motion are written in vector form with $\theta=\{\theta_1 \; \theta_2\}^T$:

$$\ddot{\theta} = \mathbf{f}(\dot{\theta}, \theta) + \mathbf{q}_1(\dot{\theta}, \theta, \Omega\tau) + \Omega\mathbf{q}_2(\theta, \Omega\tau) + \mathbf{q}_3(\theta, \Omega\tau), \quad (7.81)$$

Two time-scales are now introduced, a fast time-scale $T_0 \equiv \Omega\tau$ and a slow time-scale $T_1 \equiv \tau$. The fast scale describes motion at the rate comparable with the excitation frequency Ω, whereas the slow scale describes motion at a rate comparable with the natural frequencies of the system.

It is assumed that the solution can be separated into a term dependent of the slow time-scale only $\mathbf{x}(T_1)=\{x_1(T_1) \; x_2(T_1)\}^T$, and a term varying with both the fast as well as the slow time-scale $\psi(T_0, T_1)=\{\psi_1(T_0, T_1) \; \psi_2(T_0, T_1)\}^T$. The fast fluctuating term is assumed to be of order $\mu \equiv 1/\Omega$ compared to the slow oscillations. The solution assumption can thus be written in the form:

$$\theta(T_0, T_1) = \mathbf{x}(T_1) + \mu\psi(T_0, T_1), \quad (7.82)$$

7.4. Articulated Pipes conveying Fluid Pulsating with High Frequency

where $\mathbf{x}$ and ψ are terms of order μ^0. It is assumed also that $\psi(T_0, T_1)$ is a 2π-periodic function of T_0 with: $\langle\psi(T_0, T_1)\rangle \equiv \frac{1}{2\pi}\int_0^{2\pi} \psi(T_0, T_1)dT_0 = 0$, i.e. a zero T_0-average.

The solution assumption (7.82) is inserted into the model equation (7.81):

$$D_1^2\mathbf{x} + \mu^{-1}D_0^2\psi + 2D_0D_1\psi + \mu D_1^2\psi = \mathbf{f}(D_1\mathbf{x} + D_0\psi + \mu D_1\psi, \mathbf{x} + \mu\psi)$$
$$+ \mathbf{q}_1(D_1\mathbf{x} + D_0\psi + \mu D_1\psi, \mathbf{x} + \mu\psi) + \mu^{-1}\mathbf{q}_2(\mathbf{x} + \mu\psi) + \mathbf{q}_3(\mathbf{x} + \mu\psi), \quad (7.83)$$

with the notation $D_i^j \equiv \partial^j/\partial T_i^j$ introduced in order to denote partial differentiation with respect to the different time-scales. Applying the averaging operator to both sides of Equation (7.83) and after some rearranging of terms [24] leads to the μ^0-order approximation of the model equations:

$$D_0^2\psi = \mathbf{q}_2(\mathbf{x}), \quad (7.84)$$

$$D_1^2\mathbf{x} = \mathbf{f}(D_1\mathbf{x}, \mathbf{x})$$
$$+ \langle \mathbf{f}(D_1\mathbf{x} + D_0\psi, \mathbf{x}) - \mathbf{f}(D_1\mathbf{x}, \mathbf{x}) + \mathbf{q}_1(D_1\mathbf{x} + D_0\psi, \mathbf{x}) + \psi \cdot \partial\mathbf{q}_2(\mathbf{x})/\partial\mathbf{x}\rangle, \quad (7.85)$$

In the presented form, the averaging bracket represents the additional terms in the equation governing the slow variables that arise due to the pulsating component of the flow.

Solving Equation (7.84) for ψ, inserting into Equation (7.85) and rearranging, the equations governing the slow variables x_1 and x_2 finally become:

$$4\ddot{x}_1 + \tfrac{3}{2}\left(\ddot{x}_2 \cos(x_2 - x_1) - \dot{x}_2^2 \sin(x_2 - x_1)\right) + 2c\dot{x}_1 - c\dot{x}_2$$
$$+ 2x_1 - x_2 + 3g\sin x_1 + \beta u_0^2 \sin(x_2 - x_1) \quad (7.86)$$
$$+ \beta u_0\left(2\dot{x}_2 \cos(x_2 - x_1) + \dot{x}_1\right) + \beta\left(V_1 + \tfrac{1}{2}p^2 u_0^2 \sin(x_2 - x_1)\right) = 0,$$

$$\ddot{x}_2 + \tfrac{3}{2}\left(\ddot{x}_1 \cos(x_2 - x_1) + \dot{x}_1^2 \sin(x_2 - x_1)\right) + c\dot{x}_2 - c\dot{x}_1$$
$$+ x_2 - x_1 + g\sin x_2 + \beta u_0 \dot{x}_2 + \beta V_2 = 0, \quad (7.87)$$

where V_1 and V_2 are

$$\begin{Bmatrix} V_1 \\ V_2 \end{Bmatrix} = \frac{\beta p^2 u_0^2}{(18\cos 2(x_2 - x_1) - 46)^3} \begin{Bmatrix} -18257\sin(x_2 - x_1) - 5152\sin 2(x_2 - x_1) + \\ 9546\sin 2(x_2 - x_1) - \end{Bmatrix}$$
$$\begin{matrix} 11379\sin 3(x_2 - x_1) + 1008\sin 4(x_2 - x_1) - 2889\sin 5(x_2 - x_1) + 243\sin 7(x_2 - x_1) \\ 2664\sin 4(x_2 - x_1) + 162\sin 6(x_2 - x_1) \end{matrix} \Bigg\}.$$

(7.88)

When results based upon numerical integration of the original model Equations (7.78) and (7.79) and the autonomous Equations (7.86) and (7.87) are compared, it is noted that even for large values of p and large pipe angles the autonomous equations yield accurate results. The small fluctuations are discarded and the slow oscillations at the rate comparable with the natural frequencies are fully captured. Also, numerical integration of the simpler equations is significantly faster compared to integration of the non-autonomous equations, since the time-steps now just need to be sufficiently small to capture the slow oscillations instead of oscillations at the rate of the excitation frequency.

It should be noted however, that the transformation only captures solutions which can be written on the form (7.82). Performing sample numerical integration and simulations with many different parameter combinations has not revealed any dynamic behaviour which cannot be separated as proposed.

The effects of the added rapid flow pulsations are now studied by examining the slow components of motion governed by Equations (7.86) and (7.87).

7.4.3 Linear Stability of the Hanging Position

In this section the linear stability of the downward hanging position of the pipes is investigated. The effect of the added flow pulsations on the stability of this position with respect to onset of flutter is analyzed by examining the eigenvalues of the corresponding linearized autonomous model equations.

Linearizing the equations near the hanging position given as $(x_1, x_2) = (0, 0)$ yields:

$$\mathbf{M}\ddot{\mathbf{x}} + (\mathbf{C}_{static} + \mathbf{C}_{fluid})\dot{\mathbf{x}} + (\mathbf{K}_{static} + \mathbf{K}_{fluid} + \mathbf{K}_{puls})\mathbf{x} = 0, \quad (7.89)$$

7.4. Articulated Pipes conveying Fluid Pulsating with High Frequency

where **M** is the mass matrix, **C** the damping matrix, and **K** the stiffness matrix [26].

The pulsating flow component contributes with an asymmetric component to the stiffness matrix for all values of β. The effect of pulsation can thus be either stabilizing or destabilizing for the downward hanging position. To this first approximation, no contribution from the pulsation to the damping matrix arises. Second-order terms proportional to the inverse of the pulsation frequency, will however enter the damping and also the stiffness matrix in a second-order approximation.

With assumed time-dependent form of the solution $\mathbf{x}(\tau)=\mathbf{z}\cdot\exp(\lambda\tau)$, the following characteristic equation for the eigenvalue λ is obtained:

$$a_0\lambda^4 + a_1\lambda^3 + a_2\lambda^2 + a_3\lambda + a_4 = 0, \qquad (7.90)$$

where

$$\begin{aligned}
a_0 &= 49, a_1 = 28(9c + 2\beta u_0), \\
a_2 &= 252 + 28c^2 + 196g + 14\beta u_0(2\beta u_0 + 10c - 5u_0) + \\
&\quad (76\beta - 35)\beta p^2 u_0^2, \\
a_3 &= 56c + 140cg + 140\beta u_0 + 112\beta u_0 g + 12c\beta^2 p^2 u_0^2 + \\
&\quad (2(20\beta - 7)p^2 - 28)\beta^2 u_0^3, \\
a_4 &= 28 + 140g + 84g^2 - 28\beta u_0^2 g + 12\beta^2 p^2 u_0^2 + 2(20\beta - 7)g\beta p^2 u_0^2.
\end{aligned} \qquad (7.91)$$

Onset of flutter

The downward hanging position will be stable only if all λ satisfying Equation (7.90) have negative real parts. For the case of $u_0=0$, this is so for positive damping $c>0$. Thus, applied disturbances will decay, and the two pipes will swing to rest in the downward hanging position. If $u_0\neq 0$, for some value of u_0 a single eigenvalue or a pair of complex conjugate eigenvalues will pass the imaginary axis into the positive real half-plane, and an initially applied disturbance will be magnified. The hanging position will thus be unstable. If a pair of complex eigenvalues crosses the imaginary axis the instability will be dynamic (flutter). If a single eigenvalue passes through the origin, the instability is static (divergence). For this system both types of instabilities may occur depending on system parameters. As mentioned previously the study will be restricted to cover only the case of flutter.

For onset of flutter, a pair of complex conjugate eigenvalues crosses the imaginary axis into the right half-plane, i.e., $\lambda = \pm i\omega$ where $\omega \neq 0$ is the frequency of flutter. We obtain the following condition for onset of flutter:

$$a_1 a_2 a_3 - a_0 a_3^2 - a_1^2 a_4 = 0, \qquad (7.92)$$

with the corresponding frequency is found from $\omega^2 = a_3/a_1$. Equation (7.92) is recognized as part of the Routh-Hurwitz criteria. Evaluating a_3/a_1 it can be shown that $\omega_p > \omega_{p=0}$ for $\beta > 7/20$, whereas for $\beta < 7/20$ the flutter frequency will decrease with added pulsation, if the damping coefficient c is sufficiently small.

In the following a stability diagram is presented showing the boundaries between linear stability of the downward hanging position and flutter oscillations, based on Equation (7.92). Figure 7.17 shows the stability boundaries for four different values of the relative pulsation amplitude p. For mass ratios larger than $\beta \approx 0.58$, the effect of the added flow pulsations is stabilizing, i.e. the critical mean fluid speed u_{cri} for onset of flutter is increased. Below this value of β the added pulsation acts marginally destabilizing. Changing the frequency of excitation has no effect on the stability properties, provided that the frequency is sufficiently high for the μ^0-order approximation of the autonomous equations to be valid. Numerical integration of the full model equations

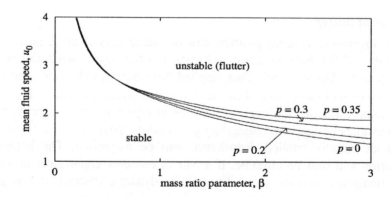

Figure 7.17. Stability diagram showing borders between linear stability of the downward hanging position and flutter oscillations for the case of constant fluid speed and a pulsating fluid for three different values of the relative pulsation amplitude p. Parameter-values: $c=0.1$, $g=0.2$. From Jensen [26].

(3.78) and (3.79) shows that for $\Omega > \approx 10$, Equation (3.92) gives reliable results for sufficiently low values of p.

For magnitudes of flow pulsations higher than $p \approx 0.35$ the results based on Equation (3.92) are inaccurate in the region of high mass ratios. Numerical integration of the full equations shows the hanging position to be unstable even for very small initial disturbances also when the theoretical results predict the hanging position to be stable. A possible reason could be that the basin of attraction for the hanging position for large values of p becomes so small that the presence of fast fluctuations is sufficient to push the system beyond the limit of this basin.

Changing the amount of viscous damping will affect the stability curves only quantitatively, no qualitative changes can be noted. For different ratios of gravity to elastic forces, the curves are also expected to be qualitatively similar, given that this ratio is so small that the pipes will not diverge.

7.4.4 Periodic Motion of the Autonomous System

The stability analysis presented in the previous section revealed when stability of the downward hanging position was lost. No information was obtained concerning the post-critical behaviour or about the global stability of this position. A quantitative analysis of periodic oscillations of the pipes is presented in the following. Small-amplitude local oscillations are analyzed using the method of multiple scales, and large-amplitude oscillations are examined using a path-following algorithm based on numerical integration of the approximating autonomous set of model equations.

Small-Amplitude Periodic Oscillations

A fifth-order Taylor expansion of the autonomous model equations is carried out in order to capture the behaviour of the pipes in the vicinity of the critical flow speed. The choice to include nonlinearities up to fifth order is made, since a third-order model is shown to be insufficient to capture the essential dynamics. The nonlinear analysis is performed using the perturbation method of multiple scales. Bifurcation diagrams are provided, showing small-amplitude oscillations of the pipes in the vicinity of the critical mean fluid speed. The transition of bifurcations from super- to sub-critical is examined also.

The notation $\mathbf{y}=\{x_1\ x_2\ \dot{x}_1\ \dot{x}_2\}^T$ is introduced. The fifth-order approximation can now be written as:

$$\dot{\mathbf{y}} = \mathbf{A}\mathbf{y} + \varepsilon\mathbf{f}(\mathbf{y}) + \varepsilon\mathbf{g}(\mathbf{y}), \qquad (7.93)$$

where the bookkeeping parameter ε is introduced to indicate that the nonlinear terms are assumed to be small compared to the linear terms and $\mathbf{A}$, $\mathbf{f}(\mathbf{y})$ and $\mathbf{g}(\mathbf{y})$ are the linear system matrix and vectors containing third- and fifth-order nonlinear terms, respectively.

The local behaviour in the vicinity of the critical mean fluid speed u_0 is to be examined. Giving u_0 a small perturbation σ from the critical value, the resulting mean fluid speed becomes $u_0 = u_{\text{cri}} + \varepsilon\sigma$, where ε indicates that the perturbation $\varepsilon\sigma$ is small. Expanding $\mathbf{A}$, $\mathbf{f}(\mathbf{y})$ and $\mathbf{g}(\mathbf{y})$ in terms of ε yields:

$$\mathbf{A} = \mathbf{A}_0 + \varepsilon\sigma\mathbf{A}_1 + O(\varepsilon^2), \quad \mathbf{f}(\mathbf{y}) = \mathbf{f}_0 + O(\varepsilon), \quad \mathbf{g}(\mathbf{y}) = \mathbf{g}_0 + O(\varepsilon), \qquad (7.94)$$

with

$$\mathbf{A}_0 = \mathbf{A}\big|_{u_0=u_{\text{cri}}}, \quad \mathbf{A}_1 = \partial\mathbf{A}/\partial u_0\big|_{u_0=u_{\text{cri}}}, \quad \mathbf{f}_0 = \mathbf{f}(\mathbf{y})\big|_{u_0=u_{\text{cri}}}, \quad \mathbf{g}_0 = \mathbf{g}(\mathbf{y})\big|_{u_0=u_{\text{cri}}}. \qquad (7.95)$$

Inserting the expansions in Equation (7.94) into Equation (7.93) yields:

$$\dot{\mathbf{y}} = \mathbf{A}_0\mathbf{y} + \varepsilon(\sigma\mathbf{A}_1 + \mathbf{f}_0 + \mathbf{g}_0) + O(\varepsilon^2), \qquad (7.96)$$

An approximate solution to Equation (7.96) is now obtained by use of the perturbation method of multiple scales. The method of analysis follows the outline of Thomsen [57], where a similar analysis was conveniently carried out for the follower-loaded double pendulum in matrix form.

The two time scales $T_1=\tau$ and $T_2=\varepsilon\tau$ are introduced, with the slow time scale T_2 describing the slow modulation of amplitudes and phases. To obtain an ε^0-order approximate solution to the solution of Equation (7.61), an ε^1-order uniformly valid expansion is introduced:

$$\mathbf{y} = \mathbf{y}_0(T_1, T_2) + \varepsilon\mathbf{y}_1(T_1, T_2) + O(\varepsilon^2). \qquad (7.97)$$

Inserting the expansion (7.97) into Equation (7.96) yields to order ε^0:

$$\partial\mathbf{y}_0/\partial T_1 - \mathbf{A}_0\mathbf{y}_0 = \mathbf{0}, \qquad (7.98)$$

and to order ε^1:

7.4. Articulated Pipes conveying Fluid Pulsating with High Frequency

$$\frac{\partial \mathbf{y}_1}{\partial T_1} - \mathbf{A}_0 \mathbf{y}_1 = -\frac{\partial \mathbf{y}_0}{\partial T_2} + \sigma \mathbf{A}_1 \mathbf{y}_0 + \sum_{j,k,l=1}^{4} \mathbf{b}_0 x_{0j} x_{0k} x_{0l} + \sum_{j,k,l,m,n=1}^{4} \mathbf{c}_0 x_{0j} x_{0k} x_{0l} x_{0m} x_{0n},$$
(7.99)

where $\mathbf{b}_0$ and $\mathbf{c}_0$ denote $\mathbf{b}_{jkl}$ and $\mathbf{c}_{jklmn}$ evaluated for $u_0 = u_{cri}$.
The solution to the ε^0-order Equation (7.98) is:

$$\mathbf{y}_0 = Y(T_2)\mathbf{u} e^{i\omega_{cri} T_1} + \overline{Y}(T_2)\overline{\mathbf{u}} e^{-i\omega_{cri} T_1},$$
(7.100)

where Y is a complex function of T_2, a bar denotes the complex conjugate, and $i\omega_{cri}$ and $\mathbf{u} = \{u_1 \; u_2 \; u_3 \; u_4\}^T$ is the purely imaginary eigenvalue and the right eigenvector, respectively, for the corresponding eigenvalue problem:

$$(\mathbf{A}_0 - i\omega_{cri}\mathbf{I})\mathbf{u} = \mathbf{0}.$$
(7.101)

Substituting Equation (7.100) into the ε^1-order Equation (7.99) yields:

$$\partial \mathbf{y}_1 / \partial T_1 - \mathbf{A}_0 \mathbf{y}_1 = \mathbf{q}_1 e^{i\omega_{cri} T_1} + \mathbf{q}_3 e^{3i\omega_{cri} T_1} + \mathbf{q}_5 e^{5i\omega_{cri} T_1} + cc,$$
(7.102)

where cc denotes complex conjugates of the preceding terms. The terms $\mathbf{q}_1$, $\mathbf{q}_3$ and $\mathbf{q}_5$ are given in [26].

The solution to Equation (7.102) will contain secular terms which grow unbounded in time unless $\mathbf{v}^T\mathbf{q}_1=0$ ([57]), where $\mathbf{v}$ is the left eigenvector of $\mathbf{A}_0$ found from $\mathbf{v}^T(\mathbf{A}_0-i\omega_{cri}\mathbf{I})=\mathbf{0}^T$, i.e., the r.h.s. of Equation (7.102) must be orthogonal to the homogenous solution. The solvability condition thus becomes:

$$\partial Y/\partial T_2 - \kappa Y - \gamma Y^2 \overline{Y} - \chi Y^3 \overline{Y}^2 = 0,$$
(7.103)

where $\Re(\;)$ and $\Im(\;)$ denote real and the imaginary parts of a complex number. Now, introducing the polar notation $Y = a \cdot \exp(i\varphi)$ and separating Equation (7.103) into a real and an imaginary equation, the solvability condition can be written as:

$$\frac{\partial a}{\partial T_2} = \Re(\kappa)a + \Re(\gamma)a^3 + \Re(\chi)a^5,$$
(7.104)

$$a\frac{\partial \varphi}{\partial T_2} = a\left(\Im(\kappa) + \Im(\gamma)a^2 + \Im(\chi)a^4\right).$$
(7.105)

The stationary solutions are found by setting $\partial a/\partial T_2 = 0$. The non-zero positive solution for a then becomes:

$$a = \sqrt{\frac{-\Re(\gamma) \pm \sqrt{\Re^2(\gamma) - 4\Re(\chi)\Re(\kappa)}}{2\Re(\chi)}}, \qquad (7.106)$$

which exists for parameter combinations that render the radicals positive. With a determined from Equation (7.106), $\varphi(T_2)$ can be found by integrating Equation (7.105).

With a given by Equation (7.106) and $\varphi(T_2)$ found from Equation (7.105), the ε^0-order approximate solution to Equation (7.93) is:

$$x_i \approx x_{0i} = 2\Re(Au_i e^{i\omega_{cri}T_1}) = 2a|u_i|\cos(\omega_{cri}\tau + \varphi(\tau) + \theta_i), \quad i = 1,\ldots,4 \quad (7.107)$$

where $\tan\theta_i = \Im(u_i)/\Re(u_i)$.

In the following, sample response curves based on the approximate solution (7.107) are presented. The results exemplify the effect of pulsating flow on the local nonlinear bifurcation behaviour of the pipes. The results based on the fifth-order expansion of the autonomous model equations are compared to results based on a third-order expansion, found by setting $\Re(\chi)=\Im(\chi)=0$ in Equations (7.104) and (7.105), as well as to numerical integration of the fifth-order autonomous equations and of the full model equations.

Figure 7.18 shows such response curves for four different values of the magnitude of the pulsating flow component, p, with all other parameters kept constant. In all four pictures the stable zero solution, corresponding to the downward hanging position, loses stability by a Hopf bifurcation for a certain critical value of the mean fluid speed u_0. The two lowest values of p (Figure 7.18a,b) show supercritical Hopf bifurcations, whereas for the two higher values of p (Figure 7.18c,d) the bifurcations are subcritical. In the two supercritical cases the fifth- and third-order expansions yield qualitatively similar results, both predicting a stable post-critical solution branch. For $p=0.3$ (Figure 7.18b), the third-order expansion predicts far too large an amplitude of oscillation. For the subcritical Hopf bifurcations the results are qualitatively different. The fifth-order expansion correctly predicts stable post-critical solution branches, whereas the third-order model predicts that only unstable pre-critical solutions exist. This shows that relying on a nonlinear model including nonlinearities only up to third order might be misleading. The unstable pre-critical solutions predicted by the third-order model suggest possible existence of large-amplitude pre-critical solutions caused by the repelling unstable solution branches. Also, the lack of post-critical periodic solutions suggests non-periodic and possibly chaotic post-critical behaviour. Examining the fifth-order model, the existence of such solutions is not indicated, but of course not ruled out either.

The numerical results based on the fifth-order expansion of the autonomous model equations, indicated with discrete box-markers in Figure 7.18, show that the approximate analytical solution adequately models the fifth-order expansion. The limited validity of the fifth-order model is indicated when the results are compared to numerical integration of the full model equations, shown with discrete circle-markers. The integration is based on the non-autonomous equations but are identical to results based on the full autonomous equations. It is seen that for amplitudes of oscillations above about 10% of a full pipe rotation, significant discrepancies exist.

Figure 7.18 indicated that the Hopf bifurcation, encountered when the mean fluid speed is increased beyond the critical value, turns from being supercritical to subcritical for a certain value of the magnitude of pulsation p. In Figure 7.19 is shown the type of Hopf bifurcation for

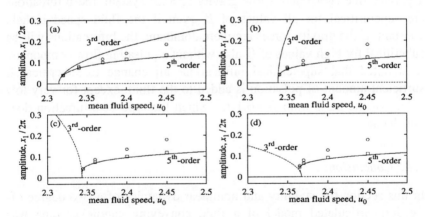

Figure 7.18. Mean fluid speed response curves for four different values of p showing results for the fifth-order and the third-order expansion, and numerical integration of the full equations and the autonomous fifth-order model, (a) $p=0.25$, (b) $p=0.30$, (c) $p=0.31$, (d) $p=0.35$. Parameter-values: $\beta=1$, $\Omega=50$, $c=0.1$, $g=0.2$. Solid lines: stable solutions, dashed lines: unstable solutions, ○ : numerical integration of (3.78) and (3.79), □ : numerical integration of (3.93). From Jensen [26].

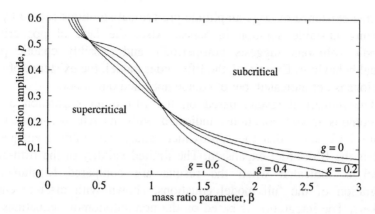

Figure 7.19. Bifurcation diagram showing the transition curves between super- and sub-critical Hopf-bifurcations for four different values of the relative strength of gravity compared to the elastic forces. Parameter-values: $c=0.1$. From Jensen (1999a)[26].

values of the mass ratio parameter β and the magnitude of fluid speed pulsation p, for four values of the gravity parameter g. It is noted that for a gravity free ($g=0$) and a low gravity ($g=0.2$) system, the bifurcations are supercritical for all values of β, even if the fluid speed is held constant. If the gravity forces are more dominant, the bifurcations will be subcritical for high values of β, also without a pulsating flow component. It is noted that supercritical bifurcations all change to subcritical if sufficient amount of pulsation is added to the fluid, except for very low values of the mass ratio $\beta < \approx 0.1$ for which the type of bifurcation does not change.

7.4.5 Conclusions

In this section the stability and nonlinear dynamics of a two degree-of-freedom articulated model of a fluid conveying cantilever pipe was analyzed, with fast excitation added to the system via a high-frequency pulsating component of the fluid speed.

An approximate analytical solution for local periodic flutter oscillations, based on a fifth-order Taylor expansion of the autonomous equations, was derived. The derived equation adequately described qualitatively and quantitatively the bifurcation behaviour, and proved superior to a third-order model. It was shown that added fluid speed pulsation could turn supercritical bifurcations subcritical.

It was demonstrated how the added pulsation increased the linear stability of the hanging position, but also that the global stability of the stabilized position was limited due to nearby repelling unstable solutions and coexisting large-amplitude nonlinear solutions.

References

1. Babitsky, V. I., & Veprik, A. M. Damping of beam forced vibration by a moving washer. *Journal of Sound and Vibration* 166(1), 77-85, 1993.
2. Bajaj, A. K. Interactions between self and parametrically excited motions in articulated pipes. *Journal of Applied Mechanics* 51, 423-429, 1984.
3. Bajaj, A. K. Nonlinear dynamics of tubes carrying a pulsatile flow. *Dynamics and Stability of Systems* 2, 19-41, 1987.
4. Benjamin, T. B. Dynamics of a system of articulated pipes conveying fluid. I. Theory. *Proceedings of the Royal Society (London)* A261, 457-486, 1961.
5. Blekhman I. I, Method of direct separation of motions in problems on the action of vibration on nonlinear dynamic systems, *Izv. AN SSSR, MTT*, 6 (1976) 13–27 (in Russian).
6. Blekhman I. I., Development of a conception of direct separation of motions in nonlinear mechanics, in *Modern Problems of Theor. and App. Mechanics: proceedings of IV All Union Congress on Theor. and App. Mechanics* (Naukova Dumka, Kiev, 1978) 148–168 (in Russian).
7. Blekhman, I. I. *Vibrational mechanics - nonlinear dynamic effects, general approach, applications.* Singapore: World Scientific, 2000.
8. Blekhman, I. I., & Malakhova, O. Z. Quasi-equilibrium positions of the Chelomei pendulum. *Soviet Physics Doklady* 31(3), 229-231, 1986.
9. Bohn, M. P. & Herrmann, G. The dynamic behavior of articulated pipes conveying fluid with periodic flow rate. *American Society of Mechanical Engineers Journal of Applied Mechanics* 41, 55-62, 1974.
10. Bolotin, V. V. *Nonconservative problems of the theory of elastic stability.* Oxford: Pergamon Press, 1963.
11. Carr, J. *Applications of centre-manifold theory.* New York: Springer-Verlag, 1981.
12. Champneys, A. R., & Fraser, W. B. The 'Indian rope trick' for a parametrically excited flexible rod: linearized analysis. *Proceedings of the Royal Society of London A* 456(1995), 553-570, 2000.
13. Chelomei, V. N. Mechanical paradoxes caused by vibrations. *Soviet Physics Doklady* 28(5), 387-390, 1983.
14. Chen, S. S. Dynamic Stability of a tube conveying fluid. *Proc. ASCE J. Eng. Mech. Div.* 97, 1469-1485, 1971.
15. Feeny, B. F., & Moon, F. C. Quenching stick-slip chaos with dither. *Journal of Sound and Vibration* 237(1), 173-180, 2000.
16. Fidlin, A. On asymptotic properties of systems with strong and very strong high-frequency excitation. *Journal of sound and vibration* 235(2), 219-233, 2000.
17. Fidlin, A., & Thomsen, J. J. Predicting vibration-induced displacement for a resonant friction slider. *European Journal of Mechanics A/Solids* 20, 155-166, 2001.

18. Flügge, W. *Handbook of engineering mechanics*. New York: McGraw-Hill, 1962.
19. Guckenheimer, J. & Holmes, P. *Nonlinear Oscillations, Dynamical Systems, and Bifurcations of Vector Fields*. New York: Springer-Verlag, 1983.
20. Hansen, M. H. Effect of high-frequency excitation on natural frequencies of spinning disks. *Journal of Sound and Vibration* 234(4), 577-589, 2000.
21. Herrmann, G. & Jong, I. C. On Nonconservative Stability Problems of Elastic Systems with Slight Damping. *Journal of Applied Mechanics* 33, 125-133, 1966.
22. Herrmann, G., Nemat-Nasser, S. & Prasad, S. N. *Models demonstrating instability of nonconservative mechanical systems*. Technical report No. 66-4, National Aeronautics and Space Administration, 1966.
23. Jensen, J. S. Transport of continuos material in vibrating pipes. In, EUROMECH 2nd European Nonlinear Oscillation Conference. Prague: Czech Technical University, Prague, 211-214, 1996.
24. Jensen, J. S. Fluid transport due to nonlinear fluid-structure interaction. *Journal of Fluids and Structures* 11, 327-344, 1997.
25. Jensen, J. S. Non-linear dynamics of the follower-loaded double pendulum with added support-excitation. *Journal of Sound and Vibration* 215(1), 125-142, 1998.
26. Jensen, J. S. Articulated pipes conveying fluid pulsating with high frequency. *Nonlinear Dynamics* 19, 171-191, 1999a.
27. Jensen, J. S. *Non-trivial effects of fast harmonic excitation*. Ph. D. Dissertation, Department of Solid Mechanics, Technical University of Denmark. *DCAMM Report*, S83, 1999b.
28. Jensen, J. S. Effects of high-frequency bi-directional support-excitation of the follower-loaded double pendulum. In E. Lavendelis, & M. Zakrzhevsky (eds.), *Klüwer series: Solid Mechanics and its Applications, Vol. 37*, IUTAM/IFToMM Symposium on Synthesis of nonlinear Dynamical Systems, August 1998. Riga: Klüwer (Dordrecht), 169-178, 2000a.
29. Jensen, J. S. Buckling of an elastic beam with added high-frequency excitation. *Int. Journal of Non-linear Mechanics* 35, 217-227, 2000b.
30. Jensen, J. S., Tcherniak, D. M., & Thomsen, J. J. Stiffening effects of high-frequency excitation: experiments for an axially loaded beam. *ASME Journal of Applied Mechanics* 67(2), 397-402, 2000c.
31. Kaas-Petersen, C., *PATH-User's Guide*, Department of Applied Mathematical Studies, University of Leeds, England, 1989.
32. Kapitza, P. L. Dynamic stability of a pendulum with an oscillating point of suspension (in Russian). *Zurnal Eksperimental'noj i Teoreticeskoj Fiziki* 21(5), 588-597, 1951.

33. Kirgetov, A. V. On the stability of quasi-equilibrium positions of Chelomey's pendulum. *Izv. AN SSSR. Mekhanika Tverdogo Tela* 21(6), 57-62, 1986.
34. Krylov, V., & Sorokin, S. V. Dynamics of elastic beams with controlled distributed stiffness parameters. *Smart Materials and Structures* 6, 573-582, 1997.
35. Kurbatov, A. M., Chelomey, S. V., & Khromushkin, A. V. On Chelomey's pendulum. *Izv. AN SSSR. Mekhanika Tverdogo Tela* 21(6), 63-65, 1986.
36. Langthjem, M. A. & Sugiyama Y. Dynamic stability of columns subjected to follower loads: a survey. *Journal of Sound and Vibration* 238(5), 809-851, 2000.
37. Levi, M. Geometry and physics of averaging with applications. *Physica D* 132, 150-164, 1999.
38. Li, G. X. & Païdoussis, M. P. Stability, double degeneracy and chaos in cantilevered pipes conveying fluid. *International Journal of Non-linear Mechanics* 29(1), 83-107, 1994.
39. Menyailov, A. I., & Movchan, A. V. Stabilization of a pendulum-ring system under conditions of vibration of the base. *Izv. AN SSSR. Mekhanika Tverdogo Tela* 19(6), 35-40, 1984.
40. Miranda, E. C., & Thomsen, J. J. Vibration induced sliding: theory and experiment for a beam with a spring-loaded mass. *Nonlinear Dynamics* 16(2), 167-186, 1998.
41. Nayfeh, A. H., & Mook, D. T. *Nonlinear oscillations.* New York: John Wiley, 1979.
42. Nayfeh, A. H. *Method of Normal Forms.* New York: John Wiley, 1993.
43. Païdoussis, M. P. & Issid, N. T. Dynamic stability of pipes conveying fluid', *Journal of Sound and Vibration* 33, 267-294, 1974.
44. Païdoussis, M. P. & Li, G. X. Pipes conveying fluid: A model dynamical problem. *Journal of Fluids and Structures* 7, 137-204, 1993.
45. Païdoussis, M. P. & Issid, N. T. Experiments on parametric resonance of pipes containing pulsatile flow. *Journal of Applied Mechanics* 43, 198-202, 1976.
46. Païdoussis, M. P. & Semler, C. Nonlinear dynamics of a fluid-conveying cantilevered pipe with an intermediate spring support. *Journal of Fluids and Structures* 7, 269-298, 1993
47. Panovko, Y. G., & Gubanova, I. I. *Stability and oscillations of elastic systems; Paradoxes, fallacies and new concepts.* New York: Consultants Bureau, 1965.
48. Perko, L. *Differential Equations and Dynamical Systems.* New York: Springer-Verlag, 1991.
49. Rousselet, J. and Herrmann G. Flutter of articulated pipes at finite amplitude. *Journal of Applied Mechanics* 44, 154-158, 1977.

50. Sudor, D. J., & Bishop, S. R. Inverted dynamics of a tilted pendulum. *European Journal of Mechanics A/Solids* 18, 517-526, 1999.
51. Tcherniak, D. M. The influence of fast excitation on a continuous system. *Journal of Sound and Vibration* 227(2), 343-360, 1999.
52. Tcherniak, D. M. Using fast vibration to change the nonlinear properties of mechanical systems. In E. Lavendelis, & M. Zakrzhevsky (eds.), *Klüwer series: Solid Mechanics and its Applications, Vol. 37*, IUTAM/IFToMM Symposium on Synthesis of nonlinear Dynamical Systems, August 1998. Riga: Klüwer (Dordrecht), 227-236, 2000.
53. Tcherniak, D., & Thomsen, J. J. Slow effects of fast harmonic excitation for elastic structures. *Nonlinear Dynamics* 17(3), 227-246, 1998.
54. Thomsen, J. J. Chaotic Dynamics of the Partially Follower-loaded Elastic Double Pendulum. *Journal of Sound and Vibration* 188(3), 385-405, 1995.
55. Thomsen, J. J. Vibration induced sliding of mass: non-trivial effects of rotatory inertia. In, EUROMECH 2nd European Nonlinear Oscillation Conference. Prague: Czech Technical University, Prague, 455-458, 1996a.
56. Thomsen, J. J. Vibration suppression by using self-arranging mass: effects of adding restoring force. *Journal of Sound and Vibration* 197(4), 403-425, 1996b.
57. Thomsen, J. J. *Vibrations and stability, order and chaos*. London: McGraw-Hill, 1997.
58. Thomsen, J. J. Using fast vibrations to quench friction-induced oscillations. *Journal of Sound and Vibration* 228(5), 1079-1102, 1999.
59. Thomsen, J. J. Vibration-induced displacement using high-frequency resonators and friction layers. In E. Lavendelis, & M. Zakrzhevsky (eds.), *Klüwer series: Solid Mechanics and its Applications, Vol. 37*, IUTAM/IFToMM Symposium on Synthesis of nonlinear Dynamical Systems, August 1998. Riga: Klüwer (Dordrecht), 237-246, 2000.
60. Thomsen, J. J. Theories and experiments on the stiffening effect of high-frequency excitation for continuous elastic systems. *DCAMM Report*, October, Technical University of Denmark, 2001a.
61. Thomsen, J. J. Some general effects of strong high-frequency excitation: stiffening, biasing, and smoothening. *Journal of Sound and Vibration*, (accepted for publication), 2001b.
62. Thomsen, J. J., & Tcherniak, D. M. Chelomei's pendulum explained. *Proceedings of the Royal Society of London A* 457(2012), 1889-1913, 2001.
63. Wolf, A., Swift, J. B., Swinney, H. L. & Vastano, J. A.. Determining Lyapunov Exponents from a Time Series. *Physica* 16D, 285-317, 1985.

Chapter 8

On the Theory of the Indian Magic Rope

I. I. Blekhman[1], H. Dresig[2], E. Shishkina[3]

[1]*Institute for Problems of Mechanical Engineering,*
Russian Academy of Sciences
and
Mekhanobr-Tekhnika Corp.
22 Liniya 3, V.O., 199106, St. Petersburg, Russia

[2]*TU Chemnitz, Institute of Mechanics,*
Str. der Nationen 62, D – 09107, Chemnitz, Germany

[3]*Institute for Problems of Mechanical Engineering,*
Russian Acad. of Sciences
61, Bolshoy pr. V.O., St.Petersburg, 199178, Russia

8.1 Preliminary Remarks

One of the remarkable effects, caused by the action of vibration on the nonlinear mechanical systems, is the transformation of the equilibrium positions of the system. Positions, unstable under vibration, may become stable and vice versa, the former equilibrium positions may disappear and the new ones may appear, etc.

The classical investigations in this direction refer to the behavior of the pendulum with a vibrating axis of the suspension (publications by A.Stephenson [1] and P.L.Kapitsa [2]). An outgrowth of that is the work by V.N.Chelomey on the possibility of raising the stability of elastic systems by means of vibration [3] and the later works by S. V. Chelomey [4,5]. An interesting effect of this type is the behavior of the so-called Indian magic rope: under the action of the vertical vibration a "soft" rope seems to acquire some additional rigidity and takes a stable vertical position (Fig. 8.1). The effect shows itself rather vividly. In some popular papers one can even see an affirmation (a little bit questionable) that a small monkey can climb such a stabilized vertical rope.

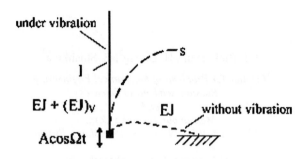

Figure 8.1. The scheme of the system.

The effect under consideration attracted the attention of a member of investigators. S. Otterbein examined an n-unit pendulum, "standing on head", with a vertically vibrating axis of suspension of the lower unit [6]. It was found that with the increase of the number of units n the region of stability decreases and at $n \to \infty$ it disappears. As it was shown by H. Weis [7], that phenomenon does not take place when one introduces the angular stiffness c_φ in the hinges: the region of stability remains finite even at $c_\varphi \to \infty$. Thus we have here an effect, well known in mechanics, of a discontinuous change in the behaviour of the system when a certain parameter becomes zero. Further we will show that in the investigation given below the region of stability of the vertical position of the rod also remains finite when its bending rigidity EJ becomes zero.

In [6] and [7] the investigation was performed on the basis of the classical theory of Floquet-Lyapunov which included the formation of the monodromy matrix. The works [8, 9 and 10] were also devoted to this problem.

Here we will show that a clear physical explanation and approximate mathematical description of the effect can be obtained in a most simple way by the approach of vibrational mechanics and using the method of direct separation of motions [11].

The effect under consideration might be used in the textile industry [12] and in cosmic technology. It is also interesting in connection with the idea of creating dynamic materials, which was proposed quite recently (see [11, 13, 14] and Part IV of this book).

8.2 Equation of Oscillations and Its Consideration

We will start from the differential equation of the bending of the rod in view of the action of the longitudinal forces (see, say, [15]):

$$EI\frac{\partial^4 u}{\partial s^4} = q + \frac{\partial}{\partial s}(N\frac{\partial u}{\partial s}) \tag{8.1}$$

where u is the flexure (transverse shift) of the rod (rope), s is the longitudinal coordinate, EI is the bending rigidity, q is the transverse distributed load, N is the longitudinal force. Counting off u from the vertical position and the coordinate s upward from the lower end of the rod, oscillating according to the law

$$s_0 = A\cos\Omega t \tag{8.2}$$

where A is the amplitude and Ω is the oscillation frequency, we will have

$$q = -\rho F\frac{\partial^2 u}{\partial t^2}, \quad N = -\rho F(g - A\Omega^2 \cos\Omega t)(l-s) \tag{8.3}$$

Here ρ is the density of the material of the rod, F is its cross section area, g is the free fall acceleration, l is the length of the rod.

In view of equalities (8.3) the equation of motion (8.1) will have the form

$$EI\frac{\partial^4 u}{\partial s^4} = -\rho F\frac{\partial^2 u}{\partial t^2} - \rho F(g - A\Omega^2 \cos\Omega t)\left[-\frac{\partial u}{\partial s} + (l-s)\frac{\partial^2 u}{\partial s^2}\right] \tag{8.4}$$

or, after the transfer to the dimensionless values

$$\frac{\partial^2 u}{\partial t_1^2} = (1-\eta^2\gamma\cos\tau)\left[\frac{\partial u}{\partial \sigma}+(\sigma-1)\frac{\partial^2 u}{\partial \sigma^2}\right] - \kappa\frac{\partial^4 u}{\partial \sigma^4} \qquad (8.5)$$

where

$$\sigma = s/l, \quad \eta = \Omega/\omega, \quad \gamma = A/l, \quad \omega = \sqrt{g/l}, \quad t_1 = \omega t,$$
$$\tau = \eta t_1 = \Omega t, \quad k = EI/\rho F g l^3 \qquad (8.6)$$

8.3 Solving the Problem by Method of Direct Separation of Motions

Let us assume that

$$\eta \gg 1, \quad \gamma \ll 1, \qquad (8.7)$$

i.e. the vibration frequency Ω is much greater than the frequency ω, and the oscillation amplitude A is much less than the length of the rod l. To solve this problem we will use the approach of vibrational mechanics and method of direct separation of motions (see [11] and chapter 2 of this book). Considering t_1 to be the "slow" and τ - the "fast" time, we will assume that

$$u = U(t_1) + \psi(t_1,\tau), \quad \langle u \rangle = U, \quad \langle \psi \rangle = 0 \qquad (8.8)$$

where U is the main slow component of motion- and ψ is the fast one (small as compared to U); the pointed brackets denote the averaging for the period $T = 2\pi/\Omega$ over the fast time τ.

To find the functions U and ψ, the following equations are obtained

$$\frac{\partial^2 U}{\partial t_1^2} = \frac{\partial U}{\partial \sigma}+(\sigma-1)\frac{\partial^2 u}{\partial \sigma^2} - \kappa\frac{\partial^4 U}{\partial \sigma^4} - \eta^2\gamma\left\langle\cos\tau\left[\frac{\partial \psi}{\partial \sigma}+(\sigma-1)\frac{\partial^2 \psi}{\partial \sigma^2}\right]\right\rangle$$

(8.9)

8.3. Solving the Problem by Method of Direct Separation of Motions

$$\frac{\partial^2 \psi}{\partial t_1^2} = -\eta^2 \gamma \cos\tau \left[\frac{\partial U}{\partial \sigma} + (\sigma-1)\frac{\partial^2 U}{\partial \sigma^2} \right] + \frac{\partial \psi}{\partial \sigma} + (\sigma-1)\frac{\partial^2 \psi}{\partial \sigma^2} - \kappa \frac{\partial^4 \psi}{\partial \sigma^4}$$

$$-\eta^2 \gamma \cos\tau \left[\frac{\partial \psi}{\partial \sigma} + (\sigma-1)\frac{\partial^2 \psi}{\partial \sigma^2} \right] + \eta^2 \gamma \left\langle \cos\tau \left[\frac{\partial \psi}{\partial \sigma} + (\sigma-1)\frac{\partial^2 \psi}{\partial \sigma^2} \right] \right\rangle$$

(8.10)

Equation (8.9) is called (to a certain extent arbitrarily) the equation of slow motions, while equation (8.10) is called the equation of fast motions. In accordance with the method used, equation (8.10) can be solved approximately. When solving this equation we will consider the function U and its derivatives to be independent of t_1 (to be "frozen") and will neglect the relatively small terms in that equation (mind that we have assumed that $\psi \ll U$). Then to determine ψ we will have the approximate equation

$$\frac{\partial^2 \psi}{\partial t_1^2} = -\eta^2 \gamma \cos\tau \left[\frac{\partial U}{\partial \sigma} + (\sigma-1)\frac{\partial^2 U}{\partial \sigma^2} \right] \qquad (8.11)$$

whose periodic solution will have the form

$$\psi = \gamma \left[\frac{\partial U}{\partial \sigma} + (\sigma-1)\frac{\partial^2 U}{\partial \sigma^2} \right] \cos\tau \qquad (8.12)$$

About the extent of sufficiency of such quite approximate solution see below.

Substituting expression (8.12) into equation (8.9), and having done the averaging in view of the equality $\langle \cos^2 \tau \rangle = 1/2$, we come to the following equation of slow motions (the main equation of vibrational mechanics):

$$\frac{\partial^2 U}{\partial t_1^2} = \frac{\partial U}{\partial \sigma} + [(\sigma-1) - \gamma^2\eta^2]\frac{\partial^2 U}{\partial \sigma^2} - \underline{2\gamma^2\eta^2(\sigma-1)}\frac{\partial^3 U}{\partial \sigma^3} - \left[\kappa + \underline{\frac{1}{2}\gamma^2\eta^2(\sigma-1)^2}\right]\frac{\partial^4 U}{\partial \sigma^4} \quad (8.13)$$

8.4 Analysis of the Result. Physical Explanation of the Effect of the Indian Rope

In equation (8.13) we have underlined the additional (as compared to equation (8.5)) coefficients, reflecting the action of vibration. We see, in particular, that instead of the usual dimensionless rigidity κ the equation contains a greater rigidity

$$k_* = k + k_v \quad (8.14)$$

where the additional ("vibrational") rigidity

$$k_v = \frac{1}{2}\left(\frac{A}{l}\right)^2\left(\frac{\Omega}{\omega}\right)^2\left(\frac{s}{l}-1\right)^2 \quad (8.15)$$

increases quadratically from the free (upper) end of the rope. Equality (8.14) can be written for the dimensional rigidity in the following way

$$(EI)_* = EI + (EI)_v \quad (8.16)$$

where

$$(EI)_v = \frac{1}{2}\rho F[A\Omega(l-s)]^2 \quad (8.17)$$

is the additional rigidity.

So we can say that the physical explanation of the effect under consideration lies in the fact that the effective rigidity of the rope (rod) increases due to vibration.

8.4. Physical Explanation of the Effect

It should be noted that there is a qualitative correspondence of expression (8.17) to the expression for the additional angular rigidity, which seems to be appearing due to the harmonic vibration of the axis of suspension of the mathematical pendulum. In accordance with [2, 11] (see also chapter 4 of this book) we obtain for this rigidity the expression

$$(c_v)_p = \frac{3}{2} m (A\Omega)^2 \tag{8.18}$$

while in our case the value is

$$(c_v)_r = \frac{(EI)_v}{l}\bigg|_{s=0} = \frac{1}{2} m (A\Omega)^2 \tag{8.19}$$

Here $m = \rho F l$ is the mass of the rope.

It should be emphasized that according to equation (8.13) the stability of the rod is determined by two dimensionless parameters $k = EI / \rho F g l^3$ and $q^2 = \eta^2 \gamma^2 = (\Omega A)^2 / gl$, while in the static case there is but one parameter k.

Let us consider the stability of the rope. In this case (see, say [15, 16]) it is sufficient to consider the static stability, i.e. to consider the equation in full derivatives

$$-\frac{d^2}{d\sigma^2}\left(k_* \frac{d^2 U}{d\sigma^2}\right) + \frac{d}{d\sigma}\left((\sigma - 1)\frac{dU}{d\sigma}\right) = 0 \tag{8.20}$$

which is obtained from (8.13) at $\partial^2 U / \partial t_1^2 = 0$

Equation (8.20) coincides with an equation describing the behaviour of a column with rigidity equal to k and loaded with a gravity force. At the same time one end of the column should be clamped, another one - free. We shell use Bubnov-Galerkin method to obtain the boundary of an area of stability for this column.

The main equation of Bubnov-Galerkin method is the following [16]:

$$\int_0^1 \left(\frac{d}{d\sigma}\left[k_* \frac{d^2U}{d\sigma^2} \right] + (1-\sigma)\frac{dU}{d\sigma} \right) \frac{d\eta}{d\sigma} d\sigma = 0 \qquad (8.21)$$

According to the method the function η should satisfy both geometric and static boundary conditions. Assuming $\eta = 6\sigma^2 - 4\sigma^3 + \sigma^4$, $U = c(6\sigma^2 - 4\sigma^3 + \sigma^4)$ and substituting U and η to (8.21) we obtain the relationship between k and q which gives the boundary of an area of stability

$$k = -0.357q^2 + 0.125 \qquad (8.22)$$

The diagram of stability of the rope in the plane of the parameters k and q is shown in Figure 8.2a where the value $k = 0.125$ at $q = 0$, corresponds to the absence of vibration (to the static stability of the column). Figure 8.2b for the sake of comparison shows a diagram, obtained by means of more exact but much more complicated investigation [9]. As we can see, the qualitative character of both diagrams is the same, but there is a considerable quantitative distinction: the boundary of stability area is intersected by the axis q in the first case at the point $q = 0.592$ and in the second case at the point $q \approx 1.13$. This differences is explained by a rather approximate character of the

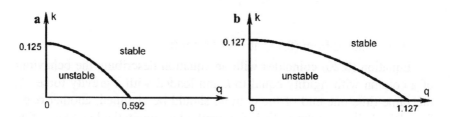

Figure 8.2. The diagram of stability of the rope upper vertical position: a - according to the approximate solution, b - according to the more exact solution [9].

8.4. Physical Explanation of the Effect

solution (8.12) of equation of fast motions (8.10) and, mainly, apparently, by the fact that the boundary conditions are satisfied but approximately - only for the slow component U, i. e. only on the average for the period $2\pi/\Omega$. Along with that, the rough solution leads, as we have seen above, to a distinct physical interpretation and to the analytical description of this effect by means of a very simple investigation. At the same time the more exact solution does not possess such clear physical illustration.

It is interesting to compare the results given in Fig. 8.2 with the condition of stability of the upper position of a simple pendulum of the length l which axis vibrates according to the same law (8.2). In accordance with Fig. 8.2 b in the case of negligibly small rigidity k we have the following condition of stability

$$(A\Omega) > 1.13\sqrt{gl} \qquad (8.23)$$

At the same time for a pendulum ([2, 6], chapter 4 of this book)

$$(A\Omega) > 1.41\sqrt{gl} \qquad (8.24)$$

The condition (8.24) is stronger than (8.23). It should be noted that the difference between these conditions is quite natural: in our case we consider a rather flexible rod with a clamped lower end while the pendulum is an absolutely rigid rod with the hinged end.

Let us consider as an example the rope of the length $l = 100$ mm, which rigidity k is negligibly small. Let the oscillation frequency Ω be equal to 200 rad/s. Then from (8.23) follows that to provide the stability the amplitude A should be greater than 5.6 mm.

Thus, one can see that even for a rope without flexural rigidity its vertical position can be stabilised by means of an action of vibration. And the magnitude of vibration parameters is easy realisable in practice.

The results obtained may be regarded as referring to the extreme case of the problem of the so-called dynamic materials) (see [11, 13, 14] and Part IV of this book). With the change of the amplitude and

8. On the Indian Magic Rope

frequency of vibration we seem to obtain the seemingly different materials (with different elastic properties) (Fig. 8.3).

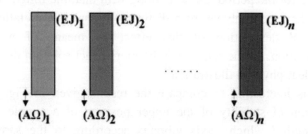

Figure 8.3. To the problem of dynamic materials.

References

1. Stephenson, A., On induced stability. *Philosophical Magazine*, **15** (1908), 233–236, **17** (1909), 765–766.
2. Kapitsa, P.L., Dynamic stability of the pendulum when the point of suspension is oscillating, *Jour. of Exper. and Theor. Physics*, **21**, 5 (1951), 588–597 (in Russian).
3. Chelomei, V. N., On the possibility of the increase of the elastic system's stability by means of vibrations, *DAN SSSR*, **110**, 3 (1956), 341–343 (in Russian).
4. Chelomei S. V., On dynamic stability of elastic systems, *DAN SSSR*, **252**, 2 (1980) 307–310 (in Russian).
5. Chelomei S. V., Dynamic stability at high-frequency parametric excitation, *DAN SSSR*, **257**, 4 (1981) 853–857 (in Russian).
6. Otterbein, S., Stabilisierung des n-Pendels und der Indische Seiltrick, *Archive for Rational Mechanics and Analysis*, **78** (1982), 381–393.
7. Weis, H. Stabilisierung der aufrecht stahenden Pendels und der Indische Seiltrick, *Vortrag an der GAMM Tagung 2000*, Götingen (2000).
8. Galan J., Fraser W.B., Acheson D.J., and Champneys A.R. The parametrically excited upside-down rod: an elastic Iointed pendulum model, Submitted to "Physica D".
9. Champneys A.R., Fraser W.B. The "Indian rope trick" for a parametrically excited flexible rod: linearized analysis. *Proc. Roy. Soc.*(2000) London, A456, p. 553-570.
10. Fraser W.B., Champneys A.R., The "Indian rope trick" for a parametrically excited flexible rod: nonlinear and subharmonic analysis, *Proc. Roy. Soc.*(2002) London, A458, p. 1353-1373.
11. Blekhman, I. I., *Vibrational Mechanics*. (World Scientific, Singapore, 2000), 509.
12. Dresig, H., *Schwingungen mechanischer Antriebsysteme*, (Springer-Verlag, Berlin-Heidelberg,2001) 420.
13. Blekhman, I.I., Lurie, K.A. On Dynamic Materials, *DAN Russia*, **371**, 2 (2000), 182–185.
14. Lurie K.A. Some recent advances in the theory of dynamic materials. *Presentation on the DCAMM Intern. Symposium "Challenges in Applied Mechanics"*, Lyngby, Denmark (2002).
15. *Strength, stability of oscillation*. (Reference book in 3 volumes, edited by I.A.Birger and Ja.G. Panovko, Moscow, 1968) (in Russian).
16. Volmir, A.S., *The stability of deformable systems* (Nauka, Moscow, 1967) 984 (in Russian).

Chapter 9

Effect of Conjugate Resonances and Bifurcations at the Biharmonic Excitation of Nonlinear Systems (By an Example of Pendulum with a Vibrating Axis)

I. I. Blekhman[1], P. S. Landa[2]

[1]*Institute for Problems of Mechanical Engineering,*
Russian Acad. of Sciences
and
Mekhanobr-Tekhnika Corp.
22 Liniya, 3, V.O., St.Petersburg, 199106, Russia

[2] *Lomonosov Moscow State University,*
Department of Physics,
119899 Moscow, Russia

9.1 Preliminary Remarks

This chapter describes the effect of conjugate resonances and bifurcations under the action of a biharmonic excitation with two essentially different frequencies on the nonlinear systems. This effect consists in the fact that resonances and bifurcations in such systems take place in both cases – when the low frequency changes with the high frequency being fixed and when the high frequency changes with the low frequency being fixed, and also when the amplitude of a high frequency action is changing with the fixed frequencies.

We consider this effect using as an example the classical problem about the behavior of a pendulum with an oscillatory axis of suspension, studying both cases - of the ordinary and of the parametric resonances. From this consideration it is clear that the effects discussed will take place in other systems as well, including those more general ones, and we can speak not only about the action of the ordinary exciting force, but also about the parametric excitation.

The investigation was performed by means of the approach of the vibrational mechanics and of the method of direct separation of motions. In the context of this approach the explanation of the described effects consists in the following. A high-frequency vibration changes essentially the parameters and properties of the system with relation to the low-frequency effects. As a result, the low-frequency effect seems in fact to be acting on another system (particularly in the case of a pendulum), on that with another effective rigidity, the latter depending essentially on the frequency and amplitude of the high-frequency effect. The results make it possible to explain, on the basis of a deterministic consideration, the mechanism of the effect widely discussed in literature, called **statistical resonance** (see, also [2]).

Thus, the effects under consideration are closely connected with the property of a high-frequency vibration to exert a major influence on the effective elastic characteristics of the body. This property was discussed in detail in the book [1] and was discussed above in this part of the book (see chapters 6 and 7, where proper references are also given).

In our joint articles [2, 3] we have considered the regularities under consideration on the example of Duffing's oscillator. The use of the approach of vibrational mechanics made it possible to establish analytically the fact of conjugacy of resonances and bifurcations in such system. It was also established that there is a good agreement with the results of the numerical investigation performed earlier [4, 5], where a part of resonances under consideration was displayed and they was called *"vibrational resonances"*. The results of these works are given briefly in the section 9.6.

At the end we say some remarks on the application of the results of investigation.

9.2 Equation of Motion

The discussed effects can already be found out in the behavior of the classical system — a pendulum with a vibrating axis of suspension. Unlike the previous investigations, here it will be assumed that the axis of suspension of the pendulum performs oscillations in two mutually

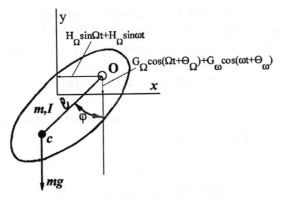

Figure 9.1. Pendulum with the biharmonical vibrating axis of suspension.

perpendicular directions according to the law (Fig. 9.1)

$$x = H_\Omega \sin \Omega t + H_\omega \sin \omega t, \quad y = G_\Omega \cos(\Omega t + \theta_\Omega) + G_\omega \cos(\omega t + \theta_\omega), \quad (9.1)$$

believing that the "fast" frequency Ω exceeds considerably (practically not less than three times) the "slow" frequency ω. The symbols H and G denote here the amplitudes, while θ denotes the phase shift of the corresponding oscillation components. The axis of the pendulum oscillates along the elliptic trajectory with each of the two frequencies.

The equation of motion of the pendulum has the form

$$I\ddot\varphi + h\dot\varphi + mgl\sin\varphi + ml\Omega^2[H_\Omega \cos\varphi \sin\Omega t - G_\Omega \sin\varphi \cos(\Omega t + \theta_\Omega)] + \qquad (9.2)$$
$$ml\omega^2[H_\omega \cos\varphi \sin\omega t - G_\omega \sin\varphi \cos(\omega t + \theta_\omega)] = 0$$

where φ is the angle of deviation of the pendulum from the lower position, I, m and l are respectively the moment of inertia, the mass and the distance from the center of gravity C to the axis of suspension of the pendulum O, g is the free fall acceleration, and h is the viscous resistance coefficient.

9.3 Equation of Slow Motions

Solving the problems in the case of one-frequency "fast" oscillations of the pendulum ($H_\omega = G_\omega = 0$) by the method of direct separation of motions is given in the book [1], in a more complicated case – in chapter 6, and in a more simple case – in section 4.8 of this book. In accordance with that method, the solution of equation (9.2) was found in the form

$$\varphi = \alpha(t) + \psi(t,\Omega), \quad \tau = \Omega t \tag{9.3}$$

where $\alpha(t)$ is the "slow" component, and ψ is the "fast" periodic one, with ψ satisfying the condition

$$\langle \psi(t,\tau) \rangle \equiv \frac{1}{2\pi} \int_0^{2\pi} \psi(t,\tau)d\tau = 0 \tag{9.4}$$

All the reasoning and calculations, given in the book [1] and in section 4.8 of this book, remain valid also with respect to equation (9.2) with the only difference that the the moments of forces with the frequency ω should also be referred to the slow moments of forces. As a result, the equation of slow motions (the main equation of vibrational mechanics) will have the form

$$I\ddot{\alpha} + h\dot{\alpha} + mgl \sin\alpha +$$
$$\frac{(ml\Omega)^2}{4I}[(G_\Omega^2 - H_\Omega^2)\sin 2\alpha + 2G_\Omega H_\Omega \cos 2\alpha \sin\theta_\Omega] = \tag{9.5}$$
$$- ml\omega^2[H_\omega \cos\alpha \sin\omega t - G_\omega \sin\alpha \cos(\omega t + \theta_\omega)]$$

When obtaining this equation, as well as the corresponding one in the book [1] and in chapter 6 and section 4.8, the following assumptions of the order of values were used:

$$\frac{p}{\omega} \sim \varepsilon, \quad \frac{ml\sqrt{G_\Omega^2 + H_\Omega^2}}{I} = \frac{\sqrt{G_\Omega^2 + H_\Omega^2}}{l_0} \sim \varepsilon, \quad \frac{h}{I\Omega} \sim \varepsilon \tag{9.6}$$

where ε is a small parameter, $p = \sqrt{mgl/I} = \sqrt{g/l_0}$, with l_0 being the equivalent length of the mathematical pendulum. The relation $\psi \sim \varepsilon\alpha$, following from those assumptions, is also taken into consideration. The main qualitative result of investigating equations (9.5) in the case of one-frequency fast oscillations of the axis amounts to the following: *due to vibration, the pendulum seems to be attracted to the big semi-axes of the elliptic trajectory of oscillations of the pendulum axis. If this attraction (in the case of vertical vibration) is strong enough, then the upper position of the pendulum, unstable in the absence of vibration, becomes stable (the main classical result).*

Let us now turn our attention to the analysis of a more complicated equation (9.5).

9.4 The Case of "Ordinary" Resonance. Conjugate Resonances and Bifurcations

This case corresponds to the vertical fast oscillations of the axis and with the horizontal slow ones. The equation (9.5) takes the form

$$I\ddot{\alpha} + h\dot{\alpha} + mgl\sin\alpha + \frac{(mlG_\Omega)^2}{4I}\Omega^2 \sin 2\alpha = -mlH_\omega\omega^2 \cos\alpha \sin\omega t \quad (9.7)$$

Let us first consider the small oscillations of the pendulum *near the lower position*. Linearizing equation (9.7), we get

$$I\ddot{\alpha} + h\dot{\alpha} + \left[mgl + \frac{(mlG_\Omega)^2}{2I}\Omega^2\right]\alpha = -mlH_\omega\omega^2 \sin\omega t \quad (9.8)$$

Hence one can see that the resonance takes place if the following equality is satisfied (disregarding the small effect of damping):

$$\omega = p_{V0} = \sqrt{p^2 + q^2\Omega^2},$$

$$p^2 = \frac{mgl}{I}, \quad q^2 = \frac{1}{2}\left(\frac{mlG_\Omega}{I}\right)^2 = \frac{1}{2}\left(\frac{G_\Omega}{l_0}\right)^2 \quad (9.9)$$

From relation (9.9) it follows that in this system the resonance (peak of the oscillation amplitude A under the change of the parameter) may take place in both cases — under the changes of the slow oscillation frequency of the axis of the pendulum ω with the fast oscillation frequency Ω being fixed (Fig. 9.2), and under the changes of the fast oscillation frequency with the slow oscillation frequency being fixed (if $\omega > p$). Such resonances whose frequencies are bound by the relation of type (9.9) may be called *conjugate resonances*. It is also remarkable that the resonance character of the dependence of the oscillation amplitude of the pendulum takes place at the change of the fast oscillation amplitude G_Ω to which the parameter q is proportional (and also if $\omega > p$). All these effects are explained by an essential dependence of the slow free oscillation frequency of the pendulum near the lower equilibrium position on the frequency Ω and on the fast vibration amplitude G_Ω. It should be also emphasized that all the described resonances correspond to the forced oscillations of the pendulum with the "slow" frequency ω.

The graph in Fig. 9.2(c) is similar to the one given when considering the effect, called the **statistical resonance** (see, e.g. [3]). From what has been said it follows that such an effect is not connected with causal jumps of the system caused by noise; it can be explained by the above consideration.

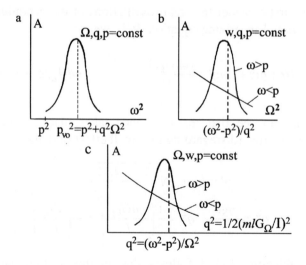

Figure 9.2. "Ordinary" conjugate resonances near the lower position of the pendulum: a — at the change of the low frequency; b — at the change of the high frequency c — at the change of the amplitude of the high frequency.

9.4. "Ordinary" Conjugate Resonances and Bifurcations

Let us consider now the small oscillations of the pendulum near the upper position $\alpha = \pi$. Assuming that in equation (9.7) $\alpha = \pi + \beta$ and linearizing it over β, we obtain the equation

$$I\ddot{\beta} + h\dot{\beta} + \left[-mgl + \frac{(mlG_\Omega \Omega)^2}{2I} \right] \beta = mlH_\omega \omega^2 \sin \omega t \qquad (9.10)$$

Let us assume that the well known condition of stability of the upper position of the pendulum under the action of the vertical harmonic vibration is fulfilled ([6, 7]; see also chapters 4 and 6)

$$ml(G_\Omega \Omega)^2 / 2gI = (G_\Omega \Omega)^2 / 2gl_0 > 1$$

which provides the positiveness of the coefficients at β in equation (9.10). Using the above-adopted symbols, we write this inequality as

$$q^2 \Omega^2 > p^2 \qquad (9.11)$$

Resonance takes place if the following equality is satisfied

$$\omega = \sqrt{q^2 \Omega^2 - p^2} \equiv p_{V_\pi}, \qquad (9.12)$$

where p_{V_π} is the slow free oscillation frequency of the pendulum near the upper equilibrium position with the fast vibration of the axis. The resonance curves, corresponding to this case are shown schematically in Fig. 9.3; they are analogous, with a certain specification, to the curves, shown in Fig. 9.2. What was said about the conjugated resonances also remains valid.

Along with that, in this case there is also an effect of conjugacy of the bifurcation points, which corresponds to the change in the character of stability of the upper position of the pendulum at $q\Omega = p$. These points can be achieved at the change in the frequency Ω with the amplitude G_Ω being constant, and also at the change of G_Ω with Ω being constant. The corresponding points are marked in Figs. 9.3 – 9.5 by the symbol *.

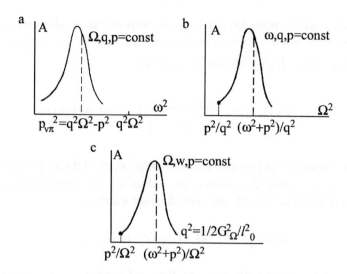

Figure 9.3. "Ordinary" conjugate resonances and bifurcations near the upper position of the pendulum: a — at the change of the low frequency; b — at the change of the high frequency c — at the change of the amplitude of the high frequency ("*" are designated the points of bifurcation.)

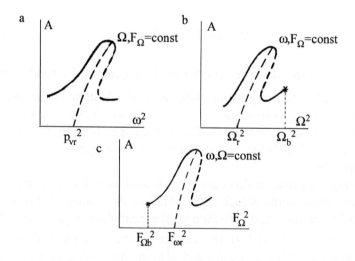

Figure 9.4. Conjugate resonances and bifurcations in the case of Duffing's oscillator: a — at the change of the low frequency; b — at the change of the high frequency c — at the change of the amplitude of the high frequency.

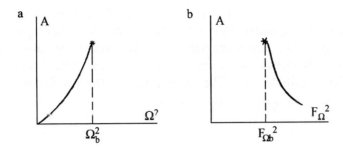

Figure 9.5. The points of bifurcations (seeming resonances) in the case of degenerate (over-damped) Duffing's oscillator: a — at the change of the high frequency; b — at the change of the amplitude of the high frequency.

9.5 The Case of Parametric Resonances

This case corresponds to the vertical fast and slow oscillations of the axis of the pendulum ($H_\Omega = 0$, $H_\omega = 0$). Equation (9.5) in this case takes the form

$$I\ddot{\alpha} + h\dot{\alpha} + mgl\sin\alpha + \frac{(mlG_\Omega)^2}{4I}\Omega^2 \sin 2\alpha - \\ mlG_\omega \omega^2 \sin\alpha \cos(\omega t + \theta_\omega) = 0 \quad (9.13)$$

Considering the small oscillations of the pendulum near the stable equilibrium position $\alpha = 0$, we linearize this equation with respect to α, then it may be written in the form

$$\ddot{\alpha} + h_1 \dot{\alpha} + p_{V0}^2[1 - \gamma \sin(\omega t + \theta_\omega)]\alpha = 0 \quad (9.14)$$

where $h_1 = h/I$, $\gamma = mlG_\omega \omega^2 / Ip_{V0}^2 = G_\omega \omega^2 / l_0 p_{V0}^2$.

The condition of the basic parametric resonance is, as is known, the equality

$$\omega = 2p_{V0} = 2\sqrt{p^2 + q^2\Omega^2} \quad (9.15)$$

which differs from (9.9) only by coefficient 2. Therefore what was said in 9.4 can be repeated here word for word: the basic parametric resonance (just like any other parametric resonance) can take place at the change of the frequency ω and also at the change of the frequency Ω and of the amplitude G_ω. The same refers to the oscillations of the pendulum near the upper positions $\alpha = \pi$.

9.6 The Biharmonic Effect on the System, Described by Duffing's Equation

The effects under study in the system, described by Duffing's differential equation

$$m\ddot{x} + h\dot{x} - cx + dx^3 = F_\Omega \cos(\Omega t + \theta) + F_\omega \cos \omega t, \qquad (9.16)$$

where $m, h, c, d, \omega, \Omega, F_\omega, F_\Omega$ are positive constants, θ is the phase shift, and, as before $\Omega \gg \omega$, were considered in the work [2]. Using the method of direct separation of motions, the solution of this equation was found in the form

$$x = X(t) + \psi(t, \Omega t), \qquad (9.17)$$

where $X(t)$ is the slow- and $\psi(t, \Omega t)$ is the fast 2π-periodic function of the fast time $\tau = \Omega t$ with a zero average over this argument. In accordance with the above-mentioned method we obtain for the functions X and ψ the following system of equations

$$m\ddot{X} + h\dot{X} - cX + dX^3 + 3dX\langle\psi^2\rangle + q\langle\psi^3\rangle = F_\omega \cos \omega t \qquad (9.18)$$

$$m\ddot{\psi} + h\dot{\psi} - c\psi + 3dX^2(\psi - \langle\psi\rangle) + 3dX(\psi^2 - \langle\psi^2\rangle) + \\ d(\psi^3 - \langle\psi^3\rangle) = F_\Omega \cos(\Omega t + \theta) \qquad (9.19)$$

where, as before, the angular brackets denote averaging for the 2π-period over the argument $\tau = \Omega t$.

9.6. The Case of Duffing's Oscillator

Just like in the case of the pendulum when solving equations of fast motions (9.19), a purely inertial approximation was used (see section 2.9) at which all the terms in the left hand side, except $m\ddot{\psi}$, are considered to be negligibly small as compared to those in the right hand side. Then the approximate periodic solution of this equation has the form

$$\psi = -\frac{F_\Omega}{m\Omega^2}\cos(\Omega t + \theta) \qquad (9.20)$$

On substituting this expression into equation (9.18) we obtain the following approximate equation of slow motions

$$m\ddot{X} + h\dot{X} + \left[\frac{3}{2}d\left(\frac{F_\Omega}{m\Omega^2}\right)^2 - c\right]X + dX^3 = A_\omega \cos\omega t \qquad (9.21)$$

This equation differs from the initial one (9.16) in particular by the fact that the coefficient at X can be both either positive or negative, while in the initial equation this coefficient is always negative.

If

$$\frac{3}{2}d\left(\frac{F_\Omega}{m\Omega^2}\right)^2 - c > 0 \qquad (9.22)$$

then at $F_\omega = 0$ equation (9.21) has only one stable equilibrium position $X=0$, while equation (9.16) at $F_\Omega = F_\omega = 0$ has three equilibrium positions. The position $x=0$ is unstable, while the other two $x = \pm\sqrt{c/q}$ are stable. (so the system under consideration is sometimes calle "bistable oscillator")

In case the condition (9.22) is not satisfied, equation (9.1) also has two stable and one unstable positions of equilibrium. Thus the relationship

$$\frac{3}{2}d\left(\frac{F_\Omega}{m\Omega^2}\right)^2 - c = 0 \qquad (9.23)$$

corresponds to bifurcation.

While the basic resonance (in the linear approximation) in case inequality (9.22) is satisfied, corresponds to the equality

$$p_V^2 = \frac{1}{m}\left[\frac{3d}{2}\left(\frac{F_\Omega}{m\Omega^2}\right)^2 - c\right] = \omega^2 \qquad (9.24)$$

i.e. to the condition of coincidence of the frequency of slow small free oscillations with that of excitation. On solving equality (9.24) with respect to Ω^2 we obtain

$$\Omega^2 = \frac{F_\Omega}{m}\sqrt{\frac{3}{2}\cdot\frac{d}{(c+m\omega^2)}} \qquad (9.25)$$

The relationships (9.23)–(9.25) bring us to the same conclusions as above: the resonance states take place at the change of the low frequency with the high frequency being fixed and also at the change of the high frequency with the low frequency being fixed (conjugated resonances); moreover, resonances may also appear at the change of the excitation amplitude F_Ω with both frequencies, low and high, being fixed. Similar conclusions are also valid with respect to the bifurcation points.

Schematically the resonance curves, similar to those in Fig. 9.2 are shown in Fig. 9.4. The values of the parameters, corresponding to the resonance relation are marked by the symbol "*r*", while the bifurcation values are marked by "*b*".

The described regularities are partially retained if we consider, as is done in many publications, the "degenrated" equation (9.16), corresponding to the so-called "*over-damped oscillator*".

$$h\dot{x} - cx + dx^3 = F_\Omega \cos(\Omega t + \theta) + F_\omega \cos\omega t \qquad (9.26)$$

Using in this case the same method and the same suppositions, we obtain

$$\psi \approx \frac{1}{h\Omega}F_\Omega \sin(\Omega t + \theta), \quad h\dot{X} + aX + dX^3 = F_\omega \cos\omega t \qquad (9.27)$$

where

$$a = -c + \frac{3}{2}d\left(\frac{F_\Omega}{h\Omega}\right)^2$$

Resonances in the ordinary sense of the word are absent here. Bifurcation relationship, however, still takes place as before and corresponds to the coefficient value $a = 0$, i.e. to the equality

$$\frac{3}{2}d\left(\frac{F_\Omega}{h\Omega}\right)^2 = c \qquad (9.28)$$

It should be noted, however, that near the indicated bifurcation value, the dependence of the amplitude of oscillations on the corresponding parameters may look similar to the resonance peaks. Schematically such dependences (at $a > 0$) are shown in Fig. 9.5. They correspond to the expression

$$X = F_\Omega \frac{1}{\sqrt{(h\omega)^2 + a^2}} \cos(\omega t + \gamma) + 0(d), \qquad (9.29)$$

obtained when finding the $2\pi/\omega$ – periodic solutions of equation (9.27) in the form of series over d.

In the articles [2, 3] the results of the approximate solution of equations are compared to those of the numerical solution of the initial equations and they were found to be in good agreement.

9.7 Some Remarks on the Application of the Results

In the context of the results of this chapter some remarkable regularities become understandable – those observed when an accident excitation with the clearly expressed high-and low- frequency parts of the spectrum acts on nonlinear oscillators: roughly speaking, we can say that the high-frequency components of the spectrum cause a change in the properties of the system with respect to the low-frequency ones, i.e. the low-frequency excitations act on another, modified system, with this change depending essentially on the intensity of the high-frequency spectrum components.

The possibility of such a conclusion is connected with the additivity of the action of high-frequency harmonic components of excitation (see chapter 6). The results stated here make it possible to explain very simply the mechanism of the above mentioned effect of statistical resonance – nonmonotonous (resonant) character of the dependence of the amplitude of the output signal in the nonlinear system on the level of a casual effect.

From the results follow also two important practical consequences:

1) the possibility of controlling the resonance states of the system with respect to the low-frequency effect by means of changing the high-frequency effect;

2) the necessity of taking into consideration the changes in the resonante frequency of the system with respect to the low-frequency effect due to the high-frequency action. That is quite necessary if the aim is to prevent the resonance situations.

References

1. Blekhman I. I., *Vibrational Mechanics* (Fizmatlit, Moscow, 1994), p. 394 (in Russian). (English translation enlarged and revised: *Vibrational Mechanics. Nonlinear Dynamic Effects, General Approach, Applications.* World Scientific, Singapore, 2000)
2. Blekhman I. I., Landa P.S., Conjugate resonances in nonlinear systems under biharmonical excitation. Vibroinduced bifurcations, *Applied Nonlinear Dynamics*, **10**, 1-2 (2002), 44-51 (in Russian)
3. Blekhman I. I., Landa P.S., Conjugate resonances and bifurcations in nonlinear systems under biharmonical excitation, *Intern. Journ. of Nonlinear Mechanics*, (in print)
4. Landa P.S. and P.V.E. McClintock, Vibrational resonance. *Phys. A: Math. Gen.*, **33**, L433 (2000).
5. Landa P.S., *Regular and Chaotic Oscillations* (Springer-Verlag, Berlin-Heidelberg, 2001).
6. Kapitsa P. L, Pendulum with vibrating axis of suspension, *Uspekhi fizicheskich nauk*, **44**, 1 (1954) 7-20 (in Russian).
7. Kapitsa P. L, Dynamic stability of the pendulum when the point of suspension is oscillating, *Jour. of Exper. and Theor. Physics*, **21**, 5 (1951) 588-597 (in Russian).

# Chapter 10

On the Investigations of the Electromechanical Systems. On the Behavior of the Conductivity Bodies of Pendulum Types in High-Frequency Magnetic Fields

I. I. Blekhman

Institute for Problems of Mechanical Engineering,
Russian Academy of Sciences
and
Mekhanobr-Tekhnika Corp.
22 Liniya 3, V.O., 199106, St. Petersburg, Russia

10.1 On Some New Results of the Theory of Electro-Mechanical Systems

Many electromechanical systems contain explicity or implicitly the elements of the type of pendulum. Though, as marked in the Preface, this also refers to the oscillatory systems in general.

This chapter pursues two aims. The first one is to draw attention to the considerable achievements in the theory of electro-mechanical systems, in particular in the theory of electruc machines, produced in recent years [1, 2, 3, 4, 5]. These results were achieved on the basis of a fruitful idea to separate the slow and fast oscillatory processes, i.e. the same classical idea which forms the basis of vibrational mechanics.

A successful use of this idea is based in this case on the circumstance that the characteristic speed of the course of mechanical processes is much lower than that of the electrical processes. Equations of slow processes (as a rule the main ones) become much simpler. This refers in particular to the classical equations of electric machines - Gorev-Park's equations. It seems that specialists in electro-mechanics have not yet fully appreciated it.

Interesting results, very important for practice, were also obtained in the theory of electric and magnetic suspensions of solid bodies [6]. One of the main of such results consists in the following. Well-known is the theorem about the impossibility of a stable equilibrium of a conducting solid body in the electro-static field under the action of electric forces alone [7]. This theorem generalizes the classic *theorem of Earnshaw* which refers to the electric charge [8, 6]. Along with that it has been shown that the high frequency voltage fluctuations in the corresponding closed circuits make it possible to overcome this "natural" instability and keep the conducting body in a state of certain equilibrium (or *quasi-equilibrium*[1]).

All the enumerated results have been obtained by the use of asymptotic methods.

The second aim of this chapter is to draw attention to the fact that a number of important results, mentioned above, (as well as some new results, referring to this range of problems) can be obtained quite easily by using the method of direct separation of motions and can be formulated in the context of the approach of vibrational mechanics.

What was said can be illustrated by two rather simple examples.

10.2 The Problem about a Passive (Resonant) Electrostatic Suspension

In his book [6] Yu. G. Martynenko gives, according to his words, a nonstrict solution of this problem on the basis of simple physical considerations, similar to the ideology of the method of direct separation of motions.

---

[1] By *quasi-equilibria* of the system we mean motions of the type of $q_j = q_j^0 + \mu\psi_j(\tau)$ where q_j are the generalized coordinates of the system, q_j^0 are the constants, μ is a small parameter, and $\psi_j(\tau)$ are periodic or almost periodic functions of τ with $\langle\psi_j(\tau)\rangle = 0$. One can also say that positions of quai-equilibrium are equilibrium positions for the equations of slow motions.

10.2. Passive Electrostatic Suspension

Here we will obtain this result by the method of direct separation of motions.

The scheme of the system under consideration is given in Fig. 10.1. Plate *1*, fixed immobile, is connected with a free plate *3* by means of an oscillatory electrical closed circuit *2*, containing the elements with the ohmic resistance R, with self-induction L and a source of the fast-alternating voltage $u = u_0 \sin \omega t$; in this closed circuit the plates form a capacitor with a capacitance $C(y) = -S/4\pi y$, where y is the coordinate of the lower plate (the axis y being directed vertically upward, so that $y < 0$) and S is the area of the plate.

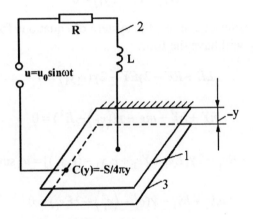

Figure 10.1. The scheme of the system.

The equations, describing the behavior of the electro-mechanical system under consideration, have the form

$$L\ddot{e} + R\dot{e} - 4\pi ey/S = u_0 \sin \omega t \qquad (10.1)$$

$$m\ddot{y} + h\dot{y} + mg - 2\pi e^2/S = 0 \qquad (10.2)$$

where e is an electric charge on the capacitor, m is the mass of the lower plate, g is a free fall acceleration, and h is the viscous friction coefficient. Here, unlike the book [6], the term $h\dot{y}$, reflecting the effect of forces of linear resistance in the mechanical part of the system, is added into Eq. (10.2).

It can be easily seen that the voltage is constant (when in the right side of Eq.(10.1) instead of $u_0 \sin \omega t$ we have $U_0 = \text{const}$), the mobile plate in full agreement with the Earnshaw theorem does not have any stable positions of equilibrium.

Following the method of direct separation of motions (see chapter 2) we will assume that

$$e = E(t) + e_1(t, \omega t), \qquad y = Y(t) + y_1(t, \omega t) \qquad (10.3)$$

where E and Y are the slow components, while e_1 and y_1 are the fast ones, with

$$\langle e_1 \rangle = 0, \quad \langle y_1 \rangle = 0 \qquad (10.4)$$

In view of latter equalities, the system of equations for the functions E, Y, e_1 and y_1 will have the form

$$L\ddot{E} + R\dot{E} - 2\gamma EY - 2\gamma \langle e_1 y_1 \rangle = 0 \qquad (10.5)$$

$$m\ddot{Y} + h\dot{Y} + mg - \gamma(\langle e_1^2 \rangle - E^2) = 0 \qquad (10.6)$$

$$L\ddot{e}_1 + R\dot{e}_1 - 2\gamma[Ey_1 + Ye_1 + e_1 y_1 - \langle e_1 y_1 \rangle] = u_0 \sin \omega t \qquad (10.7)$$

$$m\ddot{y}_1 + h\dot{y}_1 - \gamma(e_1^2 - \langle e_1^2 \rangle + 2Ee_1) = 0 \qquad (10.8)$$

where $\gamma = 2\pi/S$.

As was repeatedly mentioned above, using this method one can limit himself to an approximate solution of the equations of fast motions (10.7) and (10.8) without introducing any large error into equations of slow motions. Taking into consideration this circumstance, we consider the last term in Eq.(10.8) to be small and write down this equation as

$$m\ddot{y}_1 + h\dot{y}_1 = \varepsilon \Phi(E, e_1) \qquad (10.9)$$

where $\varepsilon \Phi(E, e_1) = \gamma(e_1^2 - \langle e_1^2 \rangle + 2Ee_1)$ and ε is a small parameter. Then in the initial approximation (at $\varepsilon = 0$) and considering the value Y to be constant ("frozen"), we will find the periodic solution of these equations satisfying conditions (10.4):

10.2. Passive Electrostatic Suspension

$$y_1 = 0, \quad e_1 = \frac{u_0 S \sin(\omega t + \varphi)}{\sqrt{(LS\omega^2 + 4\pi Y)^2 + (RS\omega)^2}} \quad (10.10)$$

where φ is a certain constant (phase shift).

After the operation of averaging in view of (10.10) as well as of $y_1 = 0$ the equation (10.5) has the solution $E = 0$, the equation of slow motions (10.6) takes the form

$$m\ddot{Y} + h\dot{Y} + mg - f(Y) = 0 \quad (10.11)$$

where

$$f(Y) = \frac{\pi S u_0^2}{(LS\omega^2 + 4\pi Y)^2 + (RS\omega)^2} \quad (10.12)$$

Equation (10.11) coincides with Eq. (1.14) of the book [6]. A simple analysis of this equation results in the following conclusions (see Fig. 10.2):

1) Positions of the quasi-equilibrium of the lower plate $Y = Y_0$ will be defined by the equality

$$mg = f(Y_0) \quad (10.13)$$

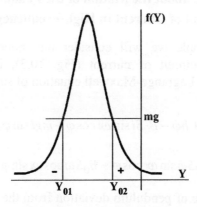

Figure 10.2. Positions of the quasi-equilibrium of the plate ($Y = Y_{01}$ is unstable, $Y = Y_{02}$ is stable).

With regard to character of the function $f(Y)$ and provided the following condition is satisfied

$$mg < f_{max} = \pi u_0^2 / S(R\omega)^2 \qquad (10.14)$$

we obtain two such positions $Y_0 = Y_{01}$ and $Y = Y_{02}$.

2) The condition of stability of the positions of equilibrium is the inequality

$$f'(Y_0) < 0 \qquad (10.15)$$

Therefore one of the positions (in the context of Fig. 10.2 it is Y_{02}) is stable, and the other position Y_{01} is unstable.

The book [6] contains a rather complicated solution of the problem in which the system (10.1), (10.2) is reduced to a singularly disturbed system of the six order. The above given result is obtained by Yu.G. Martynenko in the initial approximation. In the next approximation the author obtains instability which, according to him, is the result of the disregard of forces of mechanical resistance to the oscillations of the mobile plate. Since in the above solution this resistance was taken into account, and was not assume to be small, a stability of quasi-equilibrium was in this case ensured.

10.3 The Problem about the Motion of the Pendulum with the Closed Circuit of Current in High-Frequency Magnetic Field

As a second example we will consider the motion of a pendulum containing closed circuit of current (Fig. 10.3), in a homogeneous magnetic field. The Lagrange-Maxwell equation of such a system has the form

$$J\ddot{\varphi} + h\dot{\varphi} - B_0 Si \sin\omega t \cos\varphi + mgl \sin\varphi = 0 \qquad (10.16)$$

$$L(i) + B_0 S\dot{\varphi} \sin\omega t \cos\varphi + B_0 S\omega \cos\omega t \sin\varphi + Ri = 0 \qquad (10.17)$$

Here φ is the angle of pendulum deviation from the lower position, i is the current in the closed circuit, m is the mass of the pendulum, J is its moment of inertia with respect to the axis of suspension, L is the

10.3. Motion of the Pendulum with the Closed Circuit of Current

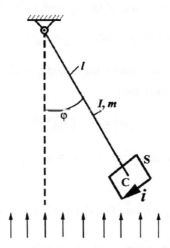

Figure 10.3. Pendulum with the closed circuit of current in a hogh-frequency magnetic field.

coefficient of self-induction, R is the ohmic resistance of the current in the closed circuit, B_0 is the amplitude of the outer magnetic field, ω is its frequency, S is the area of the closed circuit, h is the coefficient of mechanical resistance to the oscillations of the pendulum, g is the free fall acceleration. The frequency ω is believed to be much greater than the frequency of the small free oscillations of the pendulum $p = \sqrt{mgl/J}$: On introducing dimensionless values

$$i_u = i/i_*, \quad (i_* = B_0 S/L), \quad \varepsilon = p/\omega, \quad p = \sqrt{mgl/J}, \quad h_1 = h/J\omega,$$
$$\tau = \omega t, \quad \gamma = (B_0 S)^2 / Lmgl, \quad r = R/L\omega$$

(10.18)

we write down Eqs. (10.16) and (10.17) in the form

$$\varphi'' + h_1 \varphi' - \varepsilon^2 \gamma i_u \sin\tau \cos\varphi + \varepsilon^2 \sin\varphi = 0 \qquad (10.19)$$

$$i_u' + \varphi' \sin\tau \cos\varphi + \cos\tau \sin\varphi + r i_u = 0 \qquad (10.20)$$

Searching for the solution of the system (10.19), (10.20) in the form

$$\varphi = \alpha(t) + \psi(t,\omega t), \quad i_u = I_u(t) + i_1(t,\omega t) \qquad (10.21)$$

where α and I_u are the slow components, with ψ and i_1 being the fast ones, satisfying the conditions

$$\langle\psi\rangle = 0, \quad \langle i_1\rangle = 0, \qquad (10.22)$$

we obtain for the variables α, I_u, ψ and i_1 the following system of equations

$$\alpha'' + h_1\alpha' - \varepsilon^2\gamma\langle(I_u + i_1)\sin\tau\cos(\alpha+\psi)\rangle + \varepsilon^2\langle\sin(\alpha+\psi)\rangle = 0 \quad (10.23)$$

$$I_u' + \langle(\alpha'+\psi')\sin\tau\cos(\alpha+\psi)\rangle + \langle\cos\tau\sin(\alpha+\psi)\rangle + rI_u = 0 \quad (10.24)$$

$$\psi'' + h_1\psi' - \varepsilon^2 = 0 \qquad (10.25)$$

$$i_1' + (\alpha'+\psi')\sin\tau\cos(\alpha+\psi) - \langle(\alpha'+\psi')\sin\tau\cos(\alpha+\psi)\rangle$$
$$+ \cos\tau\sin(\alpha+\psi) - \langle\cos\tau\sin(\alpha+\psi)\rangle + ri_1 = 0 \qquad (10.26)$$

The approximate (at $\varepsilon = 0$ and at the constants α', α, I_u) periodic solution of Eqs. (10.25), (10.26), satisfying conditions (10.22), will be

$$\psi = 0, \quad i_1' = \frac{1}{1+r^2}[(r\alpha'\cos\alpha - \sin\alpha)\sin\tau + (\alpha'\cos\alpha - r\sin\alpha)\cos\tau] \qquad (10.27)$$

On substituting these expressions into Eq. (10.23) and on performing the operation of averaging, we obtain the following equation of slow motions of the pendulum

$$\alpha'' + \left[h_1 - \varepsilon^2\frac{\gamma\cos^2\alpha}{2(1+r^2)}\right]\alpha' + \varepsilon^2\left[1 + \frac{\gamma\cos\alpha}{2(1+r^2)}\right]\sin\alpha = 0 \quad (10.28)$$

or, passing partly to the initial variables according to (10.18),

10.3. Motion of the Pendulum with the Closed Circuit of Current

or, passing partly to the initial variables according to (10.18),

$$J\ddot{\alpha}+\left[h-\frac{mgl}{\omega}\frac{r}{2(1+r^2)}\right]\dot{\alpha}+mgl\left[1+\frac{\gamma}{2(1+r^2)}\cos\alpha\right]\sin\alpha=0 \quad (10.29)$$

The equation obtained differs from the equation of the oscillation of the pendulum by the second terms in square brackets. The terms of Eq. (10.29), corresponding to them, are quite similar to vibrational moments. The main difference from the problem about the oscillations of the pendulum with a vibrating axis of suspension (see chapters 4 and 6) lies in the fact that a term, corresponding to the "negative resistance" is added into the equation. In the case of a relatively small coefficient of the mechanical resistance h the presence of this term may result in the instability of the positions of the quasi-equilibrium of the pendulum $\alpha=0$ and $\alpha=\pi$. Then near those positions appear small quasi-self-excited oscillations[2]. As for quasi-equilibria as such, what was said in chapters 4 and 6 can be repeated about them since the coefficient at $\sin\alpha$ in Eq. (10.29) differs from the corresponding coefficient of the problem in chapters 4 and 6 only by the value of the constant. In particular, just like the vibration of the pendulum axis in the classical problem, a high-frequency magnetic field can stabilize the upper position of the pendulum.

A thorough analysis of Eq. (10.29) has been made in [5] where the stationary rotation of the pendulum under the action of the magnetic field is also considered. Such regimes are also quite analogous to those of the rotation of the pendulum with a vibrating axis of suspension (see chapter 6). Just pay attention to a certain difference of the second term in the coefficient at $\dot{\alpha}$ in Eq. (10.29) and in the corresponding equation in the work [5]. We will not dwell here on the reason of this difference, which does not result in the qualitative changes in the behavior of the system.

---

[2] Quasi-self-excited oscillations can be determined like quasi-equilibriua (see the footnote at the beginning of this chapter)

References

1. Khodzhaev K. Sh., Oscillations of nonlinear electric mechanical systems, in *Handbook "Vibrations in engineering"*, **2** (Mashinostroenie, Moscow, 1979) 331–347 (in Russian).
2. Khodzhaev K. Sh, Integral criterion of stability for systems with quasi-cyclic coordinates and energy relations under the oscillations of current-carrying conductors, *App. Math. and Mech.*, **33**, 1 (1969) 76–93 (in Russian).
3. Skubov D. Yu., Khodzhaev K. Sh., System with electromagnetic dampers of oscillations, *Mekhanika Tverdogo Tela*, **2**, (1996) 64–74.
4. Khodzhaev K. Sh., Shatalov S. D., About slow motions of the driving solid body in a magnetic field, *Mekhanika Tverdogo Tela*, **2**, (1981) 175–181.
5. Artemyeva M. S., Skubov D. Yu., Dynamics of the driving bodies of pendulum type in a high-frequency magnetic field, *Mekhanika Tverdogo Tela*, **4**, (2001) 20–39.
6. Martynenko Yu. G., *Motion of a solid body in electric and magnetic fields* (Nauka, Moscow, 1988) 368 (in Russian).
7. Landau L. D., Lifshits E. M., *Electrodynamics of continuous media* (Nauka, Moscow, 1982) 620.
8. Earnshaw S., *On the nature of the molecular forces*, **7**, (Transactions Cambridge Phys. Society, 1842) 97–112.

Part III

Problems of the Theory of Selfsynchronization

Chapter 11

On General Definitions of Synchronization

I. I. Blekhman[1], A. L. Fradkov[2]

[1] *Institute for Problems of Mechanical Engineering,*
Russian Academy of Sciences
and
Mekhanobr-Tekhnika Corp.
22 Liniya 3, V.O., 199106, St. Petersburg, Russia

[2] *Institute for Problems of Mechanical Engineering,*
Russian Academy of Sciences
61, Bolshoy, V.O., 199178, St. Petersburg, Russia

11.1 Preliminary Remarks

The term *synchronization* in scientific colloquial use means coordination or agreement in time of two or several processes or objects. Particularly, it may be coincidence or closeness of the state variables for two or several systems, correlated in time change of some quantitative characteristics of the systems, etc.

In some cases the synchronous regime arises due to natural properties of the processes themself and their natural interaction. A well known example is *frequency synchronization* of oscillating or rotating bodies. Such a phenomenon is called *selfsynchronization*. In other cases to achieve synchronization one needs to introduce special actions or impose special constraints. Then we will speak about *forced* or *controlled synchronization* understood as the stage in the time history of the system required to achieve a synchronous regime.

The synchronization phenomenon has numerous applications in vibrational technologies [2, 3, 4], in electronics and telecommunications [7, 6] and other fields.

Since the middle of the 1980s there had been significant interest in the so called chaotic synchronization, when the synchronized subsystems continue to perform complex chaotic oscillations even after the synchronous mode is achieved [1, 16, 23, 15, 24]. A number of ways how to use *chaotic synchronization* for improving security and reliability of the synchronization has been proposed (see e.g. special issues of the international journals [19, 20, 18]).

Recently an increasing interest is observed in the controlled synchronization where additional actions or feedbacks are used to achieve synchronization. Earlier such problems were studied mainly for linear systems [8, 17]. Controlled synchronization allows to extend the class of the systems possessing synchronous modes and increase their stability and robustness.

Some work is going on studying the synchronization of oscillator arrays and lattices, with applications to synchronization of the biological systems, artificial and/or natural neurons, etc. [25, 10, 24].

Due to emerging interest in synchronization from different scientific communities, a paradoxical situation has been arisen: some groups of researchers are not aware of results obtained by the researchers working in other directions and, as a rule, do not refer to their papers. Consequently, achievements of one group may be not used by the other ones. Moreover, understandings of the term "synchronization" may differ significantly from one group to another one.

In order to study different problems from a unified viewpoint it seems useful to formulate a general definition of synchronization, encompassing main exsisting definitions. Perhaps, the first definition of such kind, covering both controlled and non-controlled synchronization was proposed in [11] and extended in [4, 9, 12].

In this paper a modified definition is given, extending the ideas of [11, 4, 9, 12]. A number of examples demonstrating possibilities of the proposed definition and its relation to the main existing definitions.

11.2 Evolution of the Synchronization Concept

First versions of the general definition for periodic processes were proposed in [2] (coincidence or multiplicity of the average frequences of oscillatory or rotatory motions) and in [5] (existence of an asymptotically stable invariant torus of the dimension $n - m$, where m is the degree of synchronization). In [2] it was also pointed out that synchronization can be understood as coincidence of some functionals depending on the systems coordinates (for example, time when some coordinate crosses some level, or time when some coordinate takes its extremum value).

Studies of synchronization for chaotic processes led to a number of new versions of synchronization concept: coordinate (identical) synchronization, [16, 1, 23]; generalized synchronization [27]; phase synchronization [26], master-slave synchronization [23] etc. According to [1] synchronization is understood as existence of a homeomorphism $g : \pi_1(A_c) \to \pi_2(A_c)$, such that $g\left(\pi_1(x_1(t))\right) \to \pi_2\left(x_2(t + \alpha(t))\right)$, where $A(c)$ is attractor of the interconnected system, π_1 and π_2 are projectors to the state spaces of the partial processes, $\alpha(t)$ is asymptotically constant phase shift.

The general definition of synchronization covering both controlled and non-controlled synchronization was proposed in [11] and extended in [4, 9, 12]. Some related definitions were proposed later in [14, 13]. In [21] the definition based upon the concept of the invariant manifold of dynamical system was studied which covers both coordinate and generalized synchronization Below a modified version of the definition of [11, 4, 9, 12]. is described. It allows to easily obtain different existing definitions as its special cases.

11.3 General Definition of Synchronization

Consider a number of systems or processes, the state of each at the time t being described by some vector $x^{(i)}(t)$, $i = 1, 2 \ldots, k$ where $0 \leq t < \infty$. First assume that all the processes belong to the same functional space $\mathcal{X}$.

Let a certain characteristics of the processes is defined as the time-depending family of mappings $C_t : \mathcal{X} \to \mathcal{C}$, where C is the set of possible values of C_t. The characteristics C_t will be called the *synchronization index*. It is important that the index C_t is the same for all the processes. The value of C_t may be a scalar, a vector or a matrix, as well as a function (e.g. spectrum of the process).

Let, finally, the set of vector-functions $F_i : \mathcal{C} \to \mathbb{R}^m$, $i = 1, \ldots, k$ not depending of time are given. The functions F_i are called *comparison functions*.

Definition. We will say that the synchronization of the processes $x^{(i)}(t)$, $i = 1, \ldots, k$ with respect to the index C_t and comparison functions F_i occurs if there exist real numbers τ_i, $i = 1, \ldots, k$ (*phase shifts*) such that the following relations hold for all t:

$$F_1\left(C_{t+\tau_1}[x_1]\right) = F_2\left(C_{t+\tau_2}[x_2]\right) = \ldots = F_k\left(C_{t+\tau_k}[x_k]\right) \qquad (11.1)$$

In addition, *approximate synchronization* (ε-synchronization) is understood as approximate fulfillment of the relations (11.1), with accuracy of ε:

$$\left\|F_i\left(C_{t+\tau_i}[x_i]\right) - F_j\left(C_{t+\tau_j}[x_j]\right)\right\| \le \varepsilon \qquad \forall i,j, \quad t \ge 0, \qquad (11.2)$$

Asymptotic synchronization is understood as fulfillment of (11.1) when $t \to \infty$:

$$\lim_{t \to \infty} \left\|F_i\left(C_{t+\tau_i}[x_i]\right) - F_j\left(C_{t+\tau_j}[x_j]\right)\right\| = 0. \qquad (11.3)$$

Finally, if some averaging operator $\langle \cdot \rangle_t$ on $0 \le s \le t$ is specified, the *synchronization in the average* is defined as fulfillment of relations

$$\langle Q_s \rangle_t < \varepsilon, \qquad (11.4)$$

for all $t \ge 0$, where Q_t is a certain scalar function (*desynchronization measure*), evaluating the deflection from the synchronous mode. Often the averaging operator is specified as an integral operator $\langle Q_s \rangle_t = \frac{1}{t}\int_0^t Q_s ds$, while the desynchronization measure Q_t is defined as mean square deviation from the synchronous mode:

$$Q_t = \sum_{i,j=1}^{k} \left\|F_i\left(C_{t+\tau_i}[x_i]\right) - F_j\left(C_{t+\tau_j}[x_j]\right)\right\|^2. \qquad (11.5)$$

Remark 1. Sometimes it is convenient to write (11.1) in the form

$$F_i\left(C_{t+\tau_i}[x_i]\right) - F_k\left(C_{t+\tau_k}[x_k]\right) = 0 \quad (i = 1, 2, \ldots, k-1). \qquad (11.6)$$

The introduced definition encompasses the main existing description forms of the synchronous behavior of the processes. Let us consider some examples.

11.4 Examples

Example 1. Frequency (Huygens) synchronization. This most well known kind of synchronization is defined for the processes possessing well defined frequencies ω_i, e.g. periodic (oscillatory or rotatory) processes. Introduce the characteristics (synchronization index) C_t as the average velocity over the interval $0 \leq s \leq t$ i.e. $C_t = \omega_t = <\dot{x}>_t$. Since the frequency synchronization criterion is defined by relations $\omega_t = n_i \omega^*$, where for some integer n_i (sybchronization multiplicities), ω^* is the so called *synchronous frequency*, it is natural to introduce comparison functions as $F_i(\omega_t) = \omega_t/n_i$. For the case $n_i = 1, i = 1, \ldots, k$ the simple (multiplicity 1) synchronization is obtained.

Example 2. Extremal synchronization. Extremal synchronization is understood as a kind of behavior of scalar processes when the processes take their extremum (i.e. maximum or minimum) values simultaneously or with a certain time-shift [11, 9]. The synchronization index in this case is defined as the time of passing through the last extremum: $C_t = t^*(t)$ ($i = 1, 2, \ldots, k$), where $t^*(t)$ is the time when the process $x(t)$ takes its last extremum value in the interval $0 \leq s \leq t$, In this case the intervals between the first extremum times of $x^i(t)$ and $x^1(t)$ can play the role of the time-shifts τ_i. For vector-valued processes extremal synchronization of any component of the vector $x^{(i)}(t)$ or of any scalar function of $x^{(i)}(t)$ can be considered.

Synchronization of such kind is important for a number of chemical or biological systems.

Example 3. Phase synchronization. Systems of phase synchronization (phase locking) are well known in electronics and telecommunications [7, 6]. However, conventional engineering applications deal with periodic processes with constant or periodically changing frequencies. In the 1990s there were growing activities in the synchronization of chaotic processes. It led to appearance of new definitions of phase and phase synchronization siutable for chaotic processes [26]. It is natural to introduce the phase of a chaotic process considering its behavior between time of crossings of a certain surface (Poincare section). The synchronization index can then be introduced as the value of the phase φ_t of the process $x(t)$ that belongs to the interval from 0 to 2π and defined as follows:

$$C_t[x] = \varphi_t = 2\pi \frac{t - t_n}{t_{n+1} - t_n} + 2\pi n, \qquad t_n \leq t < t_{n+1}, \qquad (11.7)$$

where t_n is the time of the nth crossing of the trajectory with the Poincaré surface [26].

For $k = 2$ the choice of comparison function $F_1(\varphi_t) = F_2(\varphi_t) = \varphi_t$, yields the inphase synchronization. Otherwise, one may choose $F_1(\varphi_t) = \varphi_t, F_2(\varphi_t) = \varphi_t + \pi$, that corresponds to the antiphase synchronization.

A slightly more general concept of synchronization may be obtained if the synchronization index is chosen as follows: $C_t = t_*(t)$, where $t_*(t)$ is the latest time not exceeding t, of crossing the Poincaré section see [11]. Such a concept encompasses the cases that differ from phase synchronization because no physically reasonable notion of *phase* can be introduced owing to significant irregularity of the processes. For example, let the Poincaré surface be chosen so that its equation defines zero value of time derivative of a certain scalar function of the process state. Then the corresponding synchronization concept is just the extremal synchronization (see above).

Example 4. Coordinate synchronization. In the middle of the 1980s the definition of synchronization for nonperiodic processes as coincidence of the coordinates of the interacting subsystems was introduced [16, 1]. This definition has become especially popular after publishing the paper by L.Pecora and T.Carroll concerning master-slave synchronization of chaotic systems [23]. Obviously, the coordinate synchronization fits the above definition with the following synchronization index $C_t(x_i) = x_i(t)$, where $x_i(t)$ stands for the value of the state vector of the ith subsystem at the time instant t. Comparison functions can be taken identical: $F_i(x) = x$, $i = 1, \ldots, k$.

Example 5. Generalized (partial) coordinate synchronization. The coordinate synchronization from the previous example is often called *full* or *identical* to underline coincidence of all phase coordinates of the systems. In practice a more general case may take place, when only a part of all phase coordinates or certain functions of them $G_i(x_i)$ coincide. The corresponding definition was introduced in [27] and termed *generalized synchronization*. Obviously, the generalized synchronization fits the above scheme under the choice $C_t(x_i) = x_i(t)$, $F_i(x) = G_i(x)$, $i = 1, \ldots, k$.

Other examples. The proposed definition allows to formalize different properties of processes intuitively related to synchronization by means of proper choice of the synchronization index and comparison functions. For example, in order to define coordinate synchronization of processes that change in a correlated manner but with different amplitudes one may use the normalized synchronization index: $C_t[x] = \dfrac{x(t)}{\max\limits_{0 \leq s \leq t} |x(s)|}$.

If one of the two T-periodic processes is corrupted by a irregular noise, in order to reveal possible synchronization one may use a moving average based index: $C_t[x] = \dfrac{2}{T} \int\limits_{t-T/2}^{t} x(s)ds$.

A different way of using moving average allows to eliminate trends. E.g. introducing the synchronization index of the form $C_t[x] = x(t) - \frac{1}{T}\int_{t-T}^{t} x(s)ds$ allows to dascribe synchronous behavior up to a linear trend.

11.5 Discussion

Further generalizations. The above general definitions can be further generalized. For example, the time-shifts $\tau_i, i = 1, \ldots, k$ are not constant in many practical problems. It seems reasonable to extend the definition of synchronization in order to capture the problems where time-shifts are not constant but tend to some constant values instead (so called "asymptotic phases"). In such cases one may replace time-shift operators for each process by time change operators defined as follows:

$$(\sigma_{\tau_i})\, x(t) = x(t'_i(t)),$$

where $t'_i : T \to T$, $i = 1, \ldots, k$ are some homeomorphisms such that

$$\lim_{t\to\infty} (t'_i(t) - t) = \tau_i. \qquad (11.8)$$

Note that in the paper [1] a milder condition $\lim_{t\to\infty} (t'_i(t)/t) = 1$, was proposed instead of (11.8). Since the definition of the paper [1] admits infinitely large time shifts, it may describe processes that intuitively do not seem synchronized.

The definition of the synchronization may be modified to include the cases when the processes x_i belong to different functional spaces $\mathcal{X}_i$ (e.g. $x_1(t) \in \mathbb{R}^3$ is described by a Lorenz model while $x_2(t) \in \mathbb{R}^2$ obeys to a Van der Pol equation). To this end we introduce so called *precomparison functions* $F'_i : \mathcal{X}_i \to \mathcal{X}$, which map all the processes into a single space. Based on such a transformation the equalities defining the synchronous mode take the following form:

$$F_1\left(C_{t+\tau_1}[F'(x_1)]\right) = F_2\left(C_{t+\tau_2}[F'(x_2)]\right) = \ldots$$
$$= F_k\left(C_{t+\tau_k}[F'(x_k)]\right) \qquad \forall t \geq 0. \qquad (11.9)$$

Note that for the case of two processes ($k = 2$) the precomparison functions generate a "rectifying" transformation $\mathcal{F}$, introduced in the paper

[13] as follows $\mathcal{F} = (F_1', F_2')$. In this case if we choose the synchronization index as in the Example 5: $C_t(x_i) = x_i(t)$, and take one of the postcomparison functions as identity we arrive at the definition of the paper [13]. Moreover, in this case the second comparison function plays the role of synchronization function of the paper [13]. Summarizing, we have shown that the definition of [13] describes a more general concept than the generalized synchronization of the Example 5, yet still less general than the definition introduced in this paper.

Conclusions. It is worth noticing in the conclusion that the introduced definition not only provides the terminology and the conceptual tools for discussion of different synchronization-like properties, but also allows to achieve additional purposes.

Particularly, it allows to address the question of applicability area for the term "synchronization". For example, according to the above definition the property defined as reduction of the fractal dimension of the overall process with respect to the sum of the dimensions of its components (see [22]) is not a synchronization property, because it is defined via characteristics (dimensions) that do not depend on the behavior of the processes in time. In this case the terms "ordering" or "synergy" would seem more preferable.

Similarly, it seems not appropriate to term two processes as synchronized if they are strongly correlated: $|\varrho(x_1, x_2)| > 1 - \varepsilon$, where

$$\varrho(x_1, x_2) = \frac{\langle x_1 \cdot x_2 \rangle}{\sqrt{\langle x_1^2 \rangle \cdot \langle x_2^2 \rangle}}.$$

is the correlation coefficient.

Finally, introduction of the scalar desynchronization measure (e.g. (11.5)) opens the possibility for systematic design of controlled synchronization systems where the synchronization mode is created or modified artificially. Such design may be based upon the so called speed-gradient algorithms

$$u(t) = -\gamma \nabla_u \dot{Q}_t, \quad \gamma > 0,$$

where $u(t)$ is the vector of the controlling parameters or variables [15, 12].

References

[1] Afraimovich V.S., N.N. Verichev and M.I. Rabinovich (1987), "Synchronization of oscillations in dissipative systems," *Radiophysics and Quantum Electronics*, Plenum Publ. Corp. pp. 795-803.

[2] Blekhman I.I. (1971), *Synchronization of dynamical systems*, (Nauka), Moscow, 894p (in Russian).

[3] Blekhman I.I. (1988), *Synchronization in science and technology*, (ASME Press), New-York.

[4] Blekhman I.I., *Vibrational Mechanics*, World Scientific, 2000 (in Russian – 1994).

[5] Gurtovnik A.S., Neimark Yu.I. On synchronization of dynamical systems. Appl. Math. Mech. 1974, No 5.

[6] Leonov G.A., Smirnova V.B. Mathematical problems of phase synchronization theory. St.Petersburg: Nauka, 2000 (in Russian).

[7] Lindsey W. (1972), *Synchronization systems in communications and control*, (Prentice-Hall), NJ.

[8] Miroshnik I.V., Ushakov A.V. Design of the algorithm of synchronous control of the quasisimilar systems. Automation and Remote Control, 1977, No. 11.

[9] Blekhman, I.I. and Fradkov, A.L., Eds. (2001) *Control of Mechatronic Vibrational Units*. St.Petersburg, Nauka, 278 p. (in Russian).

[10] Belykh V.N., Belykh I.V., Mosekilde E. Cluster Synchronization Modes in an Ensemble of Coupled Chaotic Oscillators. Phys. Rev. E 63, 036216, 1-4 (2001).

[11] Blekhman, I.I., A.L. Fradkov, H. Nijmeijer and A.Yu. Pogromsky, (1997), "On self-synchronization and controlled synchronization", *Systems & Control Letters*, vol. 31, pp. 299–305.

[12] Blekhman I.I., Fradkov A.L., Tomchina O.P., Bogdanov D.E. Self-synchronization and controlled synchronization: general definition and example design. Mathematics and Computers in Simulation, 2002, V.58, Issue 4-6, pp.367-384.

[13] Boccaletti S., Pecora L., Pelaez A. Unifying framework for synchronization of coupled systems. Phys. Rev. E, v. 63, 2001, 066219.

[14] Brown R., Kocarev L. A unifying definition of synchronization for dynamical systems. Chaos, v. 10, No 2, 2000, pp.344–349.

[15] Fradkov A.L., Pogromsky A.Yu. Introduction to control of oscillations and chaos. Singapore; World Scientific, 1998.

[16] Fujisaka H., Yamada T. Prog.Teor.Phys. V.69, 1983, pp.32–47.

[17] Helmke U., Pratzelwolters D., Schmid S., Adaptive synchronization of interconnected linear systems. IMA J Math.Control. 8(4) 1991, pp.397-408.

[18] IEEE Trans. on Circuits and Systems (2001). V.48. No.12. Special issue on applications of chaos in modern communication systems/ Eds. L.Kocarev, G.M.Maggio, M.Ogorzalek, L.Pecora, K.Yao.

[19] IEEE Trans. on Circuits and Systems (1997). V.44. No.10. Special issue "Chaos control and synchronization"/ Eds. M.Kennedy, M.Ogorzalek.

[20] International Journal of Circuit Theory and Applications (1999) Special issue: Communications, Information Processing and Control Using Chaos./ Eds. M.Hasler, J.Vandewalle. V.27, No.6.
[21] Josić K. Invariant manifolds and synchronization of coupled dynamical systems. Phys. Rev. Lett., V. 80, 1998, 3053–3056.
[22] Landa P.S. and M.G. Rosenblum, (1993), "Synchronization and chaotization of oscillations in coupled self-oscillating systems," *Appl. Mech. Rev.*, vol. **46**(7), pp. 414–426.
[23] Pecora, L.M. and T.L. Carroll, (1990), "Synchronization in chaotic systems," *Phys. Rev. Lett.*, vol. **64**, pp. 821–823.
[24] Pikovsky A, Rosenblum M, and J. Kurths. Synchronization. A Universal Concept in Nonlinear Sciences. Cambridge University Press, 2001.
[25] Rabinovich, MI, Abarbanel, HDI, Huerta, R, Elson, R, Selverston, AI Self-regularization of chaos in neural systems: Experimental and theoretical results. IEEE Transactions On Circuits And Systems I, V.44, 1997, 997-1005.
[26] Rosenblum M G, Pikovsky A S., Kurths J. Phase synchronization of chaotic oscillators. Phys. Rev. Lett. V.76, 1996, 1804-1807.
[27] Rulkov N F., Sushchik M, Tsimring LS., Abarbanel H D I., Generalized synchronization of chaos. Phys. Rev. E, V.51, 1995, 980.

Chapter 12

A Guide to Solving Certain Self-Synchronization Problems

L. Sperling, H. Duckstein

*Institut für Mechanik, Otto-von-Guericke-Universität Magdeburg,
Universitätsplatz 2, 39106 Magdeburg, Germany*

12.1 Preliminary Remarks

Self-synchronized unbalanced rotors are frequently employed in many fields of mechanical engineering and civil engineering. While many approaches to the approximate analysis have been developed, published and demonstrated in a number of monographs and papers ([1], [2]), engineers who do not specialize in this field often face problems when it comes to solving specific tasks in connection with the non-linear phenomenon of self-synchronization. Therefore, the present chapter develops a formal, algorithmic guide to deriving the conditions for the existence and stability of self-synchronized motions of unbalanced rotors on supported mechanical systems consisting of rigid bodies and linear-elastic springs. This guide is developed on the basis of the method of direct separation of motion ([3]) employing the concept of vibrational moments and harmonic influence coefficients. The paper is not aiming at presenting entirely new approaches or results. However, the paper generalizes the type of the unbalanced rotor compared to the type assumed in relevant literature, in that dynamically unbalanced rotors are included as well (also refer to [8], [9]).

12.2 Unbalanced Rotors on an Oscillatory System

A system is considered which consists of m unbalanced rotors on an oscillatory system. With regard to the rotors, the latter is also referred to as the supporting system. Figure 12.1 shows the i-th unbalanced rotor of any shape, having a mass m_i and an inertia matrix $\mathbf{I}_i$ (refer to Eq. (12.2)). The basis for the matrix representation of the vectors is the rotor-related inertial frame $O_{Ii}, \vec{e}_{ix}, \vec{e}_{iy}, \vec{e}_{iz}$. O_{Ii} and O_i represent the pivot point of the i-th rotor in equilibrium position of the oscillatory system and in any position, respectively. The pivot point is the point of intersection of the rotor axis and the plane of motion of the mass center C_i of the rotor. At this point the rotor is rigidly connected to the supporting system. The vectors $\vec{r}_i$ and $\vec{\psi}_i$ describe the small translatory and rotatory vibrations, respectively, of the oscillatory system at the pivot point O_{Ii}. $O_i, \vec{e}_{ix}^*, \vec{e}_{iy}^*, \vec{e}_{iz}^*$ is an oscillatory system fixed frame. The rotor has one relative degree of freedom corresponding to the coordinate φ_i. $O_i, \vec{e}_{ir}^*, \vec{e}_{is}^*, \vec{e}_{iz}^*$ is a rotor-fixed frame.

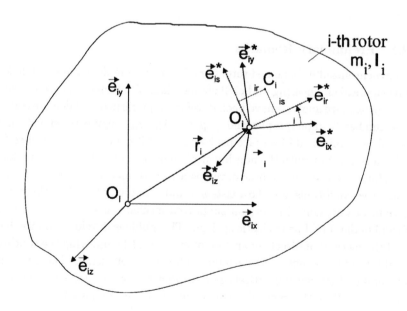

Fig. 12.1 Unbalanced rotor and reference systems

12.2. Unbalanced Rotors on an Oscillatory System

Let

$$\mathbf{r}_i = [r_{ix}\ r_{iy}\ r_{iz}]^T,\ \boldsymbol{\psi}_i = [\psi_{ix}\ \psi_{iy}\ \psi_{iz}]^T,\ i = 1, ..., m \qquad (12.1)$$

be the column matrices of the vector $\vec{r}_i$ and $\vec{\psi}_i$ with reference to the rotor-related inertial frame and

$$\mathbf{I}_i = \begin{bmatrix} I_{irr} & I_{irs} & I_{irz} \\ I_{irs} & I_{iss} & I_{isz} \\ I_{irz} & I_{isz} & I_{izz} \end{bmatrix},\ i = 1, ..., m \qquad (12.2)$$

the inertia matrix of the rotors with reference to the rotor-fixed frame. Assuming that the oscillatory system is a two- or three-dimensional supported multibody system comprising only rigid bodies and linear-elastic springs, the equations of motion are

$$\mathbf{M\ddot{q}} + \mathbf{Kq} = \mathbf{Q} \qquad (12.3)$$

with symmetrical positively definite mass and stiffness matrices:

$$\mathbf{M} = \mathbf{M}^T > 0,\ \mathbf{K} = \mathbf{K}^T > 0. \qquad (12.4)$$

The column matrix

$$\mathbf{q} = [q_1\ ...\ q_n]^T \qquad (12.5)$$

comprises the generalized coordinates of the oscillatory system. Let us assume that the vibrations of the pivot points be expressed in terms of the corresponding Jacobian matrices $\mathbf{J}_{ir}$ and $\mathbf{J}_{i\psi}$ in linear dependence on $\mathbf{q}$ which is always possible for small vibrations:

$$\mathbf{r}_i = \mathbf{J}_{ir}\mathbf{q},\ \boldsymbol{\psi}_i = \mathbf{J}_{i\psi}\mathbf{q},\ i = 1, ..., m. \qquad (12.6)$$

Then, the global column matrix of the pivot point vibrations

$$\mathbf{p} = [\mathbf{r}_1^T ... \mathbf{r}_m^T \boldsymbol{\psi}_1^T ... \boldsymbol{\psi}_m^T]^T = [\mathbf{p}_1^T ... \mathbf{p}_m^T \mathbf{p}_{m+1}^T ... \mathbf{p}_{2m}^T]^T \qquad (12.7)$$

can be expressed with the help of the global Jacobian matrix

$$\mathbf{J} = [\mathbf{J}_{1r}^T ... \mathbf{J}_{mr}^T \mathbf{J}_{1\psi}^T ... \mathbf{J}_{m\psi}^T]^T \qquad (12.8)$$

in terms of the column matrix of the generalized coordinates as follows:

$$\mathbf{p} = \mathbf{Jq}. \qquad (12.9)$$

Assuming that the column matrix **Q** of the generalized forces in (12.3) only comprises the excitations caused by the unbalanced rotors, the relationship between **Q** and the global column matrix of the inertia forces and moments

$$\mathbf{P} = [\mathbf{f}_1^T ... \mathbf{f}_m^T \mathbf{l}_1^T ... \mathbf{l}_m^T]^T \tag{12.10}$$

becomes

$$\mathbf{Q} = \mathbf{J}^T \mathbf{P}. \tag{12.11}$$

In the following, let us assume that the rotor masses be small compared to the representative mass parameters of the oscillatory system. Then, all terms which are dependent on the coordinates of the oscillatory system are approximately negligible in the inertia forces and moments. Thus, we obtain

$$f_{ix} = m_i(\varepsilon_{ix}\dot{\varphi}_i^2 + \varepsilon_{iy}\ddot{\varphi}_i), \quad f_{iy} = m_i(\varepsilon_{iy}\dot{\varphi}_i^2 - \varepsilon_{ix}\ddot{\varphi}_i), \quad f_{iz} = 0, \tag{12.12}$$

$$l_{ix} = I_{iyz}\dot{\varphi}_i^2 - I_{ixz}\ddot{\varphi}_i, \quad l_{iy} = -I_{ixz}\dot{\varphi}_i^2 - I_{iyz}\ddot{\varphi}_i, \quad l_{iz} = -I_i\ddot{\varphi}_i \approx 0 \tag{12.13}$$

with the abbreviations

$$\begin{aligned}\varepsilon_{ix} &= \varepsilon_{ir}\cos\varphi_i - \varepsilon_{is}\sin\varphi_i, & \varepsilon_{iy} &= \varepsilon_{ir}\sin\varphi_i + \varepsilon_{is}\cos\varphi_i, \\ I_{ixz} &= I_{irz}\cos\varphi_i - I_{isz}\sin\varphi_i, & I_{iyz} &= I_{irz}\sin\varphi_i + I_{isz}\cos\varphi_i,\end{aligned} \tag{12.14}$$

$$I_i = I_{izz} + m_i\varepsilon_i^2, \quad \varepsilon_i^2 = \varepsilon_{ir}^2 + \varepsilon_{is}^2. \tag{12.15}$$

It is noteworthy that, provided the bearing of the i-th rotor is damped, l_{iz} should be complemented to include an appropriate moment. Neglecting this moment in the equations of motion of the oscillatory system as well as the damping of vibrations, and also neglecting the term $I_i\ddot{\varphi}_i$ in (12.13) in view of the steady state of the motion studied, once the abbreviations (12.14) and (12.15) are substituted we obtain the column matrices of the inertia forces and moments

$$\mathbf{f}_i = m_i \begin{bmatrix} \varepsilon_{ir}(\dot{\varphi}_i^2\cos\varphi_i + \ddot{\varphi}_i\sin\varphi_i) + \varepsilon_{is}(-\dot{\varphi}_i^2\sin\varphi_i + \ddot{\varphi}_i\cos\varphi_i) \\ \varepsilon_{ir}(\dot{\varphi}_i^2\sin\varphi_i - \ddot{\varphi}_i\cos\varphi_i) + \varepsilon_{is}(\dot{\varphi}_i^2\cos\varphi_i + \ddot{\varphi}_i\sin\varphi_i) \\ 0 \end{bmatrix}, \tag{12.16}$$

$$\mathbf{l}_i = \begin{bmatrix} I_{isz}(\dot{\varphi}_i^2\cos\varphi_i + \ddot{\varphi}_i\sin\varphi_i) - I_{irz}(-\dot{\varphi}_i^2\sin\varphi_i + \ddot{\varphi}_i\cos\varphi_i) \\ I_{isz}(\dot{\varphi}_i^2\sin\varphi_i - \ddot{\varphi}_i\cos\varphi_i) - I_{irz}(\dot{\varphi}_i^2\cos\varphi_i + \ddot{\varphi}_i\sin\varphi_i) \\ 0 \end{bmatrix}. \tag{12.17}$$

12.2. Unbalanced Rotors on an Oscillatory System

The complete expressions of the inertia forces and moments linearized in the vibration coordinates include terms which correspond to the rotor masses in the mass matrix **M** considered centered in the pivot points ([8]). These terms should be taken into consideration if no other terms being dependent on the vibrational coordinates are present or if inconsistency is accepted to some extent. Let us now consider the equation of motion of the i-th rotor

$$I_i \ddot{\varphi}_i + B_i = L_i(\dot{\varphi}_i, \varphi_i). \tag{12.18}$$

Neglecting all those terms which are a non-linear function of the coordinates of the oscillatory system, we obtain the following expression for the moment of the reaction of the oscillatory system on the i-th rotor

$$B_i = \mathbf{e}_{is}^T (m_i \varepsilon_{ir} \ddot{\mathbf{r}}_i + I_{isz} \ddot{\boldsymbol{\psi}}_i) + \mathbf{e}_{ir}^T (-m_i \varepsilon_{is} \ddot{\mathbf{r}}_i + I_{irz} \ddot{\boldsymbol{\psi}}_i) \tag{12.19}$$

or

$$B_i = \mathbf{e}_{is}^T (m_i \varepsilon_{ir} \ddot{\mathbf{p}}_i + I_{isz} \ddot{\mathbf{p}}_{m+i}) + \mathbf{e}_{ir}^T (-m_i \varepsilon_{is} \ddot{\mathbf{p}}_i + I_{irz} \ddot{\mathbf{p}}_{m+i}) \tag{12.20}$$

with the column matrices

$$\mathbf{e}_{ir} = \begin{bmatrix} cos\varphi_i & sin\varphi_i & 0 \end{bmatrix}^T, \mathbf{e}_{is} = \begin{bmatrix} -sin\varphi_i & cos\varphi_i & 0 \end{bmatrix}^T. \tag{12.21}$$

In the vicinity of the stationary angular velocity, linearization of the driving moment of the rotor is permitted:

$$L_{ai}(\dot{\varphi}_i, \varphi_i) = L_{0i} - \overline{\overline{\beta}}_i \dot{\varphi}_i. \tag{12.22}$$

In addition, the damping moment

$$L_{di} = -\overline{\beta}_i \dot{\varphi}_i \tag{12.23}$$

is taken into account. Thus, we obtain for the total moment in Eq. (12.18)

$$L_i(\dot{\varphi}_i, \varphi_i) = L_{0i} - \beta_i \dot{\varphi}_i \tag{12.24}$$

where

$$\beta_i = \overline{\overline{\beta}}_i + \overline{\beta}_i. \tag{12.25}$$

Now, the equation of motion of the i-th rotor takes the following form:

$$I_i \ddot{\varphi}_i + \beta_i \dot{\varphi}_i + B_i = L_{0i}. \tag{12.26}$$

12.3 Application of the Method of Direct Separation of Motion

In general, there is no closed solution to Eq. (12.26). Our approach focuses on establishing the conditions of existence and stability of self-synchronized motions. In a first approximation these motions are characterized by angular velocities and relative phases of the rotors which are constant on an average. To solve this task the method of direct motion separation has proven very useful and descriptive. It was developed by Kapitsa ([6]) and generalized later by Blekhman who has applied it to a number of problems including problems of self-synchronization ([3], [4], also refer to [10]and chapter 2 of this book). The relationship between this method and the averaging method presented by Krylov, Bogoliubov and Mitropolskii as well as the method of multiple scales proposed by Nayfeh is described in chapter 4 of this book; see also the paper [5].

Following the basic idea of the method of direct separation of motion, we separate the motion of the rotors as follows:

$$\varphi_i(t) = \Omega t + \alpha_i(t) + \xi_i(t, \Omega t) \qquad (12.27)$$

where Ω is the synchronous angular velocity that is initially unknown, and $\alpha(t)$ is a slow component of motion:

$$\dot{\alpha}_i(t) << \Omega. \qquad (12.28)$$

It is assumed that the fast component $\xi_i(t, \Omega)$ is 2π−periodical over the so-called "fast" time Ωt. An identical assumption is made for the oscillatory motion $\mathbf{q} = \mathbf{q}(t, \Omega t)$. In addition, we assume that the average of this fast component over the period vanishes:

$$\frac{1}{2\pi} \int_0^{2\pi} \xi_i(t, \Omega t) d(\Omega t) = 0. \qquad (12.29)$$

It is another essential step of this approach that the equations of motion (12.26) are averaged over a period with regard to the "fast time" Ωt. For determining the reacting moments B_i (refer to Eqs. (12.18) to (12.21)) it is sufficient as a first approximation to assume that α_i is constant and neglect ξ_i:

$$\varphi_i^0(t) = \Omega t + \alpha_i, \quad \alpha_i = const. \qquad (12.30)$$

12.4 Harmonic Influence Coefficients and Vibrational Moments

Substituting the relation (12.30) in Eqs. (12.16) and (12.17), we obtain the centrifugal forces and moments of unbalance

$$\mathbf{f}_i^0 = \hat{f}_{ir} \begin{bmatrix} cos(\Omega t + \alpha_i) \\ sin(\Omega t + \alpha_i) \\ 0 \end{bmatrix} + \hat{f}_{is} \begin{bmatrix} -sin(\Omega t + \alpha_i) \\ cos(\Omega t + \alpha_i) \\ 0 \end{bmatrix}, \qquad (12.31)$$

$$\mathbf{l}_i^0 = \hat{l}_{ir} \begin{bmatrix} cos(\Omega t + \alpha_i) \\ sin(\Omega t + \alpha_i) \\ 0 \end{bmatrix} + \hat{l}_{is} \begin{bmatrix} -sin(\Omega t + \alpha_i) \\ cos(\Omega t + \alpha_i) \\ 0 \end{bmatrix} \qquad (12.32)$$

with the amounts

$$\hat{f}_{ir} = m_i \varepsilon_{ir} \Omega^2, \quad \hat{f}_{is} = m_i \varepsilon_{is} \Omega^2, \quad \hat{l}_{ir} = I_{isz} \Omega^2, \quad \hat{l}_{is} = -I_{irz} \Omega^2. \qquad (12.33)$$

First of all, the steady-state solution is to be determined for the equations of motion of the oscillatory system

$$\mathbf{M}\ddot{\mathbf{q}}^0 + \mathbf{K}\mathbf{q}^0 = \mathbf{Q}^0 \qquad (12.34)$$

where

$$\mathbf{Q}^0 = \mathbf{J}^T \mathbf{P}^0, \quad \mathbf{P}^0 = [\mathbf{f}_1^{0T} ... \mathbf{f}_m^{0T} \mathbf{l}_1^{0T} ... \mathbf{l}_m^{0T}]^T. \qquad (12.35)$$

To this end, we compute the steady-state solutions of the equations

$$\mathbf{M}\ddot{\mathbf{q}}^{ix} + \mathbf{K}\mathbf{q}^{ix} = (\mathbf{J}^{3i-2})^T cos\Omega t, \qquad (12.36)$$

$$\mathbf{M}\ddot{\mathbf{q}}^{iy} + \mathbf{K}\mathbf{q}^{iy} = (\mathbf{J}^{3i-1})^T cos\Omega t, \qquad (12.37)$$

$$i = 1, ..., 2m$$

where $\mathbf{J}^k$ is the k-th row of the global Jacobian matrix $\mathbf{J}$. We obtain

$$\mathbf{q}^{ix} = \hat{\mathbf{q}}^{ix} cos\Omega t, \quad \hat{\mathbf{q}}^{ix} = \mathbf{H}(\mathbf{J}^{3i-2})^T, \qquad (12.38)$$

$$\mathbf{q}^{iy} = \hat{\mathbf{q}}^{iy} cos\Omega t, \quad \hat{\mathbf{q}}^{iy} = \mathbf{H}(\mathbf{J}^{3i-1})^T \qquad (12.39)$$

with the frequency response matrix for the unit excitation

$$\mathbf{H} = (-\mathbf{M}\Omega^2 + \mathbf{K})^{-1}. \qquad (12.40)$$

From this follows the steady-state solution of the equation system (12.34)

$$\mathbf{q}^0 = \sum_{i=1}^{2m} \{\hat{g}_{ir}[\hat{\mathbf{q}}^{ix}\cos(\Omega t + \alpha_i) + \hat{\mathbf{q}}^{iy}\sin(\Omega t + \alpha_i)] \\ + \hat{g}_{is}[-\hat{\mathbf{q}}^{ix}\sin(\Omega t + \alpha_i) + \hat{\mathbf{q}}^{iy}\cos(\Omega t + \alpha_i)]\} \quad (12.41)$$

with the new abbreviations for the centrifugal forces and moments

$$\hat{g}_{ir} = \hat{f}_{ir}, \quad \hat{g}_{(m+i)r} = \hat{l}_{ir}, \quad \hat{g}_{is} = \hat{f}_{is}, \quad \hat{g}_{(m+i)s} = \hat{l}_{is} \\ i = 1, ..., m. \quad (12.42)$$

For determining the stationary pivot point motions

$$\mathbf{p}^0 = [\mathbf{r}_1^0 ... \mathbf{r}_m^0 \quad \psi_1^0 ... \psi_m^0] = \mathbf{J}\mathbf{q}^0 \quad (12.43)$$

required for the expression (12.20), the concept of the *harmonic influence coefficients* ([7], [9]) is particularly well suited. The generalized harmonic influence coefficients

$$G_{ix}^{kx} = \mathbf{J}^{3i-2}\hat{\mathbf{q}}^{kx} = \mathbf{J}^{3i-2}\mathbf{H}(\mathbf{J}^{3k-2})^T, \\ G_{ix}^{ky} = \mathbf{J}^{3i-2}\hat{\mathbf{q}}^{ky} = \mathbf{J}^{3i-2}\mathbf{H}(\mathbf{J}^{3k-1})^T, \\ G_{iy}^{kx} = \mathbf{J}^{3i-1}\hat{\mathbf{q}}^{kx} = \mathbf{J}^{3i-1}\mathbf{H}(\mathbf{J}^{3k-2})^T, \\ G_{iy}^{ky} = \mathbf{J}^{3i-1}\hat{\mathbf{q}}^{ky} = \mathbf{J}^{3i-1}\mathbf{H}(\mathbf{J}^{3k-1})^T \quad (12.44)$$

describe the translatory and the rotary motions, respectively, of the oscillatory system in the pivot point O_i attributable to a harmonic force or moment excitation with amplitudes of the value 1 at the pivot point O_k. On the basis of this definition, these coefficients may also be determined analytically or experimentally, independent of the algorithm developed here.

Due to the symmetry of **H**, the relations (12.44) directly lead to the symmetry laws of the harmonic influence coefficients

$$G_{ix}^{kx} = G_{kx}^{ix}, \quad G_{ix}^{ky} = A_{ky}^{ix}, \\ G_{iy}^{kx} = G_{kx}^{iy}, \quad G_{iy}^{ky} = G_{ky}^{iy}, \quad i = 1,...,2m, \quad k = 1,...,2m. \quad (12.45)$$

Depending on the harmonic influence coefficients, the stationary motions of the pivot points considering

$$\alpha_{m+i} = \alpha_i, \quad i = 1, ..., m \quad (12.46)$$

12.4. Harmonic Influence Coefficients and Vibrational Moments

are found in the form of

$$
\begin{aligned}
p_{ix}^0 = \sum_{k=1}^{2m} &\{\hat{g}_{kr}[G_{ix}^{kx}\cos(\Omega t + \alpha_k) + G_{ix}^{ky}\sin(\Omega t + \alpha_k)] \\
&+ \hat{g}_{ks}[-G_{ix}^{kx}\sin(\Omega t + \alpha_k) + G_{ix}^{ky}\cos(\Omega t + \alpha_k)]\},
\end{aligned}
\tag{12.47}
$$

$$
\begin{aligned}
p_{iy}^0 = \sum_{k=1}^{2m} &\{\hat{g}_{kr}[G_{iy}^{kx}\cos(\Omega t + \alpha_k) + G_{iy}^{ky}\sin(\Omega t + \alpha_k)] \\
&+ \hat{g}_{ks}[-G_{iy}^{kx}\sin(\Omega t + \alpha_k) + G_{iy}^{ky}\cos(\Omega t + \alpha_k)]\}.
\end{aligned}
\tag{12.48}
$$

Thus,

$$
\ddot{p}_{ix}^0 = -\Omega^2 p_{ix}^0, \quad \ddot{p}_{iy}^0 = -\Omega^2 p_{iy}^0. \tag{12.49}
$$

Therefore, the moments of the oscillatory system reacting on the rotors may be represented in the following form:

$$
B_i^0 = B_{i,tr}^0 + B_{i,rot}^0, \quad B_{i,rot}^0 = B_{m+i,tr}^0, \quad i = 1, ..., m \tag{12.50}
$$

where

$$
\begin{aligned}
B_{i,tr}^0 = &\hat{g}_{ir}[p_{ix}^0 \sin(\Omega t + \alpha_i) - p_{iy}^0 \cos(\Omega t + \alpha_i)] \\
&+ \hat{g}_{is}[p_{ix}^0 \cos(\Omega t + \alpha_i) + p_{iy}^0 \sin(\Omega t + \alpha_i)].
\end{aligned}
\tag{12.51}
$$

Substituting the stationary motion of the pivot points (12.47), (12.48), we obtain

$$
\begin{aligned}
B_{i,tr}^0 = \frac{1}{2} \sum_{k=1}^{2m} &\{(\hat{g}_{ir}\hat{g}_{kr} - \hat{g}_{is}\hat{g}_{ks})[(G_{ix}^{kx} - G_{iy}^{ky})\sin(2\Omega t + \alpha_i + \alpha_k) \\
&- (G_{ix}^{ky} + G_{iy}^{kx})\cos(2\Omega t + \alpha_i - \alpha_k)] \\
&+ (\hat{g}_{ir}\hat{g}_{ks} + \hat{g}_{is}\hat{g}_{kr})[(G_{ix}^{kx} - G_{iy}^{ky})\cos(2\Omega t + \alpha_i - \alpha_k) \\
&+ (G_{ix}^{ky} + G_{iy}^{kx})\sin(2\Omega t + \alpha_i + \alpha_k)] \\
&+ (\hat{g}_{ir}\hat{g}_{kr} + \hat{g}_{is}\hat{g}_{ks})[(G_{ix}^{kx} + G_{iy}^{ky})\sin(\alpha_i - \alpha_k) \\
&+ (G_{ix}^{ky} - G_{iy}^{kx})\cos(\alpha_i - \alpha_k)] \\
&- (\hat{g}_{ir}\hat{g}_{ks} - \hat{g}_{is}\hat{g}_{kr})[(G_{ix}^{kx} + G_{iy}^{ky})\cos(\alpha_i - \alpha_k) \\
&- (G_{ix}^{ky} - G_{iy}^{kx})\sin(\alpha_i - \alpha_k)]\},
\end{aligned}
\tag{12.52}
$$

which consists of 2Ω-frequent and of constant terms.

The averaging procedure mentioned after Eq. (12.29) yields the so-called vibrational moments consisting of only the constant terms,

$$V_i^0 = \frac{1}{2\pi} \int_0^{2\pi} B_i^0 d(\Omega t) = V_{i,tr}^0 + V_{i,rot}^0, \quad V_{i,rot}^0 = V_{m+i,tr}^0,$$

$$i = 1, ..., m \tag{12.53}$$

where

$$\begin{aligned} V_{i,tr}^0 &= \frac{1}{2} \sum_{k=1}^{2m} \{(\hat{g}_{ir}\hat{g}_{kr} + \hat{g}_{is}\hat{g}_{ks})[(G_{ix}^{kx} + G_{iy}^{ky})sin(\alpha_i - \alpha_k) \\ &\quad + (G_{ix}^{ky} - G_{iy}^{kx})cos(\alpha_i - \alpha_k)] \\ &\quad - (\hat{g}_{ir}\hat{g}_{ks} - \hat{g}_{is}\hat{g}_{kr})[(G_{ix}^{kx} + G_{iy}^{ky})cos(\alpha_i - \alpha_k) \\ &\quad - (G_{ix}^{ky} - G_{iy}^{kx})sin(\alpha_i - \alpha_k)]\}. \end{aligned} \tag{12.54}$$

Considering the symmetry laws of the harmonic influence coefficients (12.45) we obtain the conditions of integrability

$$\frac{\partial V_i^0}{\partial \alpha_k} = \frac{\partial V_k^0}{\partial \alpha_i}, \quad i, k = 1, ..., m. \tag{12.55}$$

Hence, a *potential function* Λ^0 must exist from which the vibrational moments may be computed as follows:

$$V_i^0 = \frac{\partial \Lambda^0}{\partial \alpha_i}. \tag{12.56}$$

It can be shown that this potential function is the averaged Lagrangian of the oscillatory system for the stationary motion:

$$\Lambda^0 = \frac{1}{\Omega t} \int_0^{2\pi} (T^0 - U^0) d(\Omega t), \quad T^0 = \frac{1}{2}(\dot{\mathbf{q}}^0)^T \mathbf{M} \dot{\mathbf{q}}^0, \quad U^0 = \frac{1}{2}(\mathbf{q}^0)^T \mathbf{K} \mathbf{q}^0. \tag{12.57}$$

Substituting solution (12.41) and considering the relations (12.38), (12.39), (12.40) and (12.44), we obtain

$$\begin{aligned} \Lambda^0 &= \frac{1}{2} \sum_{j=1}^{2m-1} \sum_{k=j+1}^{2m} \{(\hat{g}_{ir}\hat{g}_{kr} + \hat{g}_{is}\hat{g}_{ks})[-(G_{ix}^{kx} + G_{iy}^{ky})cos(\alpha_i - \alpha_k) \\ &\quad + (G_{ix}^{ky} - G_{iy}^{kx})sin(\alpha_i - \alpha_k)] \\ &\quad - (\hat{g}_{ir}\hat{g}_{ks} - \hat{g}_{is}\hat{g}_{kr})[(G_{ix}^{kx} + G_{iy}^{ky})sin(\alpha_i - \alpha_k) \\ &\quad + (G_{ix}^{ky} - G_{iy}^{kx})cos(\alpha_i - \alpha_k)]\}. \end{aligned} \tag{12.58}$$

In fact, the differentiation (12.56) yields the vibrational moments (12.53), (12.54).

12.5 Conditions for the Existence of Synchronous Motions

Considering the relations (12.27), (12.29) and (12.53) and using the so-called partial angular velocities

$$\Omega_i = \frac{L_{0i}}{\beta_i}, \quad i = 1, ..., m \tag{12.59}$$

we obtain, by averaging the Eqs. (12.26) over a period with respect to Ωt, the differential equation for the slow components

$$I_i \ddot{\alpha}_i + \beta_i \dot{\alpha}_i + V_i^0 = \beta_i(\Omega_i - \Omega), \quad i = 1, ..., m. \tag{12.60}$$

These differential equations which are less complex compared to the original system equations, may be used for the approximate numerical determination of the motion process. Determining the difference between Eqs. (12.26) and (12.60) also yields the approximate equations for the fast periodical components of the rotor motion

$$I_i \ddot{\xi}_i + \beta_i \dot{\xi}_i + B_i - V_i^0 = 0, \tag{12.61}$$

or

$$I_i \ddot{\xi}_i + \beta_i \dot{\xi}_i + B_i - B_i^0 =$$
$$-\frac{1}{2} \sum_{k=1}^{2m} \{(\hat{g}_{ir}\hat{g}_{kr} - \hat{g}_{is}\hat{g}_{ks})[(G_{ix}^{kx} - G_{iy}^{ky})sin(2\Omega t + \alpha_i + \alpha_k)$$
$$-(G_{ix}^{ky} + G_{iy}^{kx})cos(2\Omega t + \alpha_i - \alpha_k)] \tag{12.62}$$
$$-(\hat{g}_{ir}\hat{g}_{ks} + \hat{g}_{is}\hat{g}_{kr})[(G_{ix}^{kx} - G_{iy}^{ky})cos(2\Omega t + \alpha_i - \alpha_k)$$
$$+(G_{ix}^{ky} + G_{iy}^{kx})sin(2\Omega t + \alpha_i + \alpha_k)]\}.$$

In these equations it is not easily possible to express the terms $B_i - B_i^0$ by ξ_i and known quantities. In general, they also comprise parameter excitations. To ensure sufficient accuracy and reliability of the method of approximation employed for the direct separation of motion, it is essential that these motions be stable and, in a way, sufficiently small. But also in this incompletely evolved form, the equations (12.62) show the occurrence of excitations with the frequency 2Ω. In dependence of the given parameter values, the influence of the terms $B_i - B_i^0$ is often negligible. Computer simulations have confirmed that rotation at synchronous angular velocity is superimposed by periodical motions at a dominating frequency 2Ω. To derive the conditions of existence of self-synchronized motions we assume constant values α_i^* for the non-periodical components of the phases, by waiving solutions to the complete Eqs. (12.60) and, thus, we obtain the non-linear algebraic equations

$$V_i^{0*} = \beta_i(\Omega_i - \Omega^*), \quad i = 1, ..., m \qquad (12.63)$$

where

$$V_i^{0*} = V_i^0(\alpha_1 - \alpha_m = \alpha_1^* - \alpha_m^*, ..., \alpha_{m-1} - \alpha_m = \alpha_{m-1}^* - \alpha_m^*, \Omega = \Omega^*) \qquad (12.64)$$

for the synchronous angular velocity Ω^* and the related phase differences.

As it was assumed that the oscillatory system is a conservative system, the sum of all vibrational moments vanishes:

$$\sum_{i=1}^{m} V_i^0 = 0. \qquad (12.65)$$

This follows directly from the relations (12.53), (12.54) while the symmetry laws (12.45) are considered. Hence, the sum of the m equations (12.63) yields the equation

$$\sum_{i=1}^{m} \beta_i(\Omega_i - \Omega^*) = 0. \qquad (12.66)$$

The plain result obtained from this equation is the synchronous angular velocity as a weighted average of the partial angular velocities

$$\Omega^* = \frac{1}{\beta_{ges}} \sum_{i=1}^{m} \beta_i \Omega_i = \frac{1}{\beta_{ges}} \sum_{i=1}^{m} L_{0i}, \quad \beta_{ges} = \sum_{i=1}^{m} \beta_i. \qquad (12.67)$$

Subsequently, the phase differences of the synchronous motion must be determined from the remaining $m-1$ independent equations (12.63), typically producing ambiguous solutions.

Using the potential function (12.58) one finds the following alternative form of conditions of existence

$$\frac{\partial \Lambda^{0*}}{\partial \alpha_i^*} - \beta_i(\Omega_i - \Omega^*) = \frac{\partial D^*}{\partial \alpha_i^*} = 0 \qquad (12.68)$$

where

$$D = \Lambda^0 - \sum_{i=1}^{m-1} \beta_i(\Omega_i - \Omega)(\alpha_i - \alpha_m). \qquad (12.69)$$

In the special case of fully identical rotor damping parameters and identical partial angular velocities

$$\beta_1 = \beta_2 = ... = \beta_m =: \beta, \quad \Omega_1 = \Omega_2 = ... = \Omega_m =: \overline{\Omega}, \qquad (12.70)$$

the expressions (12.67) and (12.69) are reduced to

$$\Omega^* = \frac{1}{m}\sum_{i=1}^{m}\Omega_i = \overline{\Omega}, \qquad (12.71)$$

$$D = \Lambda^0. \qquad (12.72)$$

12.6 Conditions of Stability

On the basis of the prerequisites formulated earlier, we restrict our stability investigation to the constant solutions of Eqs. (12.60). Through superimposing small disturbances $\overline{\alpha}_i$,

$$\alpha_i = \alpha_i^* + \overline{\alpha}_i, \qquad (12.73)$$

$$sin\alpha_i \approx sin\alpha_i^* + \frac{dsin\alpha_i^*}{d\alpha_i^*}\overline{\alpha}_i, \quad cos\alpha_i \approx cos\alpha_i^* + \frac{dcos\alpha_i^*}{d\alpha_i^*}\overline{\alpha}_i, \qquad (12.74)$$

we obtain from Eqs. (12.60) while considering Eqs. (12.63), the following linear differential equations for small disturbances with constant coefficients:

$$I_i\ddot{\overline{\alpha}}_i + \beta_i\dot{\overline{\alpha}}_i + \sum_{k=1}^{m}\frac{\partial V_i^{0*}}{\partial \alpha_k^*}\overline{\alpha}_k = 0, \quad i = 1, ..., m. \qquad (12.75)$$

As the vibrational moments (12.64) only depend on the differences of the phases, we have

$$\sum_{k=1}^{m}\frac{\partial V_i}{\partial \alpha_k} = 0, \quad i = 1, ..., m \qquad (12.76)$$

or

$$\det \frac{\partial \mathbf{V}}{\partial \boldsymbol{\alpha}} = 0, \quad \mathbf{V} = [V_1...V_m], \quad \boldsymbol{\alpha} = [\alpha_1...\alpha_m]^T. \tag{12.77}$$

This is because the investigated system is an autonomous system and, hence, the motions are independent of a shift of the time origin. Therefore, the characteristic equation of the system (12.75) has a zero root which is not relevant to the stability of the phase differences. If all other roots exhibit negative real parts, the motion studied is characterized by an asymptotical orbital stability. Due to the zero root mentioned above, the "stiffness matrix" of the equation system (12.75) cannot be positively definite. For an asymptotical orbital stability it is necessary that this matrix is positively semi-definite. If the defect of this matrix equals 1, this condition is also sufficient.

12.7 Two Rotors

Let us now consider the special case of only two rotors which is less complex. With the phase difference

$$\alpha = \alpha_1 - \alpha_2 \tag{12.78}$$

the vibrational moments satisfy the simple relation

$$V_1^0 = -V_2^0 = \frac{d\Lambda^0}{d\alpha}. \tag{12.79}$$

Disregarding the constant terms, the potential function

$$\Lambda^0 = \frac{1}{2} \sum_{i,k=1,2;1,4;3,2;3,4} \{(\hat{g}_{ir}\hat{g}_{kr} + \hat{g}_{is}\hat{g}_{ks})[-(G_{ix}^{kx} + G_{iy}^{ky})\cos\alpha + (G_{ix}^{ky} - G_{iy}^{kx})\sin\alpha]$$

$$- (\hat{g}_{ir}\hat{g}_{ks} - \hat{g}_{is}\hat{g}_{kr})[(G_{ix}^{kx} + G_{iy}^{ky})\sin\alpha + (G_{ix}^{ky} - G_{iy}^{kx})\cos\alpha]\} + C$$

$$= \frac{1}{2}(-K_0\cos\alpha + K_1\sin\alpha) + C \tag{12.80}$$

only consists of a cosine and a sine terms. For the synchronous angular velocity it follows from Eq. (12.67)

$$\Omega^* = \frac{\beta_1 \Omega_1 + \beta_2 \Omega_2}{\beta_1 + \beta_2}. \tag{12.81}$$

From the conditions of existence of the phase difference

$$V_1^{0*} = \frac{1}{2}(K_0 sin\alpha^* + K_1 cos\alpha^*) = \frac{K}{2}cos(\alpha^* - \vartheta)$$
$$= \beta_1(\Omega_1 - \Omega^*) = \frac{\beta_1\beta_2}{\beta_1 + \beta_2}(\Omega_1 - \Omega_2) \quad (12.82)$$

with the abbreviation

$$K = \sqrt{K_0^2 + K_1^2} \quad (12.83)$$

and the phase shift ϑ corresponding to

$$tan\vartheta = \frac{K_0}{K_1} \quad (12.84)$$

we obtain the necessary condition for the existence of synchronous motions

$$\frac{K}{2} \geq \frac{\beta_1\beta_2}{\beta_1 + \beta_2}|\Omega_1 - \Omega_2|. \quad (12.85)$$

The equations for small disturbances

$$I_1\ddot{\overline{\alpha}}_1 + \beta_1\dot{\overline{\alpha}}_1 + \frac{\partial V_1^{0*}}{\partial \alpha_1^*}(\overline{\alpha}_1 - \overline{\alpha}_2) = 0,$$
$$I_2\ddot{\overline{\alpha}}_2 + \beta_2\dot{\overline{\alpha}}_2 + \frac{\partial V_1^{0*}}{\partial \alpha_1^*}(\overline{\alpha}_2 - \overline{\alpha}_1) = 0,$$
$$\left(\frac{\partial V_1^{0*}}{\partial \alpha_1^*} = -\frac{\partial V_2^{0*}}{\partial \alpha_1^*} = \frac{\partial V_2^{0*}}{\partial \alpha_2^*}\right) \quad (12.86)$$

yield the simple stability condition

$$\frac{\partial V_1^{0*}}{\partial \alpha_1^*} = \frac{\partial V_1^{0*}}{\partial \alpha^*} > 0. \quad (12.87)$$

Using the potential function Λ^0 we can combine the conditions of existence and stability into one equation

$$D^* = \Lambda^{0*} - \beta_1(\Omega_1 - \Omega^*)\alpha^* = minimum. \quad (12.88)$$

In the special case of equal rotors ($\Omega_1 = \Omega_2, \beta_1 = \beta_2$) this condition reduces to

$$\Lambda^{0*} = \frac{1}{2}\hat{f}_1\hat{f}_2[-K_0 cos\alpha^* + K_1 sin\alpha^*] + C = minimum. \quad (12.89)$$

Often the factor K_1 vanishes. We then obtain the stable phase difference 0 for positive and the stable phase difference π for negative values of the factor K_0.

12.8 Example

Finally, a simple example is considered to demonstrate the application of the approach developed. The spring-supported disk depicted in Fig. 2 represents an oscillatory system with two degrees of freedom corresponding to the coordinates x and ϑ. It is excited by a statically and a dynamically unbalanced rotors (index 1 and 2, respectively). There are only the following two non-vanishing centrifugal forces or moments of the unbalance

$$\hat{g}_{1r} = \hat{f}_{1r}, \quad \hat{g}_{4r} = \hat{l}_{2r}. \tag{12.90}$$

Hence, the potential function (12.80) reduces to

$$\Lambda^0 = \frac{1}{2}\{\hat{g}_{1r}\hat{g}_{4r}[-(G_{1x}^{4x} + G_{1y}^{4y})\cos\alpha + (G_{1x}^{4y} - G_{1y}^{4x})\sin\alpha]\} + C. \tag{12.91}$$

With regard to the column vector of the generalized coordinates

$$\mathbf{q} = [x \quad \vartheta]^T \tag{12.92}$$

the mass matrix and the stiffness matrix of the oscillatory system result as

$$\mathbf{M} = \begin{bmatrix} m & 0 \\ 0 & \Theta \end{bmatrix}, \quad \mathbf{K} = \begin{bmatrix} k_s & 0 \\ 0 & k_\vartheta \end{bmatrix}. \tag{12.93}$$

From the pivot point motions

$$\mathbf{r}_1 = [\, x - a\vartheta \quad 0 \quad 0 \,]^T, \quad \boldsymbol{\psi}_1 = [\, 0 \quad 0 \quad -\vartheta \,]^T,$$
$$\mathbf{r}_2 = [\, 0 \quad b\vartheta \quad x - a\vartheta \,]^T, \quad \boldsymbol{\psi}_2 = [\, \vartheta \quad 0 \quad 0 \,]^T \tag{12.94}$$

one obtains the global Jacobian matrix

$$\mathbf{J} = \begin{bmatrix} 1 & 0 & 0 & 0 & 0 & 1 & 0 & 0 & 0 & 0 & 0 \\ -a & 0 & 0 & 0 & b & -a & 0 & 0 & -1 & 1 & 0 \end{bmatrix}^T. \tag{12.95}$$

The rows of this matrix with the frequency response matrix

$$\mathbf{H} = (-\mathbf{M}\Omega^2 + \mathbf{K})^{-1} = \begin{bmatrix} \frac{1}{-m\Omega^2 + k_x} & 0 \\ 0 & \frac{1}{-\Theta\Omega^2 + k_\vartheta} \end{bmatrix} \tag{12.96}$$

12.8. Example

yield the required harmonic influence coefficients

$$G^{4x}_{1x} = \mathbf{J}^1\mathbf{H}(\mathbf{J}^{10})^T = \frac{-a}{-\Theta\Omega^2 + k_\vartheta}, \quad G^{4y}_{1y} = \mathbf{J}^2\mathbf{H}(\mathbf{J}^{11})^T = 0,$$
$$G^{4y}_{1x} = \mathbf{J}^1\mathbf{H}(\mathbf{J}^{11})^T = 0, \quad G^{4x}_{1y} = \mathbf{J}^2\mathbf{H}(\mathbf{J}^{10})^T = 0. \qquad (12.97)$$

It is noteworthy that the only non-vanishing influence coefficient G^{4x}_{1x} is independent of the mass m and the spring stiffness k_x. The very simple condition of existence and stability

$$\Lambda^{0*} = -\frac{1}{2}\hat{g}_{1r}\hat{g}_{4r}G^{4x}_{1x}\cos\alpha^* + C = minimum \qquad (12.98)$$

yields the condition of existence $sin\alpha^* = 0$ with the solutions 0 and π.

From the stability condition

$$G^{4x}_{1x}\cos\alpha^* > 0 \qquad (12.99)$$

we obtain the results in Table 1. This means that, with regard to the rotary motion, the value 0 is stable in the postcritical frequency range, whereas the value π is stable in the precritical frequency range.

Table 12.1 Stable and unstable synchronous motions

\multicolumn{2}{c}{$\frac{k_\vartheta}{\Theta} < \Omega^2$}		\multicolumn{2}{c}{$\frac{k_\vartheta}{\Theta} > \Omega^2$}	
$\alpha^*_1 - \alpha^*_2 = 0$	$\alpha^*_1 - \alpha^*_2 = \pi$	$\alpha^*_1 - \alpha^*_2 = 0$	$\alpha^*_1 - \alpha^*_2 = \pi$
stable	unstable	unstable	stable

From the stationary motion of the oscillatory system for any phases

$$\mathbf{q}^0 = \hat{f}_{1r}\begin{bmatrix} \frac{1}{-m\Omega^2 + k_x} \\ \frac{-a}{-\Theta\Omega^2 + k_\vartheta} \end{bmatrix}\cos(\Omega t + \alpha^*_1) + \hat{l}_{2r}\begin{bmatrix} 0 \\ \frac{1}{-\Theta\Omega^2 + k_\vartheta} \end{bmatrix}\cos(\Omega t + \alpha^*_2) \qquad (12.100)$$

we obtain for the special parameters

$$\Theta\Omega^2 > k_\vartheta, \quad \hat{l}_{2r} = a\hat{f}_{1r}, \qquad (12.101)$$

since $\alpha^* = 0$, the interesting result

$$\mathbf{q}^0 = \hat{f}_{1r} \begin{bmatrix} \dfrac{1}{-m\Omega^2 + k_x} \\ 0 \end{bmatrix} \cos(\Omega t + \alpha_1^*). \tag{12.102}$$

Therefore, in the postcritical frequency range with regard to the rotary motion, e.g. for the vanishing stiffness k_ϑ and for specific rotor parameters, the rotors only generate translatory vibrations, irrespective of the natural frequency of this translatory motion. This could be of practical use, e.g. when the right-hand side of the oscillating body is not accessible for installing a statically unbalanced rotor.

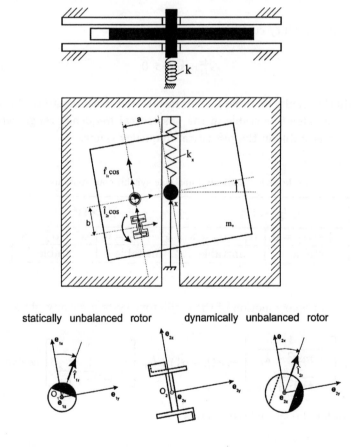

Fig. 12.2 Plane oscillatory system with a statically and a dynamically unbalanced rotors

12.9 Summary

The present chapter provides engineers with a formal algorithmic approach for practical applications in that it can be used for deriving conditions of the existence and stability of self-synchronized motions of statically and dynamically unbalanced rotors as key components of unbalanced vibrators. The mechanical supporting system excited by rotors to vibrate is assumed to be a system comprising rigid bodies and linear-elastic springs while being undamped; the rotor masses are assumed to be comparatively low with regard to the typical mass parameters of the oscillatory system.

The approach is based on the method of direct separation of motion. In addition, the concept of the vibrational moments and the harmonic influence coefficients is employed. The vibrational moments are also expressed by the averaged Lagrangian. An approximately valid equation system is also derived for the small vibrations superimposing the self-synchronized motion, the frequency of which corresponds to double the synchronous angular velocity. The simpler special case of the oscillatory system with only two rotors is considered in a separate section. Finally, the algorithm is applied to an example comprising a statically and a dynamically unbalanced rotors.

Without facing major problems, the approach may be generalized for viscously damped oscillatory systems. However, the evaluation will become much more complex. A number of results obtained with this approach, were confirmed by means of computer simulations. During these simulations also additional influences were considered, such as the terms of the forces and moments of inertia of the rotors which are a linear function of the coordinates of the oscillatory system and remained unconsidered here, a small damping of the oscillatory system, and a non-linear motor characteristic of the rotor drives.

References

[1] Blekhman I. I., *Synchronization of Dynamic Systems*, (Nauka, Moscow, 1971), p. 894 (in Russian).
[2] Blekhman I. I., *Synchronization in Science and Technology*, (ASME Press, New York, 1988) (English translation), p. 255.
[3] Blekhman I. I, Method of direct separation of motions in problems on the action of vibration on nonlinear dynamic systems, *Izv. AN SSSR, MTT*, 6 (1976) 13–27 (in Russian).
[4] Blekhman I. I., *Vibrational Mechanics* (Singapore, New Jersey, London, Hong Kong, World Scientific, 2000) 509.
[5] A. Fidlin. On the sparation of motions in systems with a large fast excitation of general form. *European Journal of Mechanics, A/Solids* 18 (1999), 527-538.
[6] Kapitsa P. L, Dynamic stability of the pendulum when the point of suspension is oscillating, *Jour. of Exper. and Theor. Physics*, 21, 5 (1951) 588–597 (in Russian).
[7] L. Sperling. Beitrag zur allgemeinen Theorie der Selbstsynchronisation umlaufender Unwuchtmassen im Nichtresonanzfall. Wissenschaftliche Zeitschrift der TH O.v.G., Magdeburg 11, 1 (1967), 63-87.
[8] L. Sperling. Selbstsynchronisation statisch und dynamisch unwuchtiger Vibratoren. Teil 1: Grundlagen,*Technische Mechanik* 14, 1 (1994), 9-24.
[9] L. Sperling. Selbstsynchronisation statisch und dynamisch unwuchtiger Vibratoren, Teil II: Ausführung und Beispiele,*Technische Mechanik* 14, 2 (1994), 85-96.
[10] L. Sperling, F. Merten, H. Duckstein. Analytical and Numerical Investigations of Rotation-Vibration- Phenomena, *Proceedings of XXV Summer School "Nonlinear Oscillations in Mechanical Systems", St.Petersburg*, (1998) 145-159.

Chapter 13

The Setting up of the Self-Synchronization Problem of the Dynamic Objects with Internal Degrees of Freedom and Methods of Its Solution

I.I.Blekhman[1], L. Sperling[2]

[1]*Institute for Problems of Mechanical Engineering,*
Russian Academy of Sciences
and
Mekhanobr-Tekhnika Corp.
22 Liniya 3, V.O., 199106, St. Petersburg, Russia

[2]*Institut für Mechanik, Otto-von-Guericke-Universität Magdeburg,*
Universitätsplatz 2, 39106 Magdeburg, Germany

13.1 Preliminary Remarks

By now only the theory of synchronization of rotating objects without any internal degrees of freedom (unbalanced rotors, oscillators, celestial bodies, considered either as mass points or as rigid bodies of simplest form) has been adequately developed (see, say, [1,2,3]). The investigations, carried out by R.F.Nagayev and P.S.Goldman [3,4] the self-synchronization of rotors with linear oscillators inside, are the only exceptions.

The investigation of this class of problems, however, seems to be very important from both – conceptual and applied standpoints. Concerning the technical systems it is possible to hope that the existence of the internal degrees of freedom in vibro-exciters will provide new means for using the effect of self-synchronization in different machines. As to the systems of celestial mechanics, it may be expected that the rotations of the system of the bodies that is being synchronized will be stabilized near the eigenfrequencies of the bodies (specifically near the central one). The appropriate model and its governing laws may be interesting for the theoretical physics, particularly as a special case of an orbit quantization.

In this chapter (see also [10]) some methods of solving problems of this class are considered, specifically the different forms of an integral criterion (extreme property) of the stability of synchronous motions, methods of small parameter and also the method of direct separation of motions. One of the forms of the integral criterion is to a certain extent generalized here, which is important for the class of problems under consideration.

In the conclusion we give an exact solution of the problem on self-synchronization of two vibro-exciters with the inner degrees of freedom, installed co-axially on a free solid body. This solution is to be compared with the results, obtained by the methods, listed above, which are valid only under certain restrictions. This solution can also be used when elaborating methods of providing stability of the synchronous motions, desirable from the point of view of applications, but unstable in the case of ordinary vibro-exciters.

13.2 Statement of the Problem

Figure 13.1,*a* shows a *"technical" version of the system* under consideration, typical of problems on the self-synchronization of vibro-exciters. For the sake of brevity we will call this version the system A. The system A comprises a certain number n of the bodies $B_{1,1}, ..., B_{1,n}$ connected with each other and with the fixed base B by elastic and damping elements – 1 and 2 respectively. The bodies $B_{I,1}, ..., B_{I,n}$ are called the *carrying bodies* and the corresponding system is called the *carrying system*. On those bodies a certain number k of unbalanced rotors (vibro-exciters) $B_{1,1}, ..., B_{k,1}$ are placed, which unlike the ordinary ones, can carry additional bodies $B_{1,2,...,k,l}$ connected with the rotors in one way or another, e.g. by elastic and damping elements (the body $B_{1,2}$ in Fig. 13.1, *a*), by hinges (the body $B_{s,2}$ in which case the main rotor seems to be provided with an additional rotor or pendulum), these bodies (or some of their points) may be able to move inside the rotors along the preset trajectories (the body $B_{2,2}$ in Fig. 13.1, *a*) etc. Finally, rotors may be connected with each other by the so-called *carried constraints*, containing certain bodies (the bodies $B_{II,1}$ and $B_{II,2}$ in Fig. 13.1, *a*). Vibro-exciters can present bodies, moving with respect to the carrying bodies along the preset closed trajectories (the body $B_{3,1}$ in Fig. 13.1, *a*). Some of the carrying bodies $B_{I,s}$ may carry several exciters, others may not have them at all. The rotors of the exciters

13.2. Statement of the Problem

are believed to be actuated by electric engines of asynchronous type, unless stated otherwise. Fig. 13.1, *a* for the sake of simplicity shows a plane scheme, while it can be spatial.

a) Technical system (System A)

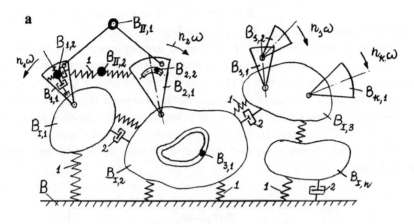

b) Planetary system (System B)

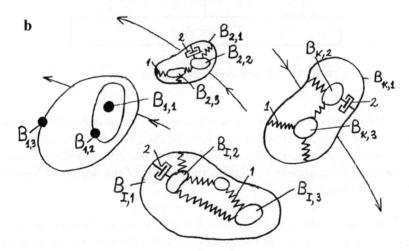

Fig. 13.1 Systems with dynamic objects with internal degrees of freedom: *a* – technical system (System A), *b* – planetary satellite system (System B)

13. Objects with Internal Degrees of Freedom

The *planetary-satellite version of the system* under consideration (called by us the system B) is shown in Fig. 13.1,b. In this case the "carrying system" is presented by the central body $B_{I,1}$ and the bodies $B_{I,2}, B_{I,3}, ..., B_{I,m}$, connected with it by elastic and damping elements. In this case objects that are being synchronized are systems of the bodies $B_{1,1}, ..., B_{1,2}; ...; B_{k,1}, ..., B_{k,r}$ which are either connected (in every system) by the elastic or damping elements (the bodies $B_{2,1}, ..., B_{2,3}$ Fig. 13.1, b) or represent a "planet with satellites" (the bodies $B_{1,1}, B_{1,,2}, B_{1,3}$ in Fig. 13.1, b).

All the material bodies in the system B are supposed to be connected by forces of gravitational interaction. Forces of the interaction between the objects – non-central bodies of the carrying system belong to the carrying constraints, while the forces of the interaction between the objects belong to the carried constraints. Every body is supposed to be moving along the closed orbit. The system B, like the system A, can also be spatial.

Figure 13.2 shows a general structural scheme of the systems under consideration

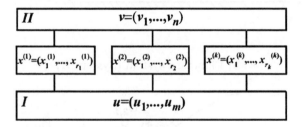

Fig. 13.2 General structural scheme of the system.

In this scheme $x^{(1)}, ..., x^{(k)}$ denote the vectors of the generalized coordinates of the objects, u denotes the vector of the generalized coordinates of the carrying system, and v denotes the vector of the generalized coordinates of the carried constraints; r_s denotes the number of the degrees of freedom of the s-th object.

The problem of the investigation of the both systems A and B consists in finding the conditions of existence and stability of motions under which the objects rotate (or revolve in their orbits) with the same or multiple mean angular velocities $n_s\omega$, while other bodies (either all of them or some of them) perform oscillations with the frequencies $n_s\omega$ where n_s are whole numbers.

13.3 Structure of the Kinetic and Potential Energy of the System. The Generalized Forces

Keeping in mind that henceforward we are going to use the equations of motion of the system in the form of the Lagrange equations of the 2nd type and the integral criterion of the stability of synchronous motions, we will examine the structure of the kinetic and potential energy of the system under consideration [3,4,5].

The "intrinsic" kinetic and potential energy of the objects have the form

$$T_s = \frac{1}{2}\dot{x}'^{(s)}A_s^{(s)}(x^{(s)})\dot{x}^{(s)}, \quad \Pi_s = \Pi_s(x^{(s)}) \qquad (13.1)$$

where $A^{(s)} = A'^{(s)}$ is a symmetric $r_s \times r_s$- matrix of the inertia, $'$ denoting the operation of transposition.

Now we will formulate supposition about the type of system of the constraint between the objects.

Due to the presence of the carrying constraints, the objects, connected with them, acquire additional mobility, so their common kinetic and potential energy are presented as

$$T^* = \sum_{s=1}^{k} T_s + \Delta T^*, \quad \Pi = \sum_{s=1}^{k} \Pi_s + \Delta \Pi^* \qquad (13.2)$$

where

$$\Delta T^* = \sum_{s=1}^{k} \dot{x}'^{(s)} A_{sm}(x,u)\dot{u} + \frac{1}{2}\dot{u}' A_m(x,u)\dot{u}, \quad \Delta \Pi^* = u'C(x) \qquad (13.3)$$

are the additional kinetic and potential energies of the objects.

By x we imply a set of all the vectors $x^{(s)}$; A_{sm} and A_m denote certain $r_s \times m$ and $m \times m$- matrixes respectively.

The intrinsic kinetic and potential energies of the carrying constraints are considered to be presentable in the form

$$T^{(I)} = \frac{1}{2}\dot{u}'M_m(u)\dot{u}, \quad \Pi = \frac{1}{2}u'C_m u \qquad (13.4)$$

The presence of the carried constraints does not lead to the increase in mobility of the objects, but their presence may be connected (though not necessarily) with the increase of the number of degrees of freedom of the system as a whole. The kinetic and potential energies of the carried constraint are considered presentable in the form

$$T^{(II)} = \frac{1}{2}\dot{v}'N_n^{(v)}(x,u,v)\dot{v} + \sum_{s=1}^{n}\dot{x}'^{(s)}N_{sn}^{(x,v)}(x,u,v)\dot{v}$$

$$+\frac{1}{2}\sum_{s=1}^{k}\sum_{j=1}^{k}\dot{x}'^{(s)}N_{sj}^{(x)}(x,u,v)\dot{x}^{(j)}$$

$$+\dot{u}'N_{mn}^{(u,v)}(x,u,v)\dot{v} + \sum_{s=1}^{k}\dot{x}'^{(s)}N_{sm}^{(x,u)}(x,u,v)\dot{u} + \frac{1}{2}\dot{u}'N_m^{(u)}(x,u,v)\dot{u}$$

$$\Pi^{(II)} = \Pi^{(II)}(x,u,v) \tag{13.5}$$

where $N_n^{(v)}$, $N_{sn}^{(xv)}$, $N_{sj}^{(x)}$, $N_{mn}^{(u,v)}$, $N_{sm}^{(x,u)}$ and $N_m^{(n)}$ are respectively $n \times n$, $r_s \times n$, $r_s \times r_j$, $r_m \times r_n$, $r_s \times m$ and $m \times m$- matrixes.

This relatively complicated form of the latter expressions is caused by the fact that the elements of the carried system acquire an additional mobility due to the motion of the objects and of the carrying constraints.

The constraints between the objects are supposed to be weak, so the total kinetic and potential energy of the system when solving the problem by method of small parameter may be written in the form

$$T = T^* + T^{(I)} + T^{(II)} = \sum_{s=1}^{k}T_s + O(\mu)$$

$$\Pi = \Pi^* + \Pi^{(I)} + \Pi^{(II)} = \sum_{s=1}^{k}\Pi_s + O(\mu) \tag{13.6}$$

where $O(\mu)$ denotes the members of the order of small parameter μ.

Conditions imposed by such supposition on the parameters of the system are mentioned in [3,5,1]. As a rule these conditions are such that the equations of the motion of the system along the coordinates of the carrying system u are linear with an accuracy of the order of μ^2.

The generalized forces in problems on the synchronization of vibro-exciters are the differences $L_s(\dot\varphi_s) - R_s(\dot\varphi_s)$ between the moments transferred to the rotors from the electric engines L_s and the moments of the resistance forces $R_s(\dot\varphi_s)$ being the turning angles of the rotors. As for the other coordinates of the technical and planetary-satellite system, the generalized forces are, as a rule, the forces of resistance.

13.4 Integral Criterion (Extreme Property) of the Stability of Synchronous Motions

13.4.1 Introduction

The so-called *integral criterion (extreme property) of the stability of synchronous motions* in many cases considerably facilitates solving problems on synchronization and makes it possible to establish in the general form the tendency to synchronization [1,2,3,5].

In accordance with the character of idealization of the objects (*quasi- and "non-quasiconservative" idealization*) and the one accepted supposedly, this criterion can be formulated in different ways. In many cases, however, especially for the technical systems, the results are mainly the same.

We will state here the adequate formulations for both types of idealization.

13.4.2 Systems with almost uniform rotations, non-quasiconservative idealization of rotation

In this case it is supposed [1] that the equations of motion of the system can be presented in the form *

$$I_s\ddot\varphi_s + k_s(\dot\varphi_s - \sigma_s n_s \omega_s) = \mu \Phi_s \qquad (s = 1,....,k) \qquad (13.7)$$

$$E_{u_r}(L) = Q_{u_r} \qquad (r = 1,...,v) \qquad (13.8)$$

where

$$E_q = \frac{d}{dt}\frac{\partial}{\partial \dot q} - \frac{\partial}{\partial q} \qquad (13.9)$$

---

*Examination of the conclusion of the relations given below [1, 6] shows that the suppositions about the type of equations (13.7) is not obligatory, though it is important so that they should allow the solution of the type of (13.11) and (13.12) (see below).

The Euler operator, corresponding to the generalized coordinate q, I_s, k_s, ω are the positive constants, n_s are the whole positive numbers, $\sigma_s = \pm 1, \mu > 0$ is the small parameter, and the function Φ_s is determined from the condition of identity of the equations (13.7) and the corresponding group of the Lagrange equations, i.e. from the equality

$$\mu \Phi_s = I_s \ddot{\varphi}_s + k_s(\dot{\varphi}_s - \sigma_s n_s) - E_{\varphi_s}(L) + Q_{\varphi_s} \qquad (13.10)$$

The functions L, Q_{φ_s}, Q_{u_r}, Φ_s may depend on the generalized coordinates and on the velocities of the system, and also on the time t. It is also supposed that they are 2π- periodic according to φ_s and $2\pi/\omega$-periodic according to t. The functions Q_{φ_s} and Q_{u_r} (the generalized forces), and also Φ_s may also depend on small parameter μ.

The coordinates φ_s are called *rotational*, and the coordinates u_r are called *oscillatory*. It should be noted that here, unlike [1,6], φ_s are not supposed to be the only one coordinate of the object. The system is believed to have a certain number k of the generalized coordinates to which at $\mu = 0$, according to (13.7), corresponds uniform rotations:

$$\varphi_s = \varphi_s^0 = \sigma_s(n_s \omega t + \alpha_s) \qquad (13.11)$$

where α_s are the arbitrary constants. All the other coordinates of the system, including perhaps some coordinates of the objects, are believed to be "oscillatory", i.e. they correspond to the oscillations in the stationary synchronous motions under consideration. This generalization follows from the consideration of the publications, cited above. It is evident that it is essential when investigating the synchronization of objects with the inner degrees of freedom.

The generating system, corresponding to equations (13.8), i.e. the system, obtained at $\mu = 0$, and when expressions (13.11) are substituted for φ_s, is believed to allow a $2\pi/\omega$- periodic solution

$$u_r = u_r^0(t, \alpha_1, ..., \alpha_k) \quad (r = 1, ..., v) \qquad (13.12)$$

which is asymptotically stable.

13.4. Integral Criterion (Extreme Property) of the Stability

If we also suppose (though it is not obligatory in the general case) that the generalized forces Q_{u_s} and Q_{φ_s} are small enough [†], then the following is established:

The stable synchronous motions of the system correspond to the points of a rough minimum of the function (potential function)

$$D = -\Lambda = -\langle [L] \rangle \qquad (13.13)$$

where $\langle \rangle$ is the sign of averaging for the period and the square brackets imply that the function L is calculated in the generating approximation. Here we mean the minimum in magnitudes α_s in the case of the problem on the outer synchronization and in the differences $\alpha_s - \alpha_k$ in the case of the problem on the inner synchronization (i.e. on self-synchronization). The absence of the minimum, discovered by analyzing the members of the second order in the decomposition of the function D near the stationary point tells about the instability of the corresponding motion, while other cases require an additional investigation. When the values of the generalized forces Q_{φ_s} and Q_{u_r} are not small, the formulated thesis may also be valid, but only for the function D, containing a certain additional term.

Expression (13.13) for the potential function D is essentially simplified in particular if the Lagrange function of the system L can be presented in the form

$$L = L^* + L^{(I)} + L^{(II)}$$

$$L^* = \frac{1}{2} \sum_{s=1}^{k} \sum_{j=1}^{k} d_{sj} \dot{\varphi}_s \dot{\varphi}_j + \sum_{r=1}^{\nu} f_r(\dot{\varphi}_1, ..., \dot{\varphi}_k; \varphi, ..., \varphi_k) \dot{u}_r + \sum_{s=1}^{k} F_s(\varphi_s),$$

$$L^{(I)} = \frac{1}{2} \sum_{r=1}^{v} \sum_{j=1}^{v} a_{rj} \dot{u}_r \dot{u}_j - \frac{1}{2} \sum_{r=1}^{v} \sum_{j=1}^{v} b_{rj} u_r u_j$$

$$L^{(II)} = \Psi(\dot{\varphi}_1, ..., \dot{\varphi}_k; \varphi_1, ..., \varphi_k) \qquad (13.14)$$

where a_{rj}, b_{rj} and d_{rj} are constants, while f_r, F_r and Ψ are periodical over φ_1 with a period 2π.

[†] In the problems under consideration this supposition is reduced to the demand of the smallness of the resistance forces according to the oscillatory coordinates and to the proximity of the so-called partial velocities of rotation of the objects ω_s or the numbers ω_s/n_s [6].

In this case the following relations are valid

$$\frac{\partial \Lambda}{\partial \alpha_s} = \frac{\partial(\Lambda^{(II)} - \Lambda^{(I)})}{\partial \alpha_s}, \quad \langle \Lambda^{(I)} = \langle [L^{(I)}] \rangle, \quad \Lambda^{(II)} = \langle [L^{(II)}] \rangle \quad (13.15)$$

and the potential function D can be presented as

$$D = \Lambda^{(I)} - \Lambda^{(II)} \quad (13.16)$$

The terms $L^{(I)}$ and $L^{(II)}$ in the expressions (13.13) can be interpreted as the Lagrange function of the carrying and carried constraints respectively.

It should be also noted that the expression for the function $L^{(I)}$ in (13.14) corresponds to the case when the carrying system in the first approximation is linear with constant coefficients.

Finally, we will note that if the expression for the function L is presented as

$$L = \sum_{s=1}^{k} L_s + L_0, \quad (13.17)$$

i.e. to specify the Lagrange function, corresponding to the rotatory coordinates, then we will have the relations

$$\langle [L_s] \rangle = C_s, \quad \langle [L] \rangle = \langle [L_0^*] \rangle = \Lambda_0^* \quad (13.18)$$

where $C_s = \text{const}$.

Then the expression (13.13) for the function D will be presented as

$$D = -\Lambda_0 \quad (13.19)$$

(since the function D was determined with an accuracy to the constants, not depending on α_s).

Note :

The values L_s can be regarded as the Lagrangians of the system with all the motions stopped except the rotations φ_s. But φ_s do not embrace all the coordinates of the synchronized objects since they may also comprise both rotational and oscillatory coordinates.

13.4.3 Systems with the objects, idealized as quasiconservative

In the theory of synchronization of *quasiconservative objects* it is supposed [3, 5] that the intrinsic and potential energies of the objects, i.e. objects in the absence of constraints, are defined by the expressions (13.1), while

13.4. Integral Criterion (Extreme Property) of the Stability

the kinetic and potential energies of the system on the whole – by the expressions (13.2) – (13.6).

The motion of the objects in the generating approximation is defined by the equations

$$E_{x^{(s)}}(L_s) = 0 \quad L_s = T_s - \Pi_s \quad (s = 1, ..., k) \tag{13.20}$$

Each of these equations is believed to allow solution of the kind

$$x_{jo}^{(s)} = x_{jo}^{(s)}(\psi_s, c_s) = \sigma^{(s)}\left[q_j^{(s)}\psi_s + y_{jo}^{(s)}(\psi_s, c_1)\right] \quad \psi_s = \tilde{\omega}_s(c_s)t + a_s,$$

$$(s = 1, ..., k; \quad j = 1, ..., r_s) \tag{13.21}$$

where α_s and c_s are the arbitrary constants, while $y_{jo}^{(s)}$ are the periodic functions of ψ_s with a period 2π, $\sigma_j^{(s)} = \pm 1$, and $q_j^{(s)}$ is equal to zero for the oscillatory coordinates, and to unit for the rotary coordinates. The values $\tilde{\omega}_s$ should not be confused with the partial frequencies of the objects ω_s. The constants c_s are mutually unambiguously connected with the constants of the energy

$$h_s(c_s) = T_s(\dot{x}_0^{(s)}, x_0^{(s)}) + \Pi_s(x_0^{(s)}) \tag{13.22}$$

and determine the values of the frequencies $\tilde{\omega}_s$. These dependencies are considered to be essential, i.e. the derivatives $d\tilde{\omega}_s/dc_s$ are not small. Such objects are called *essentially non-isochronous*. For synchronization to be possible it is necessary that the following equations should have solutions

$$\tilde{\omega}_s(c_s) = n_s\omega \quad (s = 1, ..., k) \tag{13.23}$$

where n_s are whole positive numbers; and from these equations the constants c_s of the generating solution (13.21) are determined. As for the constants α_s (or $\alpha_s - \alpha_k$) in synchronous motions which may be stable, then if the generalized forces in the generating approximation are small enough, they are determined from the conditions of the minimum of the function

$$D = -\Lambda_0 \sigma \tag{13.24}$$

where $\Lambda_0 = \langle[L_0]\rangle$ and L_0 is the Lagrange function of the system, not including the intrinsic Lagrange functions of the objects (13.1). Via

$$\sigma = \text{sign}\, e_s(\tilde{\omega}_s) \tag{13.25}$$

where

$$e_s(\tilde{\omega}_s) = \frac{1}{\tilde{\omega}_s}\frac{dh_s[c_s(\tilde{\omega}_s)]}{d\tilde{\omega}_s} \qquad (13.26)$$

is the so-called *slope of the frequency characteristic of the object*. At $e_s > 0$ the objects are called *hardening-anisochronous*, and at $e_s < 0$ they are called *softly-anisochronous*.

So equality (13.25) assumes that the character of anisochronism of the objects is the same, i.e.

$$\text{sign}\, e_1 = ... = \text{sign}\, e_k = \sigma \qquad (13.27)$$

Just like in the case of non-conservative objects, it is believed that the solution (13.21) corresponds to the asymptotically stable periodic solutions of the system of equations of the constraint

$$\begin{aligned} u_0 &= u_0(\omega t; c_1, ..., c_k; \alpha_1, ..., \alpha_k) \\ v_0 &= v_0(\omega t; c_1, ..., c_k; \alpha_1, ..., \alpha_k) \end{aligned} \qquad (13.28)$$

However, unlike this case, the conditions of the minimum of the function D give only the necessary but not the necessary and sufficient conditions of stability of the corresponding synchronous motion. The additional conditions, providing stability for certain classes of objects, can be found, say, in [7].

Along with that, the conditions, provided by the demand of the minimum of the function D are the main ones in the sense that they determine the choice of the constants α_s (or of their differences $\alpha_s - \alpha_k$) in possible stable synchronous motions.

Like before, at the additional condition that the equations of the carrying constraints in the generating approximation present a system of linear differential equations with constant coefficients, valid (with an accuracy to an insignificant constant) is the following relation:

$$\Lambda_0 = \Lambda^{(II)} - \Lambda^{(I)} \qquad (13.29)$$

and the expression (13.24) for the potential function may be presented as

$$D = (\Lambda^{(I)} - \Lambda^{(II)})\sigma \qquad (13.30)$$

13.4.4 Comparison of the results at different types of idealization. Some conclusions

Comparing the relations, given in 13.4.2 and 13.4.3, we can conclude:

In the case of hardening anisochronous objects, such as, say, unbalanced vibro-exciters, $\sigma = 1$ and the results of both types of idealization coincide (see formulas (13.16), (13.19), (13.24) and (13.30)).

In the case of softly anisochronous objects, such as celestial bodies in orbital motions, $\sigma = -1$ and the results prove to be formally different. In the case of idealizing objects as non-quasiconservative, or, more precisely, in the case of objects with almost uniform rotations, defined by equations of type (13.7), the result in the form of integral criterion appears to be independent of the type of anisochronism of the objects.

Along with that it would not be right to speak of the inconsistency of the results of the investigation. Firstly, in each case the suppositions assumed were somewhat different. ‡ Secondly, with regard to the additional conditions of stability, the final conditions of stability may prove to be similar or even identical.

In connection with the conclusions made, it seems to be of interest to consider the self-synchronization of objects in certain simplest systems for which it is easy to find exact solutions, describing synchronous motions and for which it is not difficult to investigate their stability. Such systems are considered below.

13.5 Method of Direct Separation of Motions. Methods of Small Parameter

The Poincare and Lyapunov small parameter methods as well as asymptotic methods and the method of direct separation of motions were successfully used when solving a number of problems on the synchronization of objects without internal degrees of freedom (see, say, [1 – 8]).

These methods can also be used when solving more complicated problems, considered here. In problems on synchronization the method of direct separation of motions seems to have essential advantages both in the simplicity of solutions and in the interpretation of the results. The use of this method will also be illustrated by a concrete example (see section 13.6).

The feasibility of this method when solving problems on the synchronization of objects with the internal degrees of freedom is determined by

---

‡For the problem on self-synchronization of the ordinary unbalanced vibro-exciters the comparison of the results for the concrete system was made by R.F.Nagaev [3]. It was found that at the quasi-conservative idealization the regions of stability of synchronous motions proved to be, as should be expected, more narrow. Along with that, in a number of important technical cases these regions coincide [8].

the fact that in all such problems the generalized coordinates of the system of the motions we are interested in can be presented in the form

$$q_s = X_s(t) + \psi_s(t, \omega t) \tag{13.31}$$

where X_s is the slow component, and ψ_s is the fast one. A considerable part of the coordinates (in particular those of the carrying bodies) contain only the fast component. As a result, the dimensions of the system of slow motions prove to be much lower that those of the initial system.

13.6 Self-Synchronization of Two Identical Vibro-Exciters with the Internal Degrees of Freedom, Whose Axes Pass through the Center of Gravity of a Solid Body (Plane Motion)

13.6.1 *Description of the system*

The scheme of the system under consideration is shown in Fig. 13.3, *a*. Two identical vibro-exciters, whose axes of rotation coincide, are installed on a solid body whose mass is M^o. These axes pass through the center of gravity of the body. Inside the rotor of each exciter there is an additional mass m, connected with the rotor by springs with the stiffness c_ρ and c_ψ and the damping elements h_ρ and h_ψ in radial and circumference directions. A version is possible when the additional masses have not two, but only one degree of freedom, i.e. they can be displaced inside the rotor along a certain symmetrical curve $\rho_s = \rho_s(\psi_s)$ where ρ_s and ψ_s ($s = 1, 2$) are the polar coordinates of the mass m, (Fig. 13.3, *b* and *c*). The mass M^o may be connected with the other mass M_1^o by elastic elements with a total stiffness c_1 and with the damping elements with the combined coefficient of viscous resistance h_1.

The M_1^o is assumed to be either free or connected with the base by elastic elements with negligibly small rigidities. The rotors have "intrinsic" eccentricities ε, their masses (without taking into consideration the additional masses m) are designated by m_0. The system is assumed to perform plane motions (in the plane, perpendicular to the axes of rotation of the rotors). Partial frequencies of the exciters are considered to be the same. The position of the rotors is preassigned by the angles of rotation φ_1 and φ_2, of the masses m by the polar coordinates ρ_s and ψ_s , and of the masses M^o and M_1^o by the coordinates x, y and x_1, y_1 respectively (Fig. 13.3, *b, c*).

13.6. Simplest System

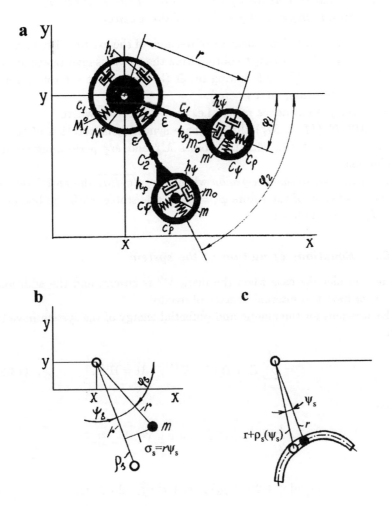

Fig. 13.3 The simplest system with two identical co-axial vibro-exciters: a – the general scheme, b – the coordinates of the additional mass m in the general case, c – the coordinates of the mass m in the case $\rho_s = \rho_s(\psi_s)$

13.6.2 Synchronization of the system in the absence of inner degrees of freedom of the exciters.

In the absence of the inner degrees of freedom of the exciters, i.e. at $m = 0$ (or at $c_\rho = c_\psi = 0$) it is easy to obtain from the previous investigations (see, say, [1, 2, 3, 5, 6, 8]) the following result (believing that the rotors rotate in the same direction, i.e. $\sigma_1 = \sigma_2$):

1) In the *pre-resonance region*, i.e. at $\omega < \sqrt{c_1/M_*}$ where $M_* = M^\circ M_1^\circ/(M^\circ + M_1^\circ)$ the synphase synchronous rotation of the rotors $\varphi_1 = \varphi_2$ is stable, with the bodies M_1° and M_2° performing circular oscillations.

2) In the *post-resonance region* when $\omega > \sqrt{c_1/M_*}$ the antiphase synchronous rotation of the rotors $\varphi_2 = \varphi_1 + \pi$ is stable, with the bodies M and M_1 remaining fixed.

13.6.3 Equations of motion of the system

Let us consider the case when the mass M_1^0 is absent, and the additional masses m have two internal degrees of freedom.

Expressions for the kinetic and potential energy of the system have the form

$$T = \sum_{s=1}^{2} T_s + \Delta T^* + T^{(I)}, \quad \Pi = \Pi^{(I)}, \qquad (13.32)$$

where

$$T_s = \frac{1}{2}[J + m_0\varepsilon^2 + m(r + \rho_s)^2 + m\sigma_s^2]\dot\varphi_s^2$$

$$+ \frac{1}{2}m[\dot\rho_s^2 + 2(r + \rho_s)\dot\varphi_s\dot\sigma_s + r^2\dot\psi_s^2 - 2\dot\rho_s\dot\varphi_s\sigma_s]$$

$$\Delta T^* = m\sum_{s=1}^{2}\{(\dot\rho_s - \sigma_s\dot\varphi_s)(\dot x\cos\varphi_s - \dot y\sin\varphi_s)$$

$$- [(r + \rho_s)\dot\varphi_s + \dot\sigma_s](\dot x\sin\varphi_s + \dot y\cos\varphi_s)\}$$

$$- m_0\varepsilon\sum_{s=1}^{2}\dot\varphi_s(\dot x\sin\varphi_s + \dot y\cos\varphi_s),$$

13.6. Simplest System

$$T^{(I)} = \frac{1}{2}(M^\circ + 2m_0 + 2m)(\dot{x}^2 + \dot{y}^2), \qquad \Pi^{(I)} = \frac{1}{2}\sum_{s=1}^{2}[c_r \rho_s^2 + c_\psi \sigma_s^2] \quad (13.33)$$

Here J is the moment of inertia of the rotor about the axis, passing through the center of gravity, and instead of ψ_s the coordinates $\sigma_s = r\psi_s$ are used (do not mistake this σ_s in formulas (13.7) and following).

The generalized forces can be presented as

$$Q_{\varphi_s} = L_s(\dot{\varphi}_s) - R_s(\dot{\varphi}_s) \approx -k(\dot{\varphi}_s - \omega) + k(\omega_s - \omega)$$

$$Q_{\rho_s} = -h_\rho \dot{\rho}_s, \quad Q_{\psi_s} = -h_\sigma \dot{\sigma}_s \qquad (13.34)$$

The symbols k, h_ρ, h_σ denote here the positive coefficients of damping, the symbol ω_s denotes the so-called partial velocities of vibro-exciters, i.e. frequencies of their rotation on a fixed base; in the problem under consideration $\omega_1 = \omega_2 = \omega$, so

$$Q_{\varphi_s} = -k(\dot{\varphi}_s - \omega) \qquad (13.35)$$

The symbols $L_s(\varphi_s)$ and $R_s(\dot{\varphi}_s)$ denote the rotating moments and the resistance moments respectively; formulas (13.34) and (13.35) represent their linear approximation, usually quite acceptable.

Expressions (13.32) – (13.35) make it possible to obtain the following equation of motion of the system, composed as the Lagrange equation

$$[J + m_0\varepsilon^2 + m(r + \rho_s)^2 + m\sigma_s^2]\ddot{\varphi}_s + k(\dot{\varphi}_s - \omega) + 2m[(r + \rho_s)\dot{\rho}_s + \sigma_s \dot{\sigma}_s]\dot{\varphi}_s$$

$$-[m_0\varepsilon + m(r + \rho_s)](\ddot{x}\sin\varphi_s + \ddot{y}\cos\varphi_s) - m\sigma_s(\ddot{x}\cos\varphi_s - \ddot{y}\sin\varphi_s)$$

$$+m[(r + \rho_s)\ddot{\sigma}_s - \sigma_s \ddot{\rho}_s] = 0$$

$$M\ddot{x} = \sum_{i=1}^{2}[m_0\varepsilon + m(r + \rho_i)](\ddot{\varphi}_i \sin\varphi_i + \dot{\varphi}_i^2 \cos\varphi_i)$$

$$-m\sum_{i=1}^{2}[\ddot{\rho}_i - \sigma_i\ddot{\varphi}_i - 2\dot{\sigma}_i\dot{\varphi}_i]\cos\varphi_i + m\sum_{i=1}^{2}[(\dot{\rho}_i - \sigma_i\dot{\varphi}_i)\dot{\varphi}_i + \ddot{\sigma}_i + \dot{\rho}_i\dot{\varphi}_i]\sin\varphi_i,$$

$$M\ddot{y} = \sum_{i=1}^{2}[m_0\varepsilon + m(r + \rho_i)](\ddot{\varphi}_i \cos\varphi_i - \dot{\varphi}_i^2 \sin\varphi_i)$$

$$+m\sum_{i=1}^{2}[\ddot{\rho}_i - \sigma_i\dot{\varphi}_i - 2\dot{\sigma}_i\dot{\varphi}_i]\sin\varphi_i + m\sum_{i=1}^{2}[(\dot{\rho}_i - \sigma_i\dot{\varphi}_i)\dot{\varphi}_i + \ddot{\sigma}_i + \dot{\rho}_i\dot{\varphi}_i]\cos\varphi_i$$

$$\ddot{\rho}_s + \beta_\rho\dot{\rho}_s + \omega_\rho^2\rho_s = \sigma_s\ddot{\varphi}_s + 2\dot{\sigma}_s\dot{\varphi}_s + (r+\rho_s)\dot{\varphi}_s^2 - (\ddot{x}\cos\varphi_s - \ddot{y}\sin\varphi_s),$$

$$\ddot{\sigma}_s + \beta_\sigma\dot{\sigma}_s + \omega_\sigma^2\sigma_s = -(r+\rho_s)\ddot{\varphi}_s - 2\dot{\rho}_s\dot{\varphi}_s + \sigma_s\dot{\varphi}_s^2 + (\ddot{x}\sin\varphi_s + \ddot{y}\cos\varphi_s)$$
(13.36)

$$(s=1, 2)$$

Here

$$\beta_\rho = h_\rho/m,\ \beta_\sigma = h_\sigma/m,\ \omega_\rho^2 = c_\rho/m,\ \omega_\sigma^2 = c_\psi/m,\ M = M^o + 2m_o + 2m$$
(13.37)

13.6.4 Stationary solutions and their stability

Equations (13.36) admit stationary solutions:

a. the *synphase* solution

$$\varphi_s = \varphi_s^0 = \omega t, \quad \rho_s = \rho^0, \quad \sigma_s = \sigma^0 = 0,$$
$$x = x^0 = -2A^0\cos\omega t, \quad y = y^0 = 2A^0\sin\omega t (s=1,2) \quad (13.38)$$

b. the *antiphase* solution

$$\varphi_1 = \varphi_1^\pi = \omega t + \pi, \quad \varphi_2 = \varphi_2^\pi = \omega t, \quad \rho_s = \rho^\pi, \quad \sigma_s = \sigma^\pi = 0 (s=1,2)$$
$$x = x^0 = 0, \quad y = y^0 = 0$$
(13.39)

Here and below we use the following designations

$$\rho^0 = \frac{\kappa(r - 2A_0)}{1 - (1 - 2q)\kappa}, \quad \rho^\pi = r\frac{\kappa}{1 - \kappa}$$
(13.40)

$$e^* = 1 - 2q, \quad q = \frac{m}{M}, \quad \kappa = \frac{\omega^2}{\omega_\rho^2}$$

$$R^0 = r + \rho^0, \quad R^\pi = r + \rho^\pi, \quad R^{*0} = R^0 - 2A^0, \quad R^{*\pi} = R^\pi - 2A^\pi,$$

$$A^0 = \frac{S^0}{M}, \quad A^\pi = \frac{S^\pi}{M}, \quad I^0 = J + m_0\varepsilon^2 + mR^{0^2}, \quad I^\pi = J + m_0\varepsilon^2 + mR^{\pi^2},$$
(13.41)

13.6. Simplest System

$$S^0 = m_0\varepsilon + mR^0, \quad S^\pi = m_0\varepsilon + mR^\pi, \quad I^{*0} = I^0 - 2MA^{0^2}, \quad I^{*\pi} = I^\pi - 2MA^{\pi^2}$$

It should be noted that in case $q \ll 1$ we have

$$\rho^0 = (r - 2A_0)\frac{\kappa}{1-\kappa} \tag{13.42}$$

and if besides that $2A_0 \ll r$, then

$$\rho^0 = \rho^\pi = r\frac{\kappa}{1-\kappa} \tag{13.43}$$

Let us first consider the stability of the *antiphase solution*. Assuming that in equations (13.36)

$$x = \xi, \quad y = \eta, \quad \varphi_1 = \omega t + \kappa_1, \quad \varphi_2 = \omega t + \pi + \kappa_2, \quad \rho_s = \rho^\pi + \gamma_s, \quad \sigma_s = \delta_s, \tag{13.44}$$

after linearizing we get the following system of equations in variations

$$I^\pi \ddot{\kappa}_1 + k\dot{\kappa}_1 + 2mR^\pi \omega \dot{\gamma}_1 + mR^\pi \ddot{\delta}_1 - S^\pi(\ddot{\xi}\sin\omega t + \ddot{\eta}\cos\omega t) = 0,$$

$$I^\pi \ddot{\kappa}_2 + k\dot{\kappa}_2 + 2mR^\pi \omega \dot{\gamma}_2 + mR^\pi \ddot{\delta}_2 + S^\pi(\ddot{\xi}\sin\omega t + \ddot{\eta}\cos\omega t) = 0,$$

$$M\ddot{\xi} = m\omega^2(\gamma_1 - \gamma_2)\cos\omega t + S^\pi\left\{\left[(\ddot{\kappa}_1 - \ddot{\kappa}_2) - \omega^2(\kappa_1 - \kappa_2)\right]\sin\omega t\right.$$

$$\left. + 2\omega(\dot{\kappa}_1 - \dot{\kappa}_2)\cos\omega t\right\} - m\left[(\ddot{\gamma}_1 - \ddot{\gamma}_2) - 2\omega(\dot{\delta}_1 - \dot{\delta}_2)\right]\cos\omega t$$

$$+ m\left[2\omega(\dot{\gamma}_1 - \dot{\gamma}_2) - \omega^2(\delta_1 - \delta_2) + \ddot{\delta}_1 - \ddot{\delta}_2\right]\sin\omega t,$$

$$M\ddot{\eta} = -m\omega^2(\gamma_1 - \gamma_2)\sin\omega t + S^\pi\left\{\left[(\ddot{\kappa}_1 - \ddot{\kappa}_2) - \omega^2(\kappa_1 - \kappa_2)\right]\cos\omega t\right.$$

$$\left. - 2\omega(\dot{\kappa}_1 - \dot{\kappa}_2)\cos\omega t\right\} + m\left[(\ddot{\gamma}_1 - \ddot{\gamma}_2) - 2\omega(\dot{\delta}_1 - \dot{\delta}_2)\right]\sin\omega t$$

$$+ m\left[2\omega(\dot{\gamma}_1 - \dot{\gamma}_2) - \omega^2(\delta_1 - \delta_2) + (\ddot{\delta}_1 - \ddot{\delta}_2)\right]\cos\omega t,$$

$$\ddot{\gamma}_1 + \beta_\rho \dot{\gamma}_1 = 2\omega\dot{\delta}_1 + 2R^\pi \omega \dot{\kappa}_1 + \left(\omega^2 - \omega_\rho^2\right)\gamma_1 - \left(\ddot{\xi}\cos\omega t - \ddot{\eta}\sin\omega t\right),$$

$$\ddot{\gamma}_2 + \beta_\rho \dot{\gamma}_2 = 2\omega\dot{\delta}_2 + 2R^\pi \omega\dot{\kappa}_2 + \left(\omega^2 - \omega_\rho^2\right)\gamma_2 + \left(\ddot{\xi}\cos\omega t - \ddot{\eta}\sin\omega t\right),$$

$$\ddot{\delta}_1 + \beta_\sigma \dot{\delta}_1 = -2\omega\dot{\gamma}_1 - R^\pi \ddot{\kappa}_1 + \left(\omega^2 - \omega_\sigma^2\right)\delta_1 + \left(\ddot{\xi}\sin\omega t + \ddot{\eta}\cos\omega t\right),$$

$$\ddot{\delta}_2 + \beta_\sigma \dot{\delta}_2 = -2\omega\dot{\gamma}_2 - R^\pi \ddot{\kappa}_2 + \left(\omega^2 - \omega_\sigma^2\right)\delta_2 - \left(\ddot{\xi}\sin\omega t + \ddot{\eta}\cos\omega\right), \quad (13.45)$$

System (13.45) with periodic coefficients is transformed into a system with constant coefficients provided we pass from ξ and η to the variables u and v, using the relations

$$\ddot{z} = \ddot{\xi} + i\ddot{\eta} = \ddot{w}e^{-i\omega t}, \quad \ddot{w} = \ddot{u} + i\ddot{v} \quad (13.46)$$

From these relations it follows

$$\ddot{\xi}\cos\omega t - \ddot{\eta}\sin\omega t = Re(\ddot{z}e^{i\omega t}) = Re\ddot{w} = \ddot{u}$$

$$\ddot{\xi}\sin\omega t + \ddot{\eta}\cos\omega t = Im(\ddot{z}e^{i\omega t}) = Im\ddot{w} = \ddot{v}$$

Simultaneously it is convenient to transfer from the variables κ_s, γ_s and δ_s to the variables z_1, z_2, z_3, y_1, y_2, and y_3 in accordance with the formulas

$$z_1 = \kappa_1 + \kappa_2, \quad z_2 = \gamma_1 + \gamma_2, \quad z_3 = \delta_1 + \delta_2,$$

$$y_1 = \kappa_1 - \kappa_2, \quad y_2 = \gamma_1 - \gamma_2, \quad y_3 = \delta_1 - \delta_2 \quad (13.47)$$

and eliminate the variables u and v.

As a result the system of equations in the variations (13.45) is transformed into two independent subsystems of equations with constant coefficients:

Subsystem 1

$$I^\pi \ddot{z}_1 + k\dot{z}_1 + 2mR^\pi \omega \dot{z}_2 + mR^\pi \ddot{z}_3 = 0$$

$$\ddot{z}_2 + \beta_\rho \dot{z}_2 - 2\omega \dot{z}_3 - 2R^\pi \omega \dot{z}_1 - (\omega^2 - \omega_\rho^2)z_2 = 0$$

$$\ddot{z}_3 + \beta_\sigma \dot{z}_3 + 2\omega \dot{z}_2 + R^\pi \ddot{z}_1 - (\omega^2 - \omega_s^2)z_3 = 0 \quad (13.48)$$

Subsystem 2

$$I^{*\pi}\ddot{y}_1 + k\dot{y}_1 + 2MA^{\pi 2}\omega^2 y_1 + 2mR^{*\pi}\omega\dot{y}_2 + mR^{*\pi}\ddot{y}_3 + 2mA^{\pi}\omega^2 y_3 = 0$$

$$e^*\ddot{y}_2 + \beta_\rho\dot{y}_2 - \left[e^*\omega^2 - \omega_\rho^2\right] y_2 - 2e^*\omega\dot{y}_3 - 2R^{*\pi}\omega\dot{y}_1 = 0$$

$$e^*\ddot{y}_3 + \beta_\sigma\dot{y}_3 - \left[e^*\omega^2 - \omega_\sigma^2\right] y_3 + 2e^*\omega\dot{y}_2 + R^{*\pi}\ddot{y}_1 + 2A^{\pi}\omega^2 y_1 = 0 \quad (13.49)$$

Characteristic equations for these two systems respectively have the form:

for subsystem 1

$$\begin{vmatrix} I^\pi\lambda^2 + k\lambda & 2mR^\pi\omega\lambda & mR^\pi\lambda^2 \\ -2R^\pi\omega\lambda & \lambda^2 + \beta_\rho\lambda - (\omega^2 - \omega_\rho^2) & -2\omega\lambda \\ R^\pi\lambda^2 & 2\omega\lambda & \lambda^2 + \beta_\sigma\lambda - (\omega^2 - \omega_\sigma^2) \end{vmatrix} = 0 \quad (13.50)$$

for subsystem 2

$$\begin{vmatrix} I^{*\pi}\lambda^2 + k\lambda + 2MA^{\pi 2}\omega^2 & 2mR^{*\pi}\omega\lambda & m(R^{*\pi}\lambda^2 + 2A^\pi\omega^2) \\ -2R^{*\pi}\omega\lambda & e^*\lambda^2 + \beta_\rho\lambda - (e^*\omega^2 - \omega_\rho^2) & -2e^*\omega\lambda \\ R^{*\pi}\lambda^2 + 2A^\pi\omega^2 & 2e^*\omega\lambda & e^*\lambda^2 + \beta_\sigma\lambda - (e^*\omega^2 - \omega_\sigma^2) \end{vmatrix} = 0 \quad (13.51)$$

Performing similar calculations and transformations for the synphase solution (13.38), we also get two subsystems:

Subsystem 1

$$I^0\ddot{y}_1 + k\dot{y}_1 + 2mR^0\omega\dot{y}_2 + mR^0\ddot{y}_3 - 2MA^{02}\omega^2 y_1 - 2mA^0\omega^2 y_3 = 0$$

$$\ddot{y}_2 + \beta_\rho\dot{y}_2 - 2R^0\omega\dot{y}_1 - 2\omega\dot{y}_3 - (\omega^2 - \omega_\rho^2)y_2 = 0$$

$$\ddot{y}_3 + \beta_\sigma\dot{y}_3 + 2\omega\dot{y}_2 + R^0\ddot{y}_1 - (\omega^2 - \omega_\sigma^2)y_3 - 2A^0\omega^2 y_1 = 0 \quad (13.52)$$

Subsystem 2

$$I^{*0}\ddot{z}_1 + k\dot{z}_1 + 2mR^{*0}\omega\dot{z}_2 + mR^{*0}\ddot{z}_3 = 0$$

$$e^*\ddot{z}_2 + \beta_\rho\dot{z}_2 - (e^*\omega^2 - \omega_\rho^2)z_2 - 2e^*\omega\dot{z}_3 - 2R^{*0}\omega\dot{z}_1 = 0$$

$$e^*\ddot{z}_3 + \beta_\sigma\dot{z}_3 - \left(e^*\omega^2 - \omega_\sigma^2\right)z_3 + 2e^*\omega\dot{z}_2 + R^{*0}\ddot{z}_1 = 0 \qquad (13.53)$$

and the corresponding characteristic equations

for subsystem 1

$$\begin{vmatrix} I^0\lambda^2 + k\lambda - 2MA^{0^2}\omega^2 & 2mR^0\omega\lambda & m\left(R^0\lambda^2 - 2A^0\omega^2\right) \\ -2R^0\omega\lambda & \lambda^2 + \beta_\rho\lambda - (\omega^2 - \omega_\rho^2) & -2\omega\lambda \\ R^0\lambda^2 - 2A^0\omega^2 & 2\omega\lambda & \lambda^2 + \beta_\sigma\lambda - (\omega^2 - \omega_\sigma^2) \end{vmatrix} = 0 \qquad (13.54)$$

for subsystem 2

$$\begin{vmatrix} I^{*0}\lambda^2 + k\lambda & 2mR^{*0}\omega\lambda & mR^{*0}\lambda^2 \\ -2R^{*0}\omega\lambda & e^*\lambda^2 + \beta_\rho\lambda - (e^*\omega^2 - \omega_\rho^2) & -2e^*\omega\lambda \\ R^{*0}\lambda^2 & 2e^*\omega\lambda & e^*\lambda^2 + \beta_\sigma\lambda - (e^*\omega^2 - \omega_\sigma^2) \end{vmatrix} = 0 \qquad (13.55)$$

It should be noted that the general order of the equations of subsystems 1 and 2 is equal to twelve, while the order of system (13.45) is equal to sixteen. The total degree of the corresponding characteristic equations is also equal to twelve. This is explained by the fact that the characteristic equation of system (13.45) has four zero roots, corresponding to the coordinates ξ and η which enter the equation only as the derivatives $\dot{\xi}$ and $\dot{\eta}$. Equations (13.50) and (13.55) have one more zero root, conditioned by the autonomy of the initial system. The existence of these five zero roots does not affect the stability of the motions that interest us.

Results of the investigation of stability of the synphase and antiphase motions of the rotors are given in Table 13.1. Investigations were made only for the cases $c_\psi \to \infty$, $\omega_\sigma \to \infty$, and $c_\rho \to \infty$, $\omega_\rho \to \infty$. In the first case in the determinants (13.50), (13.51), (13.54) and (13.55) the third line and the third column are crossed out, while in the second case the second line and the second column are crossed out.

13.6. Simplest System

As an example we will get the conditions of stability of the synphase motion for the case $\omega_\sigma \to \infty$ which follow from the consideration of subsystem 1. In this case the characteristic equation (13.54) will take the form

$$\begin{vmatrix} I^0\lambda^2 + k\lambda - 2MA^{0^2}\omega^2 & 2mR^0\omega\lambda \\ -2R^0\omega\lambda & \lambda^2 + \beta_\rho\lambda - (\omega^2 - \omega_\rho^2) \end{vmatrix} = 0 \qquad (13.56)$$

or in an expanded form

$$I^0\lambda^4 + (I^0\beta_\rho + k)\lambda^3 + [-I^0(\omega^2 - \omega_\rho^2) - 2MA^{0^2}\omega^2 + 4mR^{0^2}\omega^2 + k\beta_\rho]\lambda^2$$

$$- [k(\omega^2 - \omega_\rho^2) + 2MA^{0^2}\omega^2\beta_\rho]\lambda + 2MA^{0^2}\omega^2(\omega^2 - \omega_\rho^2) = 0 \qquad (13.57)$$

Hence we can see that the synphase motion cannot be stable at $\omega < \omega_\rho$ since in that case the constant term is negative. Neither can it be stable in the ordinary sense at $\omega > \omega_\rho$ because in this case the coefficient at λ is negative. Along with that, at $\omega > \omega_\rho$ there is so-called time stability [9]. And namely, in the absence of damping in the system ($k = 0$, $\beta_\rho = 0$) the motion is stable (not asymptotically) provided the following relation is fulfilled

$$-I^0(\omega^2 - \omega_\rho^2) - 2MA^{0^2}\omega^2 + 4mR^{0^2}\omega^2 > 0$$

or

$$\omega^2\left(1 + \frac{2MA^{0^2}}{I^0} - 4\frac{mR^{0^2}}{I^0}\right) < \omega_\rho^2 \qquad (13.58)$$

This inequality cannot be realized at $\omega > \omega_\rho$ if the term mR^{0^2}/I^0 is absent. But according to the structure of equation (13.56) this term corresponds to the gyroscopic forces. So, according to the well known theorem (see, say, [9]), the stability, taking place when inequality (13.58) is satisfied, is the so-called time stability, i.e. the stability which is destroyed in the presence of the dissipate forces, which follows from equation (13.57).

It is of interest to compare the results of the exact solution of the problem, obtained here, with the conclusion of R.F.Nagaev ([3], pp. 238 – 239) concerning the limiting case when all the inertia of the rotors is concentrated in the additional mass m, i.e. when there is the equality $mR^0/I^0 = 1$. Such case is apparently hard to realize in actual practice, it is however of fundamental interest. Using the integral criterion of stability (see section 13.4),

R.F.Nagaev comes to the conclusion that at $\omega_\sigma \to \infty$ in the systems, similar to that, at $\omega > \omega_\rho$, the synphase rotation of rotors must be stable (the vibro-exciters at $\omega > \omega_\rho$ being softly anisochronous).

The comparison of this conclusion with the data, given in the second line of the last column of Table 13.1, leads to the conclusion that the condition, obtained by R.F.Nagaev, while being necessary, is not sufficient: in all the cases it is the temporal stability that takes place, i.e. the instability in the ordinary sense.

13.6.5 *Discussion of the results of the section*

The consideration of the Table 13.1 shows that the introduction to the unbalanced rotors of the additional mass m makes it possible to control to a certain extent the stability of their synchronous motions. So, e.g. choosing the parameters of the system in accordance with the conditions $\omega > \omega_\rho$ at $\omega_\sigma \to \infty$ or $\omega > \omega_\sigma$ at $\omega_\rho \to \infty$, we can make the antiphase rotation unstable, while in the case of solid rotors such rotation is stable at any ω.

It is natural to expect that in the instability region of the antiphase rotation the synphase rotation will be stable. However, on the basis of the investigation made, that cannot be stated: at $\omega_\rho \to \infty$ such rotation is unstable at any ω, while at $\omega_\sigma \to \infty$ only the temporal stability can take place. So, the important technical problem – the provision of stability of the synphase motion of the rotors of vibro-exciters on the carrying body, either softly vibro-isolated or free, has not yet been solved. In this connection it is of interest to find, say, by numerical calculation, the character of the motion of the rotors in the region of instability of the antiphase rotation. It is also of interest to make clear whether it is possible to achieve the aim by using the rotors in which the additional masses move in the guides according to the law $\rho_s(\psi_s) = a\psi_s^2$ where the parameter can have either positive or negative values and also some other modifications of the construction of the rotor.

Table 13.1

| Supposition | Type of motion | Rotors without inner degree of freedom |||
| | | Conditions of stability |||
		Subsystem 1	Subsystem 2	General
$m = 0$ or $\omega_\rho \to \infty$, $\omega_\sigma \to \infty$	Anti-phase	—	—	Stable at any ω
	Syn-phase	—	—	Unstable at any ω

| Supposition | Type of motion | Rotors with inner degree of freedom |||
		Subsystem 1	Subsystem 2	General
$\omega_\sigma \to \infty$	Anti-phase	$\omega < \omega_\rho$	$\omega < \omega_\rho/\sqrt{e^*}$	$\omega < \omega_\rho$
	Syn-phase	1. $\omega > \omega_\rho$ 2. $\omega^2\left[1 + \dfrac{2MA^{02}}{I^0} - 4\dfrac{mR^{02}}{I^0}\right] < \omega_\rho^2$ (temporal stability)	$\omega < \omega_\rho/\sqrt{e^*}$	1. $\omega > \omega_\rho$ 2. $\omega < \omega_\rho/\sqrt{e^*}$ 3. $\omega^2\left[1 + \dfrac{2MA^{02}}{I^0} - 4\dfrac{mR^{02}}{I^0}\right] < \omega_\rho^2$ (temporal stability)
$\omega_\rho \to \infty$	Anti-phase	$\omega < \omega_\sigma$	$\omega < \omega_\sigma$	$\omega < \omega_\sigma$ $MA^\pi e^* > 2mR^{\pi*}$ (sufficient: $m_0 e > mR^\pi$)
	Syn-phase	Unstable at any ω	$\omega < \omega_\sigma/\sqrt{e^*}$	Unstable at any ω

References

[1] Blekhman I. I., *Synchronization in Science and Technology*, (ASME Press, New York, 1988, p.255.
[2] Sperling L, Selbstsynchronisation statisch und dynamisch unwuchtiger Vibratoren, *Technische Mechanic*, **14**, 1(1994),9–24; 2(1994), 85–96.
[3] Nagaev R. F., *Quasiconservative synchronizing systems*, (Nauka, St. Petersburg, 1996) (in Russian).
[4] Goldman P. S., Full Synchronization in a system of quasiconservative objects with several fast phases,*Thesis of PhD in Phys. and Math. Leningrad Polytech.Inst.*, 1985 (in Russian).
[5] Nagaev R. F., General problem on synchronization in an almost conservative system.*App.Math. and Mech.*, **29**, 5, 1965 (in Russian).
[6] Blekhman I.I., *Vibrational Mechanics*, (World Scientific, Singapore, 2000), p. 509.
[7] Nagaev R.F. and K.Sh.Khodzhaev, Synchronous motions in a system of objects with carrying constraints, *App. Math. And Mech.*, **31**, 4, 1967 (in Russian).
[8] Blekhman I.I. *Synchronization of Dynamic Systems*, (Nauka, Moscow, 1971), p. 894 (in Russian).
[9] Merkin D.R. *Introduction to the theory of the stability of motion*, (Nauka, Moscow,1976,ed. 3), p. 319 (in Russian).
[10] Blekhman I.I., L. Sperling, Synchronization of dynamic objects with internal degrees of freedom, *Proc. of the XXX Summer School "Actual Problems in Mechanics"*, St. Petersburg, Repino, 2002 (in print).

Chapter 14

On the Expansion of the Field of Applicability of the Integral Signs (Extreme Property) of Stability in Problems on the Synchronization of the Dynamic Objects with Almost Uniform Rotations

I. I. Blekhman[1], N. P. Yaroshevich[2]

[1] *Institute for Problems of Mechanical Engineering,*
Russian Academy of Sciences
and
Mekhanobr-Tekhnika Corp.
22 Liniya 3, V.O., 199106, St. Petersburg, Russia

[2] *Lutsky State Technical University*
73, Lvovskaya str., 43018, Lutsk, Ukraine

14.1 Preliminary Remark

The investigation of synchronization is greatly simplified and the results are given in a more convenient form if the so-called *integral signs of the stability of synchronous motions* are valid [1, 2, 3]. However, in problems on *multiple synchronization* and in a number of problems on the *simple synchronization* of identical unbalanced vibro-exciters in a certain rather wide class of cases, important for applications (called "not simple"), the integral signs in the form obtained by the Poincare-Lyapunov method of small parameter did not allow to find the values of the phases of rotation of the rotors of vibro-exciters in stable synchronous motions [2 – 6].

This chapter, on the basis of the method of direct separation of motions, substantiates a widened formulation of the integral signs of stability which makes it possible to consider both "simple" and "not simple" cases of problems on synchronous objects with almost uniform rotations. It was illustrated by examples that the results, obtained before by the Poincare method of small parameter and by the method of direct separation of motions can be obtained much more easily by using a widened formulation of the integral signs of stability.

14.2 On the Integral Sign (Extreme Property) of the Stability of Synchronous Motions

On the basis of using the Poincare-Lyapunov methods, the following remarkable property of the synchronous motions of the objects with almost uniform rotations and of some other dynamic objects was shown [1 3]: the stable synchronous motions correspond to the points of strict rough minimum of a certain function D – the *"potential function"*, of the so-called *generative parameters* – the phases of rotation $\alpha_1,..., \alpha_k$ (in the problem on self-synchronization – phase differences $\alpha_s\text{-}\alpha_k$, where k is the number of rotations; see below). In cases important for applications, the potential function D is the mean value of the rotation period of the Lagrange function of the system, taken with the opposite sign; in some other, more special cases – of the Lagrange function of the oscillatory part of the system, i.e. of the system with the "stopped rotations".

By using the integral signs it was proved, under rather general assumptions, that there is a *tendency to the synchronization* of a wide class of objects and a number of important applied problems has been solved [2]. The extreme property of synchronous ("resonance") motions was also established for the celestial bodies (see, e.g., [2, 7–9]).

Meanwhile there are cases when the integral signs in the above form do not make it possible to find the values of phases in stable synchronous motions. That refers, in particular, to the problems on multiple synchronization of vibro-exciters in quasi-linear systems and to some problems on the synchronization of several (more than three) identical vibro-exciters [2, 4 – 6]. In those cases, called "not simple", the function D proves to be independent of certain phases and therefore its minimum is not strict. Below the integral signs are shows to be valid provided the function D is calculated not on the basis of the generative solution, but more exactly – as far as it is necessary in order to establish its strict minimum.

14.3 The Statement of the Problem on the Synchronization of Objects with Almost Uniform Rotations

The problem named in the title can be formulated in the following way [2, 3]. We consider the system with the generalized coordinates φ_s ($s = 1,...,k$) (the *"rotational coordinates"*) and u_r ($r = 1,...,v$) (the *"oscillatory coordinates"*). We assume here that the Lagrange function of the system can be presented as

$$L = \frac{1}{2}\sum_{s=1}^{k} I_s \dot\varphi_s^2 + L_\sim(\varphi,\dot\varphi; u,\dot u; \omega t) \qquad (14.1)$$

and the nonconservative generalized forces, corresponding to rotational coordinates, as

$$Q_{\varphi_s} = -k_s(\dot\varphi_s - \sigma_s n_s \omega) + k_s(\omega_s - n_s \omega) \qquad (14.2)$$

Here I_s, k_s and ω are the positive constants; $\sigma_s = \pm 1$, n_s are positive integers; ω_s is the so-called *partial angular velocities of rotation* – angular velocities of rotation in the case, when the oscillating motions are absent ($u_r = $ const).

In the case of the problem on synchronization of rotors we can present in form (14.2) the difference between the moment $L_s(\dot\varphi_s)$, rotating the s-th rotor, and the moment of the resistance forces $R_s(\dot\varphi_s)$. It is assumed that the equations of motion of the system can be written down as

$$I_s \ddot\varphi_s + k_s(\dot\varphi_s - \sigma_s n_s \omega) = \mu \Phi_s \qquad (s = 1,...,k) \qquad (14.3)$$

$$E_{u_r}(L_\sim) = Q_{u_r} \qquad (r = 1,...,v) \qquad (14.4)$$

where $E_q = \dfrac{d}{dt}\dfrac{\partial}{\partial \dot q} - \dfrac{\partial}{\partial q}$ is Euler's operator, and Q_q is a non-conservative generalized force, corresponding to the coordinate q,

$$\mu \Phi_s = k_s(\omega_s - n_s\omega) - E_{\varphi_s}(L_\sim), \tag{14.5}$$

$\mu > 0$ is a small parameter. The functions $L_\sim$ and Q_{u_r} can depend on both – generalized coordinates and velocities, and on the time $\tau = \omega t$, being 2π – periodical over φ_s and τ; the functions $L_\sim$ and Q_{u_r} may depend on μ. As for the smoothness of the functions, suppositions are made which provide the existence of all the solutions and decompositions, considered below. The generative equations ($\mu = 0$), corresponding to equations (14.3) and (14.4), allow a set of solutions for the rotational coordinates

$$\varphi_s^0 = \sigma_s(n_s\omega t + \alpha_s), \tag{14.6}$$

answering the uniform rotations with the frequencies $\left|\dot\varphi_s^0\right| = n_s\omega$ and some arbitrary phases α_s.

The problem on synchronization consists in finding the conditions of existence and stability of the solutions of equations (14.3) and (14.4) of the type

$$\varphi_s = \sigma_s[n_s\omega t + \alpha_s + \mu\psi_s^{(p)}(\omega t, \mu)], \quad u_r = u_r^{(p)}(\omega t, \mu), \tag{14.7}$$

where $\psi_s^{(p)}$ and $u_r^{(p)}$ are the 2π-periodical functions of $\tau = \omega t$. The solution of this problem is given, in particular, in the book [2].

14.4 Solution of the Problem by Method of Direct Separation of Motions

When solving this problem by the method of direct separation of motions [3] the solution of the equations is to be found in the form

$$\varphi_s = \sigma_s[n_s\omega t + \alpha_s(t) + \psi_s(t, \omega t, \mu)], \quad u_r = u_r(t, \omega t, \mu) \tag{14.8}$$

where $\alpha_s(t)$ are the "slow" and ψ_s are the "fast" 2π-periodic components (t being the "slow" and $\tau = \omega t$ being the "fast" time, ω being

14.4. Solution by the Method of Direct Separation of Motions

the "large" parameter) with

$$\langle \psi_s(t,\omega t,\mu)\rangle = 0, \quad \langle u_r(t,\omega t,\mu)\rangle = 0, \tag{14.9}$$

and the angular brackets denote averaging over τ for the period 2π.

It is also assumed that

$$\dot{\alpha}_s \ll n_s \omega \tag{14.10}$$

The system (14.3) and (14.4) is reduced to the following system of integro-differential equations for the variables α_s, ψ_s and u_r

$$I_s \ddot{\alpha}_s = -k_s \dot{\alpha}_s + \mu \sigma_s \langle \Phi_s \rangle, \tag{14.11}$$

$$I_s \ddot{\psi}_s = -k_s \dot{\psi}_s + \mu \sigma_s (\Phi_s - \langle \Phi_s \rangle), \tag{14.12}$$

$$E_{u_r}(L) = Q_r \tag{14.13}$$

According to the method of direct separation of motions in order to obtain equations of slow motions in the first approximation, valid at least in the vicinity of the stationary regimes $\alpha_s = \text{const}$, it is sufficient to find an approximate asymptotically stable periodic solution of equations of fast motions (14.12) and (14.13) under the constant ("frozen"), $\dot{\alpha}_s, \alpha_s$ and t. We will use that solution when calculating the mean value in the right-hand parts of the equations (14.11); and we will designate this solution by ψ_s^*, u_r^* and respectively by φ_s^*. Then we arrive at the following equations of slow motions

$$I_s \ddot{\alpha}_s + k_s \dot{\alpha}_s = \mu \sigma_s \langle [\Phi_s]_* \rangle \quad (s = 1,....,k) \tag{14.14}$$

where the square brackets with the symbol "*" show, that the function enclosed in them is calculated for the solution ψ_s^* and u_r^*.

We introduce the function

$$\Lambda_{\sim *} = \langle [L_\sim]_* \rangle \tag{14.15}$$

and find the derivative of this function with respect to α_j. As a result of rather simple transformations, which include the integration by parts, in view of the equalities (14.4) and (14.8) we find

$$\frac{\partial \Lambda_{\sim *}}{\partial \alpha_j} = \frac{\partial \langle [L_{\sim}]_* \rangle}{\partial \alpha_j} = \sum_{s=1}^{k} \left\langle \left[\frac{\partial L_{\sim}}{\partial \dot{\varphi}_s}\right]_* \frac{\partial \dot{\varphi}_s^*}{\partial \alpha_j} + \left[\frac{\partial L_{\sim}}{\partial \varphi_s}\right]_* \frac{\partial \varphi_s^*}{\partial \alpha_j} \right\rangle$$

$$+ \sum_{r=1}^{v} \left\langle \left[\frac{\partial L_{\sim}}{\partial \ddot{u}_r}\right]_* \frac{\partial \ddot{u}_r^*}{\partial \alpha_j} + \left[\frac{\partial L_{\sim}}{\partial u_r}\right]_* \frac{\partial u_r^*}{\partial \alpha_j} \right\rangle = -\sum_{s=1}^{k} \left\langle [E_{\varphi_s}(L_{\sim})]_* \frac{\partial \varphi_s^*}{\partial \alpha_j} \right\rangle$$

$$- \sum_{r=1}^{v} \left\langle [E_{u_r}(L_{\sim})]_* \frac{\partial u_r^*}{\partial \alpha_j} \right\rangle = -\sigma_j \left\langle [E_{\varphi_j}(L_{\sim})]_* \right\rangle \qquad (14.16)$$

$$- \sum_{r=1}^{v} \left\langle [Q_{u_r}]_* \frac{\partial u_r^*}{\partial \alpha_j} \right\rangle - \sum_{s}^{k} \sigma_s \left\langle [E_{\varphi_s}(L_{\sim})]_* \frac{\partial \psi_s^*}{\partial \alpha_j} \right\rangle$$

The relation obtained can be considerably simplified. First of all it should be noted, that when using equalities (14.5) and (14.9), the last summand in its right-hand part can be presented as

$$\sum_{s=1}^{k} \sigma_s \left\langle [E_{\varphi_s}(L_{\sim})]_* \frac{\partial \psi_s^*}{\partial \alpha_j} \right\rangle = \sum_{s=1}^{k} \sigma_s \left\langle [k_s(\omega_s - n_s\omega) - \mu \Phi_s]_* \frac{\partial \psi_s^*}{\partial \alpha_j} \right\rangle$$

$$= -\mu \sum_{s=1}^{k} \sigma_s \left\langle \Phi_{s*} \frac{\partial \psi_s^*}{\partial \alpha_j} \right\rangle$$

(14.17)

Since, according to (14.12) and (14.9) $\psi_s^* = \psi_s^0 + \mu \psi_s^{(1)} + ...$, where $\psi_s^0 = 0$, this summand is of the order μ^2 and can, as a rule, be disregarded, except special cases when in other summands of the relation (14.16) the terms of the same order are essential. We believe such situations do not take place. This is correct when solving the problems on double synchronization, considered below, when the function Λ sufficiently find with an accuracy to the terms of the order not more than μ.

Now it should be noted, that according to (14.8) and (14.10)

$$\langle T_s \rangle = \frac{1}{2} I_s \langle \dot{\varphi}_s^2 \rangle = \frac{1}{2} I_s \left\langle [n_s \omega + \dot{\alpha}_s(t) + \dot{\psi}_s]^2 \right\rangle,$$

$$= \frac{1}{2} I_s \left\langle [n_s \omega + \dot{\psi}_s]^2 \right\rangle = \frac{1}{2} I_s \langle \dot{\psi}_s^2 \rangle + C$$

(14.18)

where C is the value independent of α_s. Therefore, again with an accuracy to the terms of the order μ^2

$$\frac{\partial \Lambda_{\sim *}}{\partial \alpha_j} = \frac{\partial \Lambda_*}{\partial \alpha_j} \qquad (14.19)$$

Finally, since according to (14.8), functions $\dot{\alpha}_s$ enters all the relations only in the combination $n_s\omega + \dot{\alpha}_s$, than due to (14.10) we can omit the symbol $*$ of all the values in equations (14.16), (14.17) and (14.19). As a result, the equations of slow motions (14.11) can be presented in the form

$$I_s\ddot{\alpha}_s + k_s\dot{\alpha}_s = -\frac{\partial D}{\partial \alpha_s} \quad (s=1,....,k) \qquad (14.20)$$

where

$$D = -\frac{\partial(\Lambda + B)}{\partial \alpha_s}, \quad \frac{\partial B}{\partial \alpha_s} = -k_s(\omega_s - n_s\omega) - \sum_{r=1}^{v}\left\langle\left[Q_{u_r}\right]\frac{\partial u_r}{\partial \alpha_s}\right\rangle, (14.21)$$

with D being the potential function, and B so-called *potential of the generalized nonconservative forces*, corresponding to the oscillatory coordinates (believing that such a potential does exist).

14.5 Extended Formulation of Integral Signs

From equation (14.20) it follows directly (see e.g. [10]) that under the assumptions made above, the extended formulation of the integral signs (extreme property) of synchronous motions of objects with almost uniform rotations is valid: *the stable synchronous motions of objects correspond to the values of the phases* $\alpha_s = $ const, *which are answered by the strict rough minimums of the potential function* $D = D(\alpha_1,...,\alpha_k)$ (in the problem on self-synchronization $D = D(\alpha_1 - \alpha_k,...,\alpha_{k-1} - \alpha_k)$ and what is meant here is the minimums of phases differences $(\alpha_s - \alpha_k)$). Then, unlike the former wording, the function D can be calculated not necessarily in the generative approximation ($\mu = 0$), but with any accuracy to μ, with two reservations:

1. In addition, it is assumed that α_s changes quite slowly as compared to ψ_s and that $\dot{\alpha}_s$ is small as compared to $n_s\omega$, i. e. $\dot{\alpha}_s \leq \dot{\psi}_s$ and $\dot{\alpha}_s \leq n_s\omega$.

2. Expressions (14.17) and (14.18) must have a higher order, than the terms, taken into considerations when calculating the functions Λ and B.

A more exact calculation of the function D makes it possible to establish the presence of strict minimums in those ("not simple") cases, when that function in the generative approximation in general cannot have such minimums.

Just like before, (see [1, 2, 3]) the expression for the function D is considerably simplified under some additional suppositions:

1. In the case, when partial velocities ω_s are equal to the corresponding multiplicity of the synchronous velocity ω, (i.e. $\omega_s = n_s\omega$), and when the non-conservative forces Q_{u_r} are negligibly small, then

$$D = -\Lambda = -\langle [L] \rangle, \qquad (14.22)$$

i.e. the potential function is the Lagrange function of the system, averaged for the period and calculated for the solution (14.7).

2. Let the system be linear with respect to the oscillatory coordinates and let the function L be presented in form

$$L = L^* + L^{(I)} + L^{(II)} \qquad (14.23)$$

where

$$L^* = \sum_{s=1}^{k} L_s(\dot{\varphi}_s, \varphi_s) + \sum_{r=1}^{v} f_r(\dot{\varphi}_1, ..., \dot{\varphi}_k, \varphi_1, ..., \varphi_k)\dot{u}_r + \sum_{s=1}^{k} F_s(\varphi_s) \qquad (14.24)$$

$$L^{(I)} = \frac{1}{2}\sum_{r=1}^{v}\sum_{j=1}^{v}(a_{rj}\dot{u}_r\dot{u}_j - b_{rj}u_ru_j), \quad L^{(II)} = \Psi(\dot{\varphi}_1, ..., \dot{\varphi}_k, \varphi_1, ..., \varphi_k)$$

$$(14.25)$$

Here a_{rj} and b_{rj} are constants, and L_s, f_r, F_r and Ψ are the functions of the listed variables, with L_s, f_r and F_r being periodical with respect to φ_s with the period 2π. Let the generalized forces Q_{u_r} be either absent or negligibly small.

Then the following relations are valid

$$\frac{\partial \Lambda}{\partial \alpha_s} = \frac{\partial \left(\Lambda^{(II)} - \Lambda^{(I)} \right)}{\partial \alpha_s} + 0(\mu),$$

(14.26)

$$\left(\Lambda^{(I)} = \left\langle \left[L^{(I)} \right] \right\rangle, \quad \Lambda^{(II)} = \left\langle \left[L^{(II)} \right] \right\rangle \right)$$

They differ from those, valid for generative solution (see equalities (2.20) in book [3]) by the fact that they are fulfilled with an accuracy up to terms of the order μ. This circumstance is easily established by means of calculations analogous to those given in the book [3].

As a result, in this case, provided the terms of the order of μ in equality (14.26) can be neglected, we can have

$$D = \Lambda^{(I)} - \Lambda^{(II)},$$

(14.27)

and in the case $\Lambda^{(II)} = 0$ we have

$$D = \Lambda^{(I)}$$

(14.28)

Thus in this case the potential function is the Lagrange function $L^{(I)}$ for only the oscillatory part of the system, averaged for the period.

It should be noted that one can arrive at the same results using the Hamilton variational principle and having modified and amplified the investigation, made by A.I. Lurie [11].

14.6 Examples, Comparison to the Results, Obtained by Other Methods

14.6.1 Double Synchronization of Two Vibro-Exciters on a Platform with One Degree of Freedom

As a simplest example let us consider a system containing two unbalanced vibro-exciters placed on the carrying body (Fig. 14.1). They are rotated by electric engines of asynchronous type. The carrying body can move with respect to the immovable base along a certain fixed direction Ou and is connected with the base by the means of linear elastic

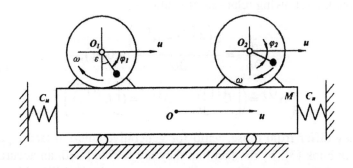

Figure 14.1. Scheme of the system with two vibro-exciters on a platform with one degree of freedom.

elements. Equations of motions (14.3) and (14.4) for the problem under consideration may be written down as (see, e.g.[2])

$$I_s \ddot{\varphi}_s + \kappa_s (\dot{\varphi}_s - \sigma_s n_s \omega) = \mu [L_s (\sigma_s n_s \omega) - R_s (n_s \omega) \\ + m_s \varepsilon_s (\ddot{u} \sin \varphi_s + g \cos \varphi_s)] \quad (s=1,2; \quad n_1 = 1, \quad n_2 = 2),$$
(14.29)

$$M\ddot{u} + c_u u = \sum_{j=1}^{2} m_j \varepsilon_j \left(\ddot{\varphi}_j \sin \varphi_j + \dot{\varphi}_j^2 \cos \varphi_j \right)$$
(14.30)

where m_s is the mass of rotor of the s-th vibro-exciter, ε_s is its eccentricity, and I_s is the moment of inertia with respect to the axis of rotation; $\kappa_s > 0$ are constant coefficients, characterizing damping; M is the mass of the carrying body in view of the mass of vibro-exciters; c_u is the rigidity of the elastic elements; g is the gravitational acceleration.

When solving the problem by the method of direct separation of motions, in view of (14.8) we pass from equations (14.29) and (14.30) to the equations of slow and fast motions (14.11) and (14.12), where

$$\Phi_s = L_s (n_s \omega) - R_s (n_s \omega) + m_s \varepsilon_s \big[\ddot{u} \sin(n_s \omega t + \alpha_s + \psi_s) \\ + g \cos(n_s \omega t + \alpha_s + \psi_s) \big]$$

14.6. Examples, Comparison to Other Results

Equation (14.13) takes the form

$$M\ddot{u} + c_u u = \sum_{j=1}^{2} m_j \varepsilon_j [\ddot{\varphi}_j \sin(n_j \omega t + \alpha_j + \psi_j) + (n_j \omega + \dot{\psi}_j)^2 \cos(n_j \omega t + \alpha_j + \psi_j)] \quad (14.31)$$

The periodic solutions of the equations of fast motions (14.12) are found in the form of a series according to the degrees μ. It should be noted that in the case of double synchronization, considered here, it is not sufficient to have in the first approximation $\psi_s = 0$. Along with that, as we will see below, it is enough to calculate ψ_s with an exactness to the terms of not higher than the first order. With that exactness we find:

$$\mu\psi_1 = -\frac{m_1^2 \varepsilon_1^2 \omega^2}{8MI_1(\omega^2 - p^2)} \sin 2(\omega t + \alpha_1) - \frac{m_1 \varepsilon_1 g}{I_1 \omega^2} \cos(\omega t + \alpha_1) -$$

$$-\frac{8 m_1 \varepsilon_1 m_2 \varepsilon_2 \omega^2}{MI_1(4\omega^2 - p^2)}\left[\frac{1}{9}\sin(3\omega t + \alpha_1 + \alpha_2) - \sin(\omega t - \alpha_1 + \alpha_2)\right],$$

$$\mu\psi_2 = \frac{m_1 \varepsilon_1 m_2 \varepsilon_2 \omega^2}{2MI_2(\omega^2 - p^2)}\left[\frac{1}{9}\sin(3\omega t + \alpha_1 + \alpha_2) + \sin(\omega t + \alpha_2 - \alpha_1)\right] -$$

$$-\frac{m_2^2 \varepsilon_2^2 \omega^2}{2MI_2(4\omega^2 - p^2)}\sin 2(2\omega t + \alpha_2) - \frac{m_2 \varepsilon_2 g}{4I_2 \omega^2} \cos(2\omega t + \alpha_2).$$

Then the solution of the equation (14.31) gets the form

$$u = -\frac{m_1 \varepsilon_1 \omega^2}{M(\omega^2 - p^2)} \cos(\omega t + \alpha_1) - \frac{4 m_2 \varepsilon_2 \omega^2}{M(4\omega^2 - p^2)} \cos(2\omega t + \alpha_2)$$

$$-\frac{2 m_1^2 \varepsilon_1^2 g \omega^2}{MI_1 \omega^2 (4\omega^2 - p^2)} \sin 2(\omega t + \alpha_1) \quad (14.32)$$

Here we have written down only the terms, affecting within the frames of this approximation, the result of calculating the Lagrange averaged function.

Substituting (14.32) into the expression for kinetic and potential energy of the oscillatory part of the system

$$T^{(1)} = \frac{1}{2}M\dot{u}^2, \quad \Pi^{(1)} = \frac{1}{2}C_u u^2,$$

and on averaging them for the period, we obtain for the case under consideration the potential function

$$D = \Lambda^{(1)} = \left\langle \left[L^{(1)} \right] \right\rangle = \left\langle \left[T^{(1)} - \Pi^{(1)} \right] \right\rangle = \frac{4m_1^2 \varepsilon_1^2 m_2 \varepsilon_2 g \omega^2}{MI_1 \left(4\omega^2 - p^2\right)} \sin(2\alpha_1 - \alpha_2)$$

As a result, the equations of slow motions, describing also the motions in the vicinity of the stationary synchronous regimes $\alpha_s = \text{const}$, can be presented in form (14.20).

The expression obtained for the function D makes it possible to write down the expressions for the vibrational moments

$$W_1 = \frac{\partial \Lambda^{(1)}}{\partial \alpha_1} = \frac{8m_1^2 \varepsilon_1^2 m_2 \varepsilon_2 g \omega^2}{MI_1 \left(4\omega^2 - p^2\right)} \cos(2\alpha_1 - \alpha_2)$$

$$W_2 = \frac{\partial \Lambda^{(1)}}{\partial \alpha_2} = -\frac{4m_1^2 \varepsilon_1^2 m_2 \varepsilon_2 g \omega^2}{MI_1 \left(4\omega^2 - p^2\right)} \cos(2\alpha_1 - \alpha_2)$$

(14.33)

These expressions exactly coincide with those, obtained by a more complicated investigation by means of the method of Poincare [4]. Consequently, all other results will also coincide.

14.6.2 Double Synchronization of Three Unbalanced Vibro-Exciters, Located Symmetrically on a Softly Vibro-Isolated Flatly Oscillating Solid Body

The system under consideration is shown in Fig. 14.2.

14.6. Examples, Comparison to Other Results

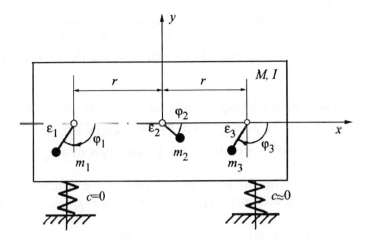

Figure 14.2. Scheme of the system with three vibro-exciters on a softly vibro-isolated flatly oscillating solid body.

Equations of motions of the system look as (see, e.g. [2, 5])

$$I_s \ddot{\varphi}_s + \kappa_s(\dot{\varphi}_s - \sigma_s n_s \omega) = \mu[L_s(\sigma_s n_s \omega) - R_s(n_s \omega)$$
$$+ m_s \varepsilon_s(\ddot{x} \sin \varphi_s + \ddot{y} \cos \varphi_s - r_s \ddot{\varphi} \cos \varphi_s + g \cos \varphi_s)] \quad (s=1,2,3)$$

$$M\ddot{x} = \sum_{j=1}^{3} m_j \varepsilon_j (\ddot{\varphi}_j \sin \varphi_j + \dot{\varphi}_j^2 \cos \varphi_j),$$

$$M\ddot{y} = \sum_{j=1}^{3} m_j \varepsilon_j (\ddot{\varphi}_j \cos \varphi_j - \dot{\varphi}_j^2 \sin \varphi_j),$$

$$I\ddot{\varphi} = \sum_{j=1}^{3} m_j \varepsilon_j r_j (\dot{\varphi}_j^2 \sin \varphi_j + \ddot{\varphi}_j \cos \varphi_j)$$

where I is the moment of inertia of the carrying body, r_s is the distance from the axis of the s-th vibro-exciter to the center of gravity of the carrying body. The outer vibro-exciters are similar and their axes are at the same distances from the center of gravity, i.e. $m_1 \varepsilon_1 = m_3 \varepsilon_2$, $I_1 = I_3$, $r_1 = r_3 = r$, $r_2 = 0$. All the vibro-exciters rotate in the same direction, the frequency of the rotation of the central vibro-exciter being twice as great as that of the outer ones, i.e. $n_1 = n_3 = 1$, $n_2 = 2$.

Applying the method of direct separation of motions in view of (14.8) it is not difficult to arrive at the equations of slow and fast motions (14.11), (14.12) and (14.13) where Φ_s for the problem under consideration is

$$\Phi_s = L_s(n_s\omega) - R_s(n_s\omega) + m_s\varepsilon_s[\ddot{x}\sin(n_s\omega t + \alpha_s) + \ddot{x}\psi_s\cos(n_s\omega t + \alpha_s)]$$
$$+ \ddot{y}\cos(n_s\omega t + \alpha_s) - \ddot{y}\psi_s\sin(n_s\omega t + \alpha_s) + r_s\ddot{\varphi}\cos(n_s\omega t + \alpha_s)$$
$$- r_s\ddot{\varphi}\psi_s\sin(n_s\omega t + \alpha_s) + g\cos(n_s\omega t + \alpha_s) - g\psi_s\sin(n_s\omega t + \alpha_s)]$$

Searching first for the periodic solutions of the equations of fast motions (14.11)(in the same approximation as in example 14.6.1), we find

$$\mu\psi_s = \frac{4m_1\varepsilon_1 m_2\varepsilon_2}{MI_1}\sin(\omega t + \alpha_2 - \alpha_s) - \frac{m_1\varepsilon_1 g}{I_1\omega^2}\cos(\omega t + \alpha_s)$$
$$+ \frac{m_1^2\varepsilon_1^2 r^2}{8II_1}[\sin 2(\omega t + \alpha_s) - \sin(2\omega t + \alpha_1 + \alpha_3)] \quad (s = 1,3),$$

$$\mu\psi_2 = -\frac{m_1\varepsilon_1 m_2\varepsilon_2}{MI_2}[\sin(\omega t + \alpha_2 - \alpha_1) + \sin(\omega t + \alpha_2 - \alpha_3)]$$
$$- \frac{m_2\varepsilon_2 g}{4I_2\omega^2}\cos(2\omega t + \alpha_2)$$

Then equations (14.13), describing the oecilltions of the carrying solid body, take the form

$$\ddot{x} = \frac{m_1\varepsilon_1\omega^2}{M}[\cos(\omega t + \alpha_1)] + \cos(\omega t + \alpha_3)] + \frac{4m_2\varepsilon_2\omega^2}{M}\cos(2\omega t + \alpha_2)$$
$$+ \frac{2m_1^2\varepsilon_1^2 g}{MI_1}[\sin 2(\omega t + \alpha_1) + \sin 2(\omega t + \alpha_3)],$$

$$\ddot{y} = -\frac{m_1\varepsilon_1\omega^2}{M}[\sin(\omega t + \alpha_1) + \sin(\omega t + \alpha_3)] - \frac{4m_2\varepsilon_2\omega^2}{M}\sin(2\omega t + \alpha_2)$$
$$+ \frac{2m_1^2\varepsilon_1^2 g}{MI_1}[\cos 2(\omega t + \alpha_1) + \cos 2(\omega t + \alpha_3)],$$

$$\ddot{\varphi} = -\frac{m_1\varepsilon_1 r\omega^2}{I}[\sin(\omega t + \alpha_1) - \sin(\omega t + \alpha_3)]$$

14.6. Examples, Comparison to Other Results

In (14.36) only those terms are written out which affect, within the frames of the approximation under consideration, the result of the calculation of the Lagrange function.

Since we assume that the elastic supports are soft, the mean value of the Lagrange function is equal to the mean value of the kinetic energy of the carrying body

$$D = \Lambda^{(1)} = \left\langle \left[T^{(1)} \right] \right\rangle = \left\langle \left[M(\dot{x}^2 + \dot{y}^2) + I\dot{\varphi}^2 \right] \right\rangle$$

$$= \left(\frac{m_1^2 \varepsilon_1^2 \omega^2}{M} - \frac{m_1^2 \varepsilon_1^2 r^2 \omega^2}{2I} \right) \cos(\alpha_1 - \alpha_3)$$

$$+ \frac{2m_1^2 \varepsilon_1^2 m_2 \varepsilon_2 g}{I_1 M} \left[\sin(2\alpha_1 - \alpha_2) + \sin(2\alpha_3 - \alpha_2) \right]$$

As a result, in view of (14.20) in order to determine the slow components (mean rotation phases of the rotors α_s) we get a system of differential equations

$$I_1 \ddot{\alpha}_1 + \kappa_1 \dot{\alpha}_1 = L_1(\omega) - R_1(\omega)$$

$$+ \frac{m_1^2 \varepsilon_1^2 \omega^2}{M} \left(1 - \frac{Mr^2}{2I} \right) \sin(\alpha_1 - \alpha_3) - \frac{4m_1^2 \varepsilon_1^2 m_2 \varepsilon_2 g}{I_1 M} \cos(2\alpha_1 - \alpha_2),$$

$$I_2 \ddot{\alpha}_2 + \kappa_2 \dot{\alpha}_2 = L_2(2\omega) - R_2(2\omega)$$

$$+ \frac{2m_1^2 \varepsilon_1^2 m_2 \varepsilon_2 g}{I_1 M} \left[\cos(2\alpha_1 - \alpha_2) + \cos(2\alpha_3 - \alpha_2) \right], \qquad (14.37)$$

$$I_3 \ddot{\alpha}_3 + \kappa_3 \dot{\alpha}_3 = L_3(\omega) - R_3(\omega)$$

$$- \frac{m_1^2 \varepsilon_1^2 \omega^2}{M} \left(1 - \frac{Mr^2}{2I} \right) \sin(\alpha_1 - \alpha_3) - \frac{4m_1^2 \varepsilon_1^2 m_2 \varepsilon_2 g}{I_1 M} \cos(2\alpha_3 - \alpha_2),$$

Equations of the slow processes of establishing the synchronous motions (14.37) exactly coincide with those, obtained in a more complicated way i.e. by using the ethod of direct separation of motions [5]. So all the results will also coincide.

It should be emphasized that in the problems of double synchronization considered here the solutions of equations (14.12) with the accuracy of the order μ proved to be sufficient. Equations of motion of the carrying body are solved in the same way as in the "simple" case of the problem on synchronization, but in view of a more exact solution of equations (14.12).

References

1. Blekhman I. I., Vertification of the integral sign of stability of motion in problems on self-synchronization of vibro-exciters, *App. Math. and Mech.*, **24**, 6 (1960) 1100–1103 (in Russian).
2. Blekhman I. I., *Synchronization in Nature and Technology* (Nauka, Moscow, 1981), p. 351 (in Russian)
3. Blekhman I. I., *Vibrational Mechanics* (World Scientific, Singapore, 2000), p. 509.
4. Barzukov O. P., Multiple synchronization in the system of weakly connected objects with one degree of freedom, *App. Math. and Mech.*, **36**, 2 (1972) 225–231 (in Russian).
5. Yaroshevich N. P. The use of the self-synchronization effect under excitation of bi-harmonious vibrations, *The problems of machine-building and the reliability of machines*, **6** (1990), 23–27.
6. Malakhova O. Z., On a special case in the theory of self-synchronization of mechanical vibro-exciters, *Izv. AN SSSR. MTT*, **1** (1990) 29–36 (in Russian).
7. Ovenden M. W., T. Feagin, and O. Graff, On the principle of least interaction action and the Laplacean satellites of Jupiter and Uranus, *Celestial Mechanics*, **8**, 3 (1974) 455–471.
8. Beletsky V. V., Extremal properties of resonance motions; in *Stability of motion. Analytical mechanics. Control of motion*, (Nauka, Moscow, 1981) 41–55 (in Russian).
9. Khentov A. On the principle of strong interaction for the resonant orbital motions of some celestial bodies, *Mathematical and computers in simulation. Trans. of IMACS*, **58**, 4-6 (Special Iss. "Chaos, Synchronization and Control"), p. 423–434.
10. Merkin D. R., *Introduction to the theory of the stability of motion* (Nauka, Moscow, 1976, ed. 3), p. 319 (in Russian).
11. Lurie A. I., Certain problems of synchronization, *Proceedings of the 5th International Conf. on Nonlinear Oscillations*, **3** (Mathematical Institute of AN USSR, Kiev, 1970) 440–455 (in Russian).

Part IV

Problems of Creating Dynamic Materials

Part II

Chapter 15

On Dynamic Materials

I. I. Blekhman

*Institute for Problems of Mechanical Engineering,
Russian Academy of Sciences
and
Mekhanobr-Tekhnika Corp.
22 Liniya 3, V.O., 199106, St. Petersburg, Russia*

15.1 Briefly on the Idea of Dynamic Materials

The idea of creating the dynamic materials - deformable solids, liquids and others - was put forth by K. A. Lurie and the author [1, 2, 3].

Dynamic materials was defined as the media with material parameters alternating in space and time. An important class of dynamic materials is represented by *dynamic composites*. These formations differ from the ordinary composites due to the explicit involvement of time as an additional fast variable participating in the microstructure that becomes spatio-temporal in this case.

The properties of dynamic materials can, naturally, be quite different from the properties of the original constituents. Changing the material parameters of the original constituents the microgeometry of the assemblage, and the character of change of those parameters in time, we can control the dynamic properties of the materials, obtaining effects, which are absolutely impossible when using ordinary materials (see below).

What has been said refers not only to the mechanical materials, characterized by inertial, elastic, dissipative and other parameters, but also to the electro-technical materials, whose main characteristics are self-induction, capacitance, etc. Aspects of the problem which are of principal importance are also connected with the relativistic effects [4, 5, 6, 7, 8].

One can imagine two ways of obtaining dynamic materials, and, accordingly, two types of such materials.

The materials of the first type are obtained by either instaneous or gradual change of the material parameters of different parts of the system (of masses, fluids, self-induction, capacitance, etc.) in the absence of relative shift of those parts. This way is called *activation* [4, 6], and the corresponding materials are called *dynamic materials of the first kind or the activated dynamic materials.*

The second way presupposes that certain parts of the system are endowed with relative motions, prearranged or excited in some way. That way will be arbitrarily called *kinetization*, and the corresponding materials will be called *dynamic materials of the second kind or the kinetic dynamic materials*.

The dynamic material of the second type can be imagined as two or several media penetrating into each other, filling up a certain domain of space, with every medium performing a certain motion (in particular fast oscillations) with respect to other media. It is quite natural that the material parameters and properties of such a material will differ essentially from the parameters and properties of the original media and that there are great potentialities to control these properties. It should be noted that dynamic materials of both the first and the second type are, as a rule, the systems with the inflow and outflow of the energy, i. e. they are essential nonconservative systems.

Among the possibilities, afforded by the use of dynamic materials, we must note a possibility of isolating a certain part of the deformable medium from the long-wave effects and also of excitating, strengthening or damping the impact momentum.

15.2 On the Development of the Idea of Creating Dynamic Materials

In the book [3] there were considered some possibilities of the implementation of dynamic materials of second kind and some potentialities provided by using these materials.

In recent years a number of constructive ideas appears on the creating the dynamic materials of both the first and the second kind. These ideas are now in the stage of patenting.

One of the simplest kind of dynamic materials and corresponding possibilities of its use is considered below in chapter 16.

Note also that the Indian magic rope, considered in chapter 8, can be considered one of the limiting cases of the dynamic materials as well.

The theory of creating the *electro-technical dynamic materials* is also developing now [8]. We will not dwell on these investigations which are irrelevant to the topic of this book.

References

[1] Blekhman I. I., Forming the properties of nonlinear mechanical systems by means of vibrations *DCAMM Report*, **616**, (Technical University of Denmark, 1999; Proceeding of the IUTAM/IFTAM Symposium on Synthesis of nonlinear Dynamical Systems, Riga, Latvia, 1998. Klüwer Academic Press, 1999).

[2] Blekhman I. I. and K. A. Lurie, Dynamic Materials as the elements of Material Design, *Paper on the EURODINAME'99 Conference "Dynamic Problems in Mechanics and Machatronics"* (Günsburg, Germany, 11-16 July, 1999).

[3] Blekhman I. I., *Vibrational Mechanics* (Singapore, New Jersey, London, Hong Kong, World Scientific, 2000) p. 509.

[4] Lurie K. A., *Applied Optimal Control Theory of Distributed Systems* (Plenum Press, New York, 1993).

[5] Lurie K. A., Effective properties of smart laminates and screening phenomenon, *Journ. Solids Structures*, **34**, 13 (1997) 1633–1643.

[6] Lurie K. A., The problem of effective parameters of a mixture of two isotropic dielectrics distributed in space-time and conservation law for wave impedance in one-dimensional wave propagation, *Proc. Royal Soc.*, A(454) (London, July, 1998) 1767–1779 .

[7] Blekhman I. I., Lurie K. A., On Dynamic Materials, *DAN Russia* **371** 2 (2000), 182–185.

[8] Lurie K. A., Some recent advances in the theory of dynamic materials. *Paper on the Intern. Symposium "'Chalenges in Applied Mechanics"'*, (Lyngby, Denmark, 2002).

Chapter 16

The Active Control of Vibrations of Composite Beams by Parametric Stiffness Modulation

S. V Grishina., O. A. Ershova, S. V. Sorokin

State Marine Technical University of St.Petersburg
Lotsmanskaya Str.,#3, St.Petersburg, 190008, Russia

16.1 Preliminary Remarks

In many technical applications, it is necessary to control the amplitudes of forced vibrations of thin walled structures, e.g., plates and shells. In the most of cases, such a control is closely linked to the control of the sound radiation from a vibrating structure and is aimed at improvements in NVH (noise, vibration and harshness) characteristics of cars, aircraft, turbine engines, etc. It is quite typical then that a spectrum of driving forces is given and not subjected to possible modifications. Thus, a reduction in the amplitudes of the structural forced response may be achieved by some 'protection from vibrations' of the structure itself. A very detailed description of the up-to-date state of affairs in the control of vibrations is given in [5]. Besides thorough discussion of the passive control of vibrations, which is performed by absorption of the vibrational energy by dissipative elements, a concept of an active control is outlined and exemplified in [5]. As is seen, the feed back and feed forward active control are considered as the most efficient tools to suppress vibrations. Specifically, the adaptive active control based on measurements of vibration/sound radiation field from primary sources with the use of some 'corrective' secondary ones is widely explored. In fact, such a control strategy relies upon the linearity of a problem, which justifies use of the superposition principle in forced vibrations of a controlled structure.

An alternative way of an active control of vibrations and radiation of sound is associated with the recent advances in material technology, which make it possible to manufacture so-called smart materials. In particular, the concept of 'dynamic materials' as a special type of smart materials designed to suppress vibrations has been suggested and elaborated in recent papers [7, 8, 9, 14, 15] This concept is in effect based on the ideas of 'vibrational rheology' suggested in [1, 16]. If a micro-inhomogeneous composite material is considered, then its mechanical 'macro-level' characteristics are usually derived by homogenisation (averaging) at the micro-level. There are many publications related to various averaging techniques for periodic arrays of elementary cells composing micro-inhomogeneous material, see for example [17]. We are not aimed at a detailed discussion of this issue, but note that in the literature these elementary cells are considered as immobile so that the 'global' mechanical properties of a smart material are uniquely defined. However, in principle it is possible to introduce some actuators, which provoke the motions (oscillations) of cells in a prescribed manner. For example, high-frequency small-amplitude vibrations may be excited in these elements of a microstructure by piezo-electric actuators [5, 18]. Then it appears that the 'global' mechanical properties of a smart material after homogenisation depend on parameters of these 'hidden motions' (e.g., the frequency and the amplitude) besides 'static' parameters of cell elements (e.g., dimensions of cells, their material properties, etc.).

In recent papers [7, 14, 15] it has been shown that a parametric control introduced as the modulation of the stiffness parameters of a structure in response to specific excitation conditions effectively prevents vibrations with large amplitudes. Due to such a parametric control very small stiffness modulation at a certain frequency extinguishes the resonant behaviour of a structure in the given excitation conditions. In addition, it is shown in [7] that suppression of propagation of flexural waves in an infinitely long plate may also be performed by the same technique.

16.2 The Governing Equations

The governing equations for vibrations of a sandwich plate in cylindrical bending are derived using Hamilton's principle. Details of derivation of a sandwich plate's theory adopted here are available in [6, 11, 13, 14].

An element of the sandwich beam consists of two symmetrical relatively thin, stiff skin plies and a thick, soft core ply. The following non-dimensional parameters are introduced to describe the internal structure of a sandwich plate: $\varepsilon = \dfrac{h_{skin}}{h_{core}}$ as a thickness parameter, $\delta = \dfrac{\rho_{core}}{\rho_{skin}}$ as a density parameter, $\gamma = \dfrac{E_{core}}{E_{skin}}$ as a longitudinal stiffness parameter and $\gamma_g = \dfrac{G_{core}}{G_{skin}}$ as a shear stiffness parameter. Hereafter, subscripts denoting parameters of skin plies are omitted. The deformation of a sandwich beam element is governed by two independent variables: a displacement of the mid-surface of the whole element w (which is the same for all plies), and a shear angle between mid-surfaces of skin plies θ.

The Hamiltonian of a vibrating composite beam (a plate undergoing cylindrical bending) in the absence of external forcing is

$$H = \frac{1}{2}\int_{t_1}^{t_2}\int_0^l \left[m(\dot{w})^2 + I_1(\dot{w}')^2 + I_2\dot{\theta}^2 - D_1\kappa_1^2 - D_2\kappa_2^2 - \Gamma\tau^2\right]dxdt \quad (16.1)$$

Here the first term $m(\dot{w})^2$ is a sum of kinetic energies of all plies in their vertical motion, while the second and the third ones correspond to rotation; the term $I_1(\dot{w}')^2$ is a sum of rotational kinetic energies of all plies in their rotations about their own axes, and the term $I_2(\dot{\theta})^2$ is the rotational kinetic energy generated by sliding (tangential motion) of skin plies. The last three terms formulate the energy of bending of each ply related to curvature of the overall bending, the energy of membrane deformation in skin plies generated by the presence of a shear angle and the energy of longitudinal shear deformation in a core ply, respectively.

The effective elastic parameters are [11]

$$D_1 = \frac{Eh^3}{12(1-v^2)}(2+\frac{\gamma}{\varepsilon^3}), \quad D_2 = \frac{Eh^3}{12(1-v^2)}(1+\frac{1}{\varepsilon})^2$$

$$\Gamma = \frac{Eh}{2(1+v)}(1+\frac{1}{\varepsilon})^2\gamma\varepsilon, \quad m = \rho h(2+\frac{1}{\varepsilon}) \quad (16.2)$$

$$I_1 = \frac{\rho h^3}{12}(2+\frac{\delta}{\varepsilon^3}) \quad I_2 = \frac{\rho h^3}{2}(1+\frac{1}{\varepsilon})^2$$

Here, $E = E_{skin}$, $h = h_{skin}$, $\rho = \rho_{skin}$, the Poisson coefficient v is assumed to be the same for all plies, the moments and forces are related to lateral displacement and shear angle as

$$M_1 = -D_1 w'', \quad M_2 = D_2\theta', \quad Q = \Gamma(\theta + w') \quad (16.3)$$

M_1 is a bending moment of the 'first kind', composed as a sum of bending moments acting in each ply, and produced by normal stresses linearly distributed in each ply in accordance with classic Bernoulli-Euler model applied to each ply individually. These stresses are related to the curvature $\kappa_1 = w''$, which is the same for all plies. M_2 is a bending moment of the 'second kind', generated by the uniform part of normal stresses acting in skin plies. These stresses are produced by the 'pseudo'-curvature $\kappa_2 = \theta'$. Shear force Q in a core ply is uniformly distributed, and it is proportional to the full shear angle in a core ply $\tau = \theta + w'$.

Stationarity conditions for the functional (1) are formulated as equations of motion

$$(D_1 w''(x,t))'' - I_1 \ddot{w}''(x,t) + m\ddot{w}(x,t) - (\Gamma(w'(x,t)+\theta(x,t)))' = q_w(x,t) \quad (16.4a)$$

$$D_2\theta''(x,t) - \Gamma(w'(x,t)+\theta(x,t)) - I_2\ddot{\theta}(x,t) = -q_\theta(x,t) \quad (16.4b)$$

Here q_w is the lateral intensity of distributed force and q_θ is the distributed intensity of shear moment.

The boundary conditions are obtained as non-integral terms in integration of the functional (16.1) by parts. They are:

$$D_1 w'' = 0 \quad \text{or} \quad w' = 0 \quad (16.5a)$$

$$I_1 \ddot{w}' + \Gamma(\theta + w') - D_2 w''' = 0 \quad \text{or} \quad w = 0 \quad (16.5b)$$

$$D_2 \theta' = 0 \quad \text{or} \quad \theta = 0 \quad (16.5c)$$

The first condition (16.5a) is related to the bending moment M_1, composed as a sum of bending moments acting in each ply. Its counterpart formulates the absence of the overall slope. Respectively, the condition (16.5b) formulates the absence of a transverse force at the edge of a beam. Its first term presents a contribution of rotational inertia ('Timoshenko'-type term), the second one accounts for shear interfacial stresses between plies and the last one is relevant to a standard Bernoulli-Euler transverse force. Alternative formulation of this boundary condition implies the absence of lateral displacement at the edge of a beam. The last set of boundary conditions (5c) is introduced by an independent variable of shear angle θ. Either a bending moment M_2, generated by the uniform part of normal stresses acting in skin plies should be absent or the shear angle is equal to zero.

16.3 Modal Analysis of Vibrations

We begin with the analysis of spectra of eigenfrequencies and eigenmodes of a beam with time- and space-independent parameters of inertia and stiffness. It is a necessary preliminary step to be made before looking at possibilities to implement the active control of vibrations by parametric stiffness modulation. The analysis here is restricted by the case of a simply supported beam, when generalized displacements may be expanded into simple series on trigonometric functions. One should expect only quantitative changes in the results obtained for any other types of boundary conditions, whereas constructing solution for a free vibrations problem becomes more cumbersome.

As it has just been discussed, no stiffness modulation is introduced, so that the bending stiffness parameter D_1 and the shear parameter Γ are taken as constants. The boundary conditions of a simple support are

$$D_1 w'' = 0 \quad w = 0 \quad D_2 \theta' = 0 \quad (16.6)$$

The displacements are expanded as

$$w(x,t)=\sum_k A_k \sin\frac{k\pi x}{l}\exp(-i\omega t) \quad \theta(x,t)=\sum_k B_k \cos\frac{k\pi x}{l}\exp(-i\omega t)$$

For each individual number k we obtain a characteristic bi-quadratic equation to find two eigenfrequencies

$$\begin{vmatrix} d_{11} & d_{12} \\ d_{21} & d_{22} \end{vmatrix}=0$$

$$\begin{aligned}d_{11} &= \frac{Eh^3}{12(1-v^2)}\left(2+\frac{\gamma}{\varepsilon^3}\right)\left(\frac{\pi k}{l}\right)^4 + \frac{Eh\varepsilon\gamma}{2(1+v)}\left(1+\frac{1}{\varepsilon}\right)^2\left(\frac{\pi k}{l}\right)^2 \\ &\quad -\frac{\rho h^3}{12}\left(2+\frac{\delta}{\varepsilon^3}\right)\left(\frac{\pi k}{l}\right)^2 \omega_k^2 - \rho h\left(2+\frac{\delta}{\varepsilon}\right)\omega_k^2 \\ d_{12} &= \frac{Eh\varepsilon\gamma}{2(1+v)}\left(1+\frac{1}{\varepsilon}\right)^2\left(\frac{\pi k}{l}\right) \\ d_{21} &= \frac{Eh\varepsilon\gamma}{2(1+v)}\left(\frac{\pi k}{l}\right) \\ d_{22} &= \frac{Eh^3}{12(1-v^2)}\left(\frac{\pi k}{l}\right)^2 + \frac{Eh\varepsilon\gamma}{2(1+v)}\left(\frac{\pi k}{l}\right) - \frac{\rho h}{2}\omega_k^2 \end{aligned} \quad (16.7)$$

A dependence of the first eigenfrequencies $\omega h/c_p$, ($c_p=\sqrt{E/(\rho(1-v^2))}$ is a velocity of sound in plate's material) calculated for $k=1$ (symmetric mode) and $k=2$ (skew-symmetric mode) on the length-to-thickness ratio l/h is shown in Figure 16.1a.

Parameters of the sandwich beam composition are $\varepsilon = 0.05$, $v = 0.3$, $\gamma_0 = 0.01$, $\delta = 0.1$. Respectively, in Figure 16.1b a dependence of the second eigenfrequencies upon l/h is plotted.

In each case, a relevant eigenvector is easily constructed by taking, for example $B_k = 1$ and calculating $A_k = -\frac{d_{12}}{d_{11}} = -\frac{d_{22}}{d_{21}}$ for a given eigenfrequency. In Figure 16.2a,b the amplitude of a non-dimensional lateral displacement A_k is plotted versus l/h for the first and the second eigenfrequency, respectively.

16.3. Modal Analysis of Vibrations

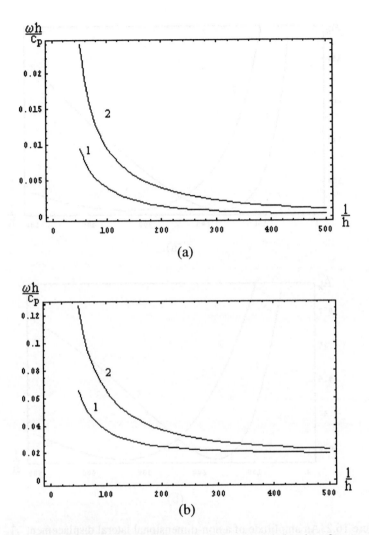

Figure 16.1. A dependence of the first (a) and the second (b) eigenfrequencies $\omega h/c_p$ calculated for $k=1$ (symmetric mode) and $k=2$ (skew-symmetric mode) on the length-to-thickness ratio l/h.

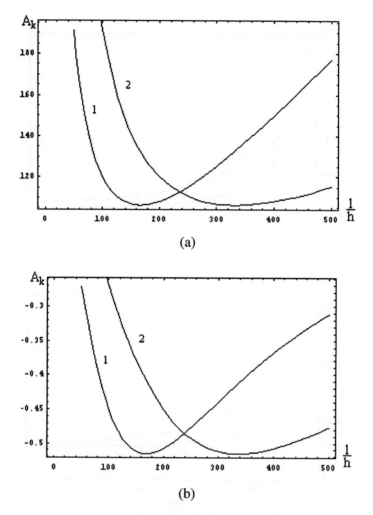

Figure 16.2. An amplitude of a non-dimensional lateral displacement A_k versus l/h for the first (a) and the second (b) eigenfrequency.

These graphs are typical and qualitatively remain unchanged for other combinations of beam's parameters. Two conclusions follow from the analysis of these graphs:

the first (low) spectrum of eigenfrequencies of vibrations is related to the modes of dominantly flexural motions of a plate, whereas the second spectrum of comparatively high eigenfrequencies is composed by dominantly shear modes;

there is a considerable difference in frequencies belonging to these spectra, specifically, the second (skew-symmetric) frequency of a 'shear' spectrum is about 20-25 times higher than the first (symmetric) frequency of a 'flexural' spectrum.

16.4 Direct Partition of Motions[1]

The method of direct partition of motions in its original general formulation is presented in [1, 16]. In recent papers [4, 20] its basic equations are rigorously derived by a use of the method of multiple scales [19, 10]. The derivation of equations of the method of direct partition of motions is briefly outlined here for a sandwich beam with parametric stiffness modulation to give a reader a key idea of this method.

We assume that flexural vibrations are excited by a force acting at comparatively low frequency ω_s:

$$q(x,t) = q_s(x)\cos\omega_s t \qquad (16.8)$$

This driving frequency may be quite close to, say, the first resonant frequency of vibrations. For simplicity, a spatial distribution of the driving force is selected to be fairly close to the shape of the first eigenmode (which is symmetric). Then one should expect large amplitudes of the forced response of a beam to develop.

To avoid large amplitudes of vibrations, a control mechanism based on the parametric stiffness modulation at comparatively high frequency ω_f is suggested. Since the driving frequency cannot be changed, the

---

[1] Here as before in chapter 7, unlike other chapters, the term "Direct partition of motions" is used instead of "Direct separation of motions".

tool for reduction of the amplitude of a forced response should only be a shift of the resonant frequency from the driving one. Such a shift of an eigenfrequency is associated with the concept of 'vibrational rheology' [1,16]. We assume the bending stiffness and the shear stiffness to be decomposed as

$$D_1(x,t) = D_{10} + D_{11}(x)\cos\omega_f t$$
$$\Gamma(x,t) = \Gamma_0 + \Gamma_1(x)\cos\omega_f t \tag{16.9}$$

Here the second terms are associated with some fairly small deviations of the stiffness of a core ply from its nominal value, i.e., the following inequalities are held:

$$D_{11}(x) \ll D_{10}, \quad \Gamma_1(x) \ll \Gamma_0 \tag{16.10}$$

These two inequalities are easily justified from the practical viewpoint: it is unrealistic to produce the stiffness modulation comparable with the nominal value of the stiffness of an uncontrolled beam.

To be in a position to apply the method of direct partition of motions, it is necessary to impose one more inequality

$$\omega_s \ll \omega_f \tag{16.11}$$

that requires a considerable difference between the rate of variability of a stiffness modulation and the rate of variability of a driving force, see [1,16].

As we have two different time scales in the input to the beam, it is reasonable to assume the response to be decomposed as

$$w(x,t) = w_s(x,t) + w_f(x,t)$$
$$\theta(x,t) = \theta_s(x,t) + \theta_f(x,t) \tag{12}$$

Then equations (4) are re-written as (we assume an absence of any shear loading, $q_\theta = 0$)

$$[(D_{10} + D_{11}(x)\cos\omega_f t)(w_s''(x,t) + w_f''(x,t))]'' -$$
$$I_1[\ddot{w}_s''(x,t) + \ddot{w}_f''(x,t)] + m[\ddot{w}_s(x,t) + \ddot{w}_f(x,t)] -$$
$$[(\Gamma_0 + \Gamma_1(x)\cos\omega_f t)(w_s'(x,t) + w_f'(x,t) + \theta_s(x,t) + \theta_f(x,t))]'$$
$$= q_s(x)\cos\omega_s t \tag{16.13a}$$

16.4. Direct Partition of Motion

$$D_2[\theta_s''(x,t)+\theta_f''(x,t)] - I_2[\ddot{\theta}_s(x,t)+\ddot{\theta}_f(x,t)] -$$
$$[(\Gamma_0 + \Gamma_1(x,t)\cos\omega_f t)(w_s'(x,t)+w_f'(x,t)+\theta_s(x,t)+\theta_f(x,t))] = 0 \qquad (16.13b)$$

To solve this system of partial differential equations, we use condition (11) and formulate 'fast' components as those depending upon two different time scales

$$w_f(x,t) = \overline{w}_f(x,t)\cos\omega_f t$$
$$\theta_f(x,t) = \overline{\theta}_f(x,t)\cos\omega_f t \qquad (16.14a)$$

In a sense, this assumption is equivalent to the approach conveniently used in the method of multiple scales, when several time variables are introduced to describe the motion of a system, i.e., $w(x,t) = w_0(x,T_0,T_1,...) + \varepsilon w_1(x,T_0,T_1,...) + ...$ with $T_0 = t$, $T_1 = \varepsilon t,...$. The function $\cos\omega_f t$ displays fast oscillations whilst the functions $\overline{w}_f(x,t)$, $\overline{\theta}_f(x,t)$ are slowly modulated amplitudes of these oscillations.

We use inequality (11) to formulate the inertial terms in equations (13) as

$$\ddot{w}_f(x,t) = -\omega_f^2 \overline{w}_f(x,t)\cos\omega_f t$$
$$\ddot{\theta}_f(x,t) = -\omega_f^2 \overline{\theta}_f(x,t)\cos\omega_f t \qquad (14b)$$

Here it is assumed that derivatives of the function $\overline{w}_f(x,t)$ are asymptotically small as compared with the right hand sides of equation (14b).

Then, collecting terms proportional to $\cos\omega_f t$ we obtain the following equations

$$D_{10}\overline{w}_f^{(4)}(x,t) + \left[D_{11}(x)w_s''(x,t)\right]'' + \omega_f^2 I_1 \overline{w}_f''(x,t) - \omega_f^2 m \overline{w}_f(x,t)$$
$$-\Gamma_0\left(\overline{w}_f''(x,t)+\overline{\theta}_f'(x,t)\right) - \left[\Gamma_1(x)(w_s'(x,t)+\theta_s(x,t))\right]' = 0$$
$$D_2\overline{\theta}_f''(x,t) + \omega_f^2 I_2 \overline{\theta}_f(x,t) - \Gamma_0\left(\overline{w}_f'(x,t)+\overline{\theta}_f(x,t)\right)$$
$$-\left[\Gamma_1(x)(\overline{w}_s'(x,t)+\overline{\theta}_s(x,t))\right] = 0 \qquad (16.15)$$

These equations are subtracted from the original equations (4). Then in accordance with the procedure of direct partition of motions [1] averaging over period of fast motions is performed that results in formulating so-called vibrational forces (terms in equations (16.16), proportional to w_f and θ_f).

$$D_0 w_s^{(4)}(x,t) + \frac{1}{2}\left[D_{11}(x)\overline{w}_f''(x,t)\right]'' - I_1 \ddot{w}_s''(x,t) + m\ddot{w}_s(x,t)$$
$$-\Gamma_0\left(w_s''(x,t) + \theta_s'(x,t)\right) - \frac{1}{2}\left[\Gamma_1(x)\left(\overline{w}_f'(x,t) + \overline{\theta}_f(x,t)\right)\right]' = q_s(x)\cos\omega_s t$$
$$D_2 \theta_s''(x,t) - I_2 \ddot{\theta}_s(x,t) - \Gamma_0\left(w_s'(x,t) + \theta_s(x,t)\right) \qquad (16.16)$$
$$-\frac{1}{2}\Gamma_1(x)\left(\overline{w}_f'(x,t) + \overline{\theta}_f(x,t)\right) = 0$$

The equations (16.16) are linear with respect to displacements and shear angles, so the forced response is now sought as

$$w_s(x,t) = \hat{w}_s(x)\cos\omega_s t$$
$$\theta_s(x,t) = \hat{\theta}_s(x)\cos\omega_s t$$

Respectively, due to the linearity of equations (16.16), the functions defining slow amplitude modulations of 'fast' motion become

$$\overline{w}_f(x,t) = \hat{w}_f(x)\cos\omega_s t$$
$$\overline{\theta}_f(x,t) = \hat{\theta}_f(x)\cos\omega_s t$$

Thus, we substitute these formulas into equations (16.15–16.16), omit time dependence expressed as $\cos\omega_s t$ and obtain a system of two ordinary differential equations for 'fast' (occurring at the frequency ω_f) motions of a sandwich beam

$$D_{10}\hat{w}_f^{(4)}(x) + \left[D_{11}(x)\hat{w}_s''(x)\right]'' + \omega_f^2 I_1 \hat{w}_f''(x) - \omega_f^2 m \hat{w}_f(x)$$
$$-\Gamma_0\left(\hat{w}_f''(x) + \hat{\theta}_f'(x)\right) - \left[\Gamma_1(x)\left(\hat{w}_s'(x) + \hat{\theta}_s(x)\right)\right]' = 0$$
$$D_2 \hat{\theta}_f''(x) + \omega_f^2 I_2 \hat{\theta}_f(x) - \Gamma_0\left(\hat{w}_f'(x) + \hat{\theta}_f(x)\right) \qquad (16.17)$$
$$-\left[\Gamma_1(x)\left(\hat{w}_s'(x) + \hat{\theta}_s(x)\right)\right] = 0$$

coupled with a system of equations for 'slow' (occurring at the frequency ω_s) motions

$$D_{10}\hat{w}_s^{(4)}(x,t) + \frac{1}{2}\left[D_{11}(x)\hat{w}_f''(x)\right]'' + \omega_s^2 I_1 \hat{w}_s''(x) - \omega_s^2 m\hat{w}_s(x)$$
$$-\Gamma_0\left(\hat{w}_s''(x) + \hat{\theta}_s'(x)\right) - \frac{1}{2}\left[\Gamma_1(x)\left(\hat{w}_f'(x) + \hat{\theta}_f(x)\right)\right]' = q_s(x) \quad (16.18)$$
$$D_2 \hat{\theta}_s''(x) + \omega_s^2 I_2 \hat{\theta}_s(x) - \Gamma_0\left(\hat{w}_s'(x) + \hat{\theta}_s(x)\right)$$
$$-\frac{1}{2}\Gamma_1(x)\left[\hat{w}_f'(x) + \hat{\theta}_f(x)\right] = 0$$

The coupling of 'slow' and 'fast' motions is generated by the variable parts of bending and shear stiffness. Thus, as expected if $D_{11}(x) = 0$ and $\Gamma_1(x) = 0$, then forced vibrations at the frequency ω_s become uncontrolled by the stiffness modulation at the frequency ω_f, see equations (16.17–16.18).

The implementation of a stiffness modulation for a beam of the sandwich composition will be discussed in Section 8. We just note here, that at the micro-level, a kind of two-phase micro-structure should be created with one of the phases able to perform 'hidden' motions with very small amplitudes. At macro-level, such a two-phase core material is treated as a homogeneous one having an effective averaged stiffness. The motions of elements of micro-structure produce some fluctuations of the effective stiffness of a core material, so that parameters D_1 and Γ are presented as

$$D_1 = \frac{Eh^3}{12(1-v^2)}\left[2 + \frac{\gamma_0 + \gamma_1(x)\cos\omega_f t}{\varepsilon^3}\right]$$

$$\Gamma = \frac{Eh}{2(1+v)}(1+\frac{1}{\varepsilon})^2 \varepsilon\left[\gamma_0 + \gamma_1(x)\cos\omega_f t\right]$$

The spatial dependence of the function $\gamma_1(x)$ may be selected rather arbitrary. Respectively, various combinations of interacting modes may be analyzed. The particular case when the stiffness modulation creates modal interaction of 'slow' symmetric resonant motions and 'fast' skew-symmetric resonant motions is considered.

These conclusions are important from the viewpoint of implementing the resonant parametric stiffness modulation on forced vibrations of a beam.

16.5 The Eigenvalue Problem for a Beam with the Resonant Parametric Stiffness Modulation

We use a two-terms approximation of a lateral displacement and a shear angle for the sandwich structure. The spatial dependence of the functions describing the 'fast' and the 'slow' vibrations is (here a non-dimensional co-ordinate $\xi = \dfrac{x}{h}$ is introduced)

$$\begin{aligned}
\hat{w}_f(\xi) &= w_{f1} \sin(\pi \tfrac{h}{l}\xi) + w_{f2} \sin(2\pi \tfrac{h}{l}\xi) \\
\hat{\theta}_f(\xi) &= \theta_{f1} \cos(\pi \tfrac{h}{l}\xi) + \theta_{f2} \cos(2\pi \tfrac{h}{l}\xi) \\
\hat{w}_s(\xi) &= w_{s1} \sin(\pi \tfrac{h}{l}\xi) + w_{s2} \sin(2\pi \tfrac{h}{l}\xi) \\
\hat{\theta}_s(\xi) &= \theta_{s1} \cos(\pi \tfrac{h}{l}\xi) + \theta_{s2} \cos(2\pi \tfrac{h}{l}\xi)
\end{aligned} \qquad (16.19)$$

These functions exactly satisfy boundary conditions (6) at the edges of a beam. A skew-symmetric (with respect to a mid-section of a beam) distribution of stiffness modulation is considered

$$\gamma_1(\xi) = \gamma_{10} \cos \pi \tfrac{h}{l} \xi \qquad (16.20)$$

Here free vibrations are considered, so the amplitude of a driving force in (18) is put to zero. Galerkin's orthogonalization of differential equations (17-18) to each of selected trial functions results in a system of eight homogeneous algebraic equations with respect to amplitudes $w_{s1}, w_{s2},\ldots$ This system of algebraic equations is split into two subsystems. The first one couples symmetric 'slow' motions (w_{s1}, θ_{s1}) to skew-symmetric 'fast' motions (w_{f2}, θ_{f2}). The second one couples skew-symmetric 'slow' motions (w_{s2}, θ_{s2}) to symmetric 'fast' motions (w_{f1}, θ_{f1}). We consider the coupling between symmetric 'slow' motions and skew-symmetric 'fast' motions because low-frequency symmetric vibrations of a baffled plate produce more efficient sound radiation than skew-symmetric vibrations [2] and from the practical viewpoint it is more important to control symmetric vibrations.

The problem of free vibrations of a beam having modulated stiffness is formulated provided that a modulation frequency ω_f and a modulation amplitude γ_{10} are treated as parameters of vibrational rheology [16], i.e., these parameters are variable. Then we may analyse a dependence of the slow resonant frequency upon both of them. The slow resonant frequencies are sought by putting to zero the following determinant

$$\begin{vmatrix} a_{11} - k_{11}\omega_{0s}^2 & a_{12} & \gamma_{10}a_{13} & \gamma_{10}a_{14} \\ a_{21} & a_{22} - k_{22}\omega_{0s}^2 & \gamma_{10}a_{23} & \gamma_{10}a_{24} \\ \gamma_{10}a_{31} & \gamma_{10}a_{32} & a_{33} - k_{33}\omega_f^2 & a_{34} \\ \gamma_{10}a_{41} & \gamma_{10}a_{42} & a_{43} & a_{44} - k_{44}\omega_f^2 \end{vmatrix} = 0 \qquad (16.21)$$

Explicit formulas for elements of this determinant are rather cumbersome and they are not displayed here. They are easily obtained through Galerkin's ortogonalization performed in two systems of differential equations (16.17–16.18) with trial functions (16.19).

Apparently, if $\gamma_{10} = 0$, then slow resonant frequencies are those of symmetric vibrations of a beam with the constant stiffness and there is no modal coupling. The stiffness modulation should be performed at the frequency fairly close to the eigenfrequency of the fast motions of a plate, $\omega_f = \mu\omega_{0f}$, with μ as a detuning parameter close to unit and ω_{0f} as the eigenfrequency of skew-symmetric vibrations of a beam. It should be pointed out that to use equations (16.17–16.18) derived by the method of direct partition of motions it is necessary to make sure that 'fast' modulation frequency is indeed much higher than a frequency of controlled 'slow' motions. This condition is actually held only if stiffness modulation is performed at the second eigenfrequency of skew-symmetric motion. As it has been discussed earlier, this second frequency is relevant to dominantly shear motions.

It is convenient to expand the determinant (16.21) as

$$\chi\omega_{0s}^4 + (\gamma_{10}^2 A_1 + \chi B_1)\omega_{0s}^2 + \chi B_0 + \gamma_{10}^2 A_0 + \gamma_{10}^4 A_2 = 0$$

Here another small parameter χ is introduced as

$$\chi = \begin{vmatrix} a_{33} - k_{33}\omega_f^2 & a_{34} \\ a_{43} & a_{44} - k_{44}\omega_f^2 \end{vmatrix} \approx (\mu - 1)\left[2k_{33}k_{44}\omega_{0f}^4 - (a_{44}k_{33} + a_{33}k_{44})\omega_{0f}^2\right]$$

since the modulation frequency is selected to be fairly close to the fast resonant one.

Parameters A_0, A_1, A_2, B_0, B_1, are expressed via components of the determinant (16.21). They are rather cumbersome and therefore we do not present them here.

Equation (22) contains two small parameters χ and γ_{10} which are independent of each other. However, it is convenient to find roots of this equation if we assume a certain link between them to be held $\chi = \sigma^k$ and $\gamma_{10}^2 = \sigma^n$, with σ selected as a formal small parameter. Then two principal cases should be distinguished: $n > k$ and $n < k$. The former one is actually related to the absence of vibration control. In such a case, two frequencies are obtained as

$$\omega_{0s} = \sqrt{-\frac{B_1}{2} \pm \sqrt{\left(\frac{B_1}{2}\right)^2 - B_0}} \qquad (16.22)$$

and they are exactly the same as the ones found in Section 3.

The second case is relevant to the precise tuning of the stiffness modulation with the fast resonant frequency. It is remarkable, that in this limit, asymptotic analysis shows that there is only one eigenfrequency, defined by a simple formula

$$\omega_{0s} = \sqrt{\frac{A_0}{A_1}} \qquad (16.23)$$

16.5. The Eigenvalue Problem for a Beam

Finally, in a particular case of $n = k$ we get

$$\omega_{0s} = \sqrt{-\frac{B_1 + A_1}{2} \pm \sqrt{\left(\frac{B_1 + A_1}{2}\right)^2 - A_0 - B_0}} \qquad (16.24)$$

In Figure 16.3, curves 1 and 3 display the dependence of the first and the second resonant frequencies of an uncontrolled beam given by equation (16.23) upon the parameter γ_0, respectively, whereas curve 2 is plotted after formula (24), i.e. for a perfectly tuned resonant stiffness modulations.

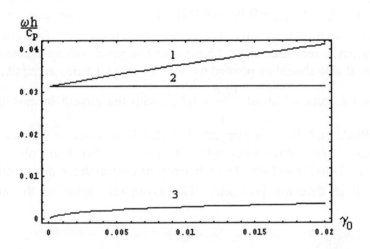

Figure 16.3. The first (curve 1) and the second (curve 3) eigenfrequencies of an uncontrolled beam and the eigenfrequency of a beam with perfectly tuned resonant stiffness modulations (curve 2) versus the parameter γ_0

The parameters of sandwich composition are selected as $\varepsilon = 0.05$, $v = 0.3$, $h/l = 0.01$, $\delta = 0.1$. As it follows from these graphs, the parametric stiffness modulation results in significant increase in the resonant frequency of the controlled beam. It is remarkable, that a controlled frequency becomes much closer to the second eigenfrequency of an uncontrolled beam, than to the first one. Another feature clearly displayed in Figure 16.3 is a very weak dependence of the controlled eigenfrequency upon parameter γ_0 as compared with the same dependence of the second eigenfrequency of an uncontrolled beam. This Figure is representative also for other combinations of sandwich parameters.

In Figure 16.4a, a dependence of the first 'slow' resonant frequency of a beam upon the parameter of stiffness modulation γ_{10} is shown for the same set of parameters of sandwich composition.

Curve 1 is plotted for detuning parameter $\chi = 1.0001$, i.e., for nearly perfectly tuned stiffness modulations. It is clear, that in these excitation conditions, a very small amplitude of stiffness is sufficient to produce a substantial increase in the first resonant frequency. Specifically, if $\gamma_{10} = 0.1\gamma_0 = 0.001$, i.e. the variable part of core material's stiffness is only 10% of its constant part, the first resonant frequency is increased about 2.5 times as compared with an uncontrolled beam. It also should be pointed out that the curve 1 tends asymptotically toward a value of about $\dfrac{\omega_r h}{c_p} \approx 0.012$ with the growth in modulation amplitude and the saturation effect is clearly observed at not too large values of modulation amplitude. The curves 2 and 3 are plotted for $\chi = 1.01$ and $\chi = 1.05$. The efficiency of control drops down with the growth in detuning parameter. The asymptotic value of the eigen-

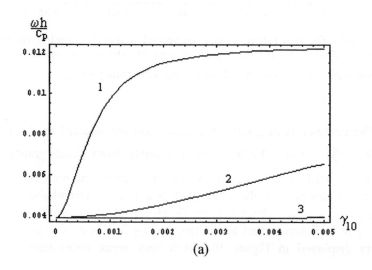

(a)

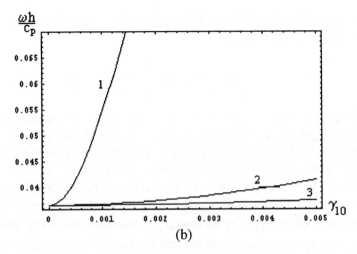

(b)

Figure 16. A dependence of the first (a) and the second (b) resonant frequencies of a beam upon the parameter of stiffness modulation γ_{10}

frequency of a beam with stiffness modulation has appeared to be the same for all the three detuning parameters (it is not shown in Figure 16.4a because too large modulation amplitudes are necessary to reach it). It should be pointed out that, for example, in the case of $\chi = 1.01$, to avoid resonant excitation it is necessary to introduce stiffness modulation of $\gamma_{10} = 0.002$, i.e. 20% of the constant part of core stiffness to have the resonant frequency increased from $\dfrac{\omega_r h}{c_p} \approx 0.0039$ to $\dfrac{\omega_r h}{c_p} \approx 0.0054$, i.e., in about 1.5 times. In Figure 16.4b, the similar graphs are plotted for the second slow resonant frequency.

Qualitatively, the role of the amplitude of stiffness modulation is the same as for the first frequency. However, it is probably not of much practical relevance since the second slow resonant frequency is rather high and the applicability conditions for the method of direct partition of motions are not fully fulfilled.

16.6 Forced Vibrations. An Influence of Internal Damping

The above analysis has addressed free vibrations of an undamped beam. Now we proceed to forced vibrations of a beam with some material losses. A spatial distribution of the driving force is taken as symmetric with respect to its mid-section

$$q_s(\xi) = Q_0 \sin\left(\pi \frac{h}{l} \xi\right) \qquad (16.25)$$

To find a forced response in such a case, an inhomogeneous system of linear algebraic equations is set up. Its left-hand side is composed by the matrix (16.21) and a right hand side is generated by orthogonalization of the spatial distribution (16.25) of a driving load to selected trial functions (16.19). A presence of material losses means that for a case of stationary vibrations a real-valued Young's module E is replaced by its counterpart $E(1-i\eta)$. The parameter η characterizes energy dissipation in stationary vibrations of a beam and for simplicity it is taken to be the same for the material of core ply and skin plies. In reality, material losses are normally associated with energy dissipation mostly in the core material, whilst energy dissipation in stiff, thin skin plies is much smaller. Although this does not complicate much the analysis, in the present paper we assume the parameter η to be some averaged material losses coefficient and do not pursue this subject any further. Thus, all the stiffness-dependent elements of the matrix (16.21) are multiplied by a complex-valued quantity $(1-i\eta)$.

In Figure 16.5, a dependence of amplitude of vibrations at the mid-span of a beam upon frequency parameter $\dfrac{\omega h}{c_p}$ of excitation is presented for perfectly tuned resonant parametric stiffness modulation and four values of internal damping parameter.

Curves 1-4 is plotted at $\eta = 0.001$, $\eta = 0.0001$, $\eta = 0.00001$ and $\eta = 0$, respectively. Parameters of the sandwich beam composition are $\varepsilon = 0.05$, $\nu = 0.3$, $h/l = 0.01$, $\gamma_0 = 0.01$, $\delta = 0.1$. We conclude from these graphs, that an influence of material losses upon forced vibrations of a sandwich beam with parametric stiffness modulations is two-folded. Firstly, similarly to a standard case of forced vibrations of an uncontrolled structure, material losses reduce amplitude of vibrations nearby resonant excitation and they produce a phase shift between displacement and a driving force. This is a 'trivial' effect of an internal damping.

16.6. Forced Vibrations

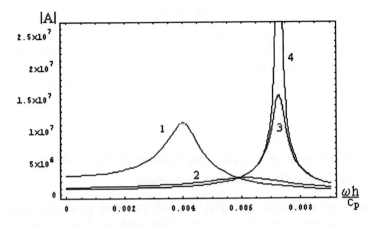

Figure 16.5. A dependence of amplitude of vibrations at the mid-span of a beam upon frequency parameter $\omega h / c_p$ of excitation

However, there is also another aspect of the presence of material losses. As discussed, to maximize an efficiency of parametric stiffness modulation, its frequency should be as close as possible to an eigenfrequency of shear skew-symmetric vibrations of a beam. The latter is detected by putting to zero determinant (16.7) and it becomes complex-valued as soon as $\eta \neq 0$. If a modulation of stiffness is initiated, then its time dependence is purely periodic, i.e. it is expressed by a real-valued frequency. Thus, it occurs somewhat out of tune with the exact value of an eigenfrequency of a beam with material losses and the efficiency of vibration suppression due to modal interaction drops down.

Both these aspects are easily recognized by inspection into graph displayed in Figure 16.5. As material losses are rather large, internal damping makes resonant amplitude smaller without shifting a resonant frequency due to stiffness modulation, see curve 1. If material losses tends to zero, then controlled resonant frequency tends to its comparatively large value determined by an amplitude of stiffness modulation, but a value of resonant amplitude in the case $\eta = 0$ tends to infinity, see curve 4. In this situation, we obtain suppression of vibrations at the 'former' resonant frequency (that actually has been a purpose of stiffness modulation), but a new higher resonant frequency is then created. The curve 3 plotted for $\eta = 0.00001$ suggests a somewhat better vibrations suppression than in the case of the absence of material losses because resonant frequency is still shifted from a frequency of an uncontrolled beam, but amplitude of vibrations and this 'new' frequency is bounded. Finally, as shown by curve 2, there exists a certain 'optimal damping' situation, when both aspects of a presence of damping 'smoothen out' resonant oscillations in the whole frequency range.

16.7 Vibrations of a Beam with Parametrically Modulated Stiffness in Heavy Fluid Loading Conditions

Control of vibrations of a fluid-loaded structure is an important issue because suppression of noise, vibrations and harshness is necessary in almost every engineering application. In literature on structure-borne sound and vibrations, this complicated problem is treated in an uncoupled formulation, i.e., when analysis and control of forced vibrations is performed for a structure with no 'feedback' from an acoustic medium. Such an approach is applicable in the so-called 'light' fluid loading conditions [2, 22]. Although analysis of radiated sound field from controlled and uncontrolled vibrating structure in such a formulation is straightforward and easy to be implemented into, say, some optimization algorithm, it could be dangerous to rely upon results obtained with a use of this theory in any case of acoustic loading. Specifically, modern composite materials are characterized as essentially light-weighted ones [13], so it is reasonable to suggest that their interaction with an acoustic medium should be formulated as structural dynamics under heavy fluid loading.

The analysis of vibrations of a fluid-loaded beam is performed here in a consistent coupled formulation, when equations of vibrations of a structure are solved simultaneously with equations of motions of an acoustic medium. The derivation of governing equations for a fluid-loaded structure by the method of direct partition of motions is not affected by the presence of an acoustic medium. In the case of a baffled plate, the stationary fluid response is given by Rayleigh integral:

– for the slow component of contact acoustic pressure

$$p_s(x) = \frac{i\rho_0 \omega_s^2}{2} \int_0^l \hat{w}_s(\xi) H_0^{(1)}(\frac{\omega_s}{c_0}|x-\xi|)d\xi \qquad (16.26a)$$

16.7. Vibrations of a Beam with Parametrically Modulated Stiffness

– for the fast component

$$p_f(x) = \frac{i\rho_0 \omega_f^2}{2} \int_0^l \hat{w}_f(\xi) H_0^{(1)}(\frac{\omega_f}{c_0}|x-\xi|)d\xi \qquad (16.26b)$$

These formulas are substituted into equations (16.17–16.18), and functions $\hat{w}_s(x)$, $\hat{w}_f(x)$ are sought in the form of two-term expansions (16.19). The system of linear algebraic equations that has been set up for analysis of an uncoupled problem is now modified by adding convolutions

$$\int_0^{1/h} \sin\left(k\kappa \frac{h}{l}\xi\right) H_0^{(1)}\left(\frac{\omega}{c_0} \frac{h}{l}|x-\xi|\right) d\xi, \quad k=1,2$$

to the matrix elements a_{11}, a_{33}.

No viscosity of fluid is taken into account for, thus fluid does not produce any shear loading in the equation (16.4b). Then the other diagonal terms in the matrix (16.21) are not affected by the presence of an acoustic medium. The real part of convolutions (16.26a,b) is relevant to an added mass effect whereas the imaginary part corresponds to a radiation damping. The roles of these two components are easily compared by calculating eigenfrequencies of vibrations of a beam with and without fluid loading.

A sandwich beam having steel skin plies in water is considered, so that density ratio is $\frac{\rho_0}{\rho_p} = 0.128$ and sound velocity ratio is $\frac{c_0}{c_p} = 0.307$.

In Figure 16.6a the first resonant frequency of 'slow' symmetric vibrations (strictly speaking, its real part) is plotted versus thickness parameter h/l with (curve 1) and without (curve 2) fluid loading. Parameters of sandwich composition are $\varepsilon = 0.05$, $v = 0.3$, $\gamma_0 = 0.01$, $\delta = 0.1$. Similarly to the case of vibrations of a beam with material losses, an eigenfrequency of a fluid-loaded beam contains both the real and the imaginary parts. It should be pointed out that a resonant frequency (which is identified as a real number at which maximum of the amplitude of a forced response is reached) is fairly close to the real part of an eigenfrequency. In Figure 16.6b, the second resonant frequency

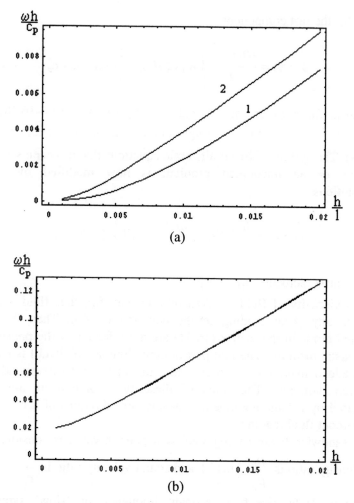

Figure 16.6. The first (a) and the second (b) resonant frequencies of 'slow' symmetric vibrations versus thickness parameter h/l with (curve 1) and without (curve 2) fluid loading

(the resonant frequency of 'fast' skew-symmetric vibrations) is plotted for both the fluid loading case and the case of the absence of a fluid loading.

It is remarkable, that these curves are hardly distinguished, and therefore they are not labeled. This effect is explained by a modal shape of vibration at this eigenfrequency. As it is shown in Section 16.3 (see

Figure 16.2b), this is a dominantly shear mode, and sliding motion of the surface of a plate is not coupled with motions of an acoustic medium. This is an important feature of dynamics of composite beams of sandwich structure, which should be utilized for possible applications. Although the resonant frequency of slow flexural motions (that is subjected to control by the parametric stiffness modulation) changes depending upon various fluid loading conditions, the frequency at which an active control should be performed is invariant to these conditions.

In Figure 16.7, a dependence of the amplitude of forced vibrations upon excitation frequency is plotted for a sandwich beam submerged into water. Parameters of sandwich composition are the same as the ones for Figure 16.6.

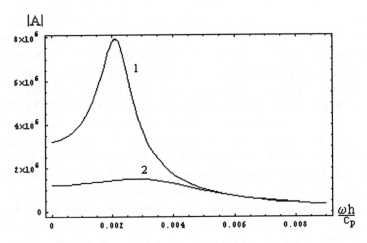

Figure 16.7. A dependence of the amplitude of forced vibrations upon excitation frequency for a sandwich beam submerged into water.

Curve 1 is relevant to the absence of stiffness modulation, curve 2 is plotted for $\eta = 0.0001$ and $\gamma_{10} = 0.2\gamma_0$. Perfectly tuned parametric stiffness modulation is considered, i.e. modulation frequency is taken to be equal 'fast' resonant frequency found from the graph in Figure 16.6b. A peak of the curve 2 is shifted to the left from the resonant peak of an uncontrolled beam (curve 1), but not as much as in Figure 16.5. This is a typical tendency also detected for other sets of parameters of a sandwich composition. Thus, an efficiency of control is less for a fluid loaded beam than in the case of vibrations in vacuum. However, we still get a favorable reduction approximately in 7 times in amplitude of vibrations at the 'former' resonant frequency, whereas amplitude of forced vibrations at the shifted resonant frequency is in around 3.5 times smaller than in an uncontrolled case, see Figure 16.7. Several computations for other sandwich composition parameters give qualitatively the same results. Those are not displayed here for brevity reasons.

16.8 Discussion of the Parametric Stiffness Modulation

As is shown, the parametric stiffness modulation may be used as a tool to control resonant vibrations of sandwich plates. However, before summing up the results reported in the previous Sections it is appropriate to formulate some remarks concerning several aspects of this way of an active control of vibrations and the method used to solve this problem.

So far, we have not discussed possible types of a micro-structure, which may be suggested to provide the necessary stiffness modulations. This aspect (which is very important and possibly crucial for the practical applicability of this method) has been tackled in [7, 14, 15]. As mentioned in [7], the effective stiffness of the plate can be changed by varying the internal structure of the core material or by having multiple pin-like inclusions whose orientation may be varied. Switched elements [21] based on piezoceramic micro-actuators [23] shape memory alloys, etc., also offer a wide opportunity for distributed stiffness control. The microscopic elements made of a shape memory alloy and embedded into the core material may be initially curved to produce certain 'basic' stiffness. Then they may be heated or cooled to modify their shape and therefore to modify their contribution to a bending stiffness. In initial configuration, these elements could create a regular honeycomb-like structure that is controlled in space and time domains. We consider control of vibrations at comparatively low frequencies of a structure and wavelengths pronouncedly exceeding the characteristic size of a microstructure. The frequency of the modulatory signal is markedly higher, because it is associated with rather small deviations of initial stiffness parameters. However, from the practical viewpoint, the necessary modulation frequency is not too high (up to 200 KHz) and lies within the working frequency range of piezoelectric actuators, which may be inserted into the microinhomogeneous material.

16.9. A Modal Formulation of the Control Problem

Apparently, an active control of any kind implies the additional energy input into the structure. Then it is also of high relevance to compare this input with, for example, a decrease in the energy of vibrating structure due to a decrease in forced amplitudes of vibrations. Of course, such a comparison may make sense only when a particular mechanism of the stiffness modulation at the micro-level is specified. Although this important aspect has not been tackled in this paper, we note that it has been thoroughly analysed in the related paper [15] for a honeycomb beam (a plate in one-dimensional cylindrical bending). Results reported in this paper with respect to this energy balance show that for precise tuning, the energy input is much less, than reduction in the energy of slow vibrations, so that this mechanism of suppression of vibrations is efficient.

In the paper [15], direct numerical integration of equations similar to (16.17), (16.21) has been made and steady-state amplitudes of forced vibrations of a controlled honeycomb beam have coincided with the predictions obtained by the method of direct partition of motions. The results of such a comparison in the case considered here are similar and we do not report them here in details for brevity.

It should also be pointed out that the suggested control mechanism is applicable for a sandwich rectangular plate with any boundary conditions. The structure of frequency spectra found in Section 3 for Navier conditions remains the same for other boundary conditions. Thus, the eigenfrequencies of dominantly shear vibrations are markedly higher than the first eigenfrequency of dominantly flexural vibrations and the 'fast' stiffness modulations may control 'slow' forced resonant motions. Respectively, the selection of a spatial distribution of the stiffness modulation should be adjusted at each particular case to provide a necessary link between two interacting modes.

16.9 A Modal Formulation of the Control Problem for a Sandwich Beam

We consider a simply supported sandwich beam. The boundary conditions (16.5) are satisfied when

$$w(x,t) = \sum_k A_k(t) \sin \frac{k\pi x}{l},$$

16. The Active Control of Vibrations of Composite Beams

We retain the first two terms in each of these expansions,

$$w(x,t) = A_1(t)\sin\frac{\pi x}{l} + A_2(t)\sin\frac{2\pi x}{l}$$
$$\theta(x,t) = B_1(t)\cos\frac{\pi x}{l} + B_2(t)\cos\frac{2\pi x}{l}, \quad (16.27)$$

and apply Galerkin's method to obtain the following system of four ordinary differential equations (here $q_w(x,t) = q\cos\omega_q t$)

$$\left[m + I_1\left(\frac{\pi}{l}\right)^2\right]\ddot{A}_1 + \left[D_{10}\left(\frac{\pi}{l}\right)^4 + \Gamma_0\left(\frac{\pi}{l}\right)^2\right]A_1 + \frac{\pi}{l}\Gamma_0 B_1$$
$$+ \gamma\cos\omega_m t\left\{\frac{1}{2}\left[4D_{11}\left(\frac{\pi}{l}\right)^4 + 2\Gamma_{11}\left(\frac{\pi}{l}\right)^2\right]A_2 + \frac{1}{2}\Gamma_{11}\frac{\pi}{l}B_2\right\} = \frac{2q}{l}\cos\omega_q t$$

(16.28a)

$$\gamma\cos\omega_m t\left[\frac{1}{2}\Gamma_{11}\frac{2\pi}{l}A_2 + \frac{1}{2}\Gamma_{11}B_2\right] + I_2\ddot{B}_1$$
$$+ \left[D_2\left(\frac{\pi}{l}\right)^2 + \Gamma_0\right]B_1 + \frac{\pi}{l}\Gamma_0 A_1 =$$

(16.28b)

$$\left[m + I_1\left(\frac{2\pi}{l}\right)^2\right]\ddot{A}_2 + \left[D_{10}\left(\frac{2\pi}{l}\right)^4 + \Gamma_0\left(\frac{2\pi}{l}\right)^2\right]A_2 + \frac{2\pi}{l}\Gamma_0 B_2 +$$
$$\gamma\cos\omega_m t\left\{\frac{1}{2}\left[4D_{11}\left(\frac{\pi}{l}\right)^4 + 2\Gamma_{11}\left(\frac{\pi}{l}\right)^2\right]A_1 + \frac{1}{2}\Gamma_{11}\frac{2\pi}{l}B_1\right\} = 0$$

(16.28c)

$$\gamma\cos\omega_m t\left[\frac{1}{2}\Gamma_{11}\frac{\pi}{l}A_1 + \frac{1}{2}\Gamma_{11}B_1\right] + I_2\ddot{B}_2$$
$$+ \left[D_2\left(\frac{2\pi}{l}\right)^2 + \Gamma_0\right]B_2 + \frac{2\pi}{l}\Gamma_0 A_2 =$$

(16.28d)

16.9. A Modal Formulation of the Control Problem

These equations are written in 'physical variables', i.e., with respect to amplitudes of lateral and shear vibrations. It is more convenient to transform them to modal equations by letting

$$A_1(t) = A_{11}(t) + A_{12}(t), \qquad A_2(t) = A_{21}(t) + A_{22}(t)$$
$$B_1(t) = \mu_{11}A_{11}(t) + \mu_{12}A_{12}(t), \qquad B_2(t) = \mu_{21}A_{21}(t) + \mu_{22}A_{22}(t) \qquad (16.29)$$

Here μ_{jk}, $j,k = 1,2$ are the modal coefficients defined at $\gamma = 0$,

$$\mu_{jk} = -\frac{\left(j\pi/l\right)\Gamma_0}{D_2\left(j\pi/l\right)^2 + \Gamma_0 - \omega_{jk}^2 I_2}$$

$$\equiv -\frac{D_{10}\left(j\pi/l\right)^4 + \Gamma_0\left(j\pi/l\right)^2 - \omega_{jk}^2\left[m + I_1\left(j\pi/l\right)^2\right]}{\left(j\pi/l\right)\Gamma_0}$$

In these formulas, ω_{jk} are the roots of a characteristic equation

$$\left[D_2\left(\frac{j\pi}{l}\right)^2 + \Gamma_0 - \omega_{jk}^2 I_2\right]$$
$$\times\left\{D_{10}\left(j\pi/l\right)^4 + \Gamma_0\left(j\pi/l\right)^2 - \omega_{jk}^2\left[m + I_1\left(j\pi/l\right)^2\right]\right\} - \left(j\pi/l\right)^2\Gamma_0^2 = 0$$

If the original variables in equations (16.28a) and (16.28c) are replaced by a use of formulas (16.29), we get

$$\left[m + I_1\left(\frac{\pi}{l}\right)^2\right](\ddot{A}_{11} + \ddot{A}_{12}) + \left[m + I_1\left(\frac{\pi}{l}\right)^2\right]\left(\omega_{11}^2 A_{11} + \omega_{12}^2 A_{12}\right)$$

$$+\gamma\cos\omega_m t\left[\left(\frac{1}{2}D_{11}\frac{4\pi^4}{l^4} + \frac{1}{2}\Gamma_{11}\frac{2\pi^2}{l^2} + \frac{1}{2}\Gamma_{11}\mu_{21}\frac{\pi}{l}\right)A_{21}\right.$$

$$\left.+\left(\frac{1}{2}D_{11}\frac{4\pi^4}{l^4} + \frac{1}{2}\Gamma_{11}\frac{2\pi^2}{l^2} + \frac{1}{2}\Gamma_{11}\mu_{22}\frac{\pi}{l}\right)A_{22}\right] = \frac{2q}{l}\cos\omega_q t$$

$$(16.30a)$$

$$\frac{\Gamma_0 I_2 \pi/l \left[-\omega_{11}^2 A_{11} - \ddot{A}_{11}\right]}{D_2 \left(\pi/l\right)^2 + \Gamma_0 - \omega_{11}^2 I_2} + \frac{\Gamma_0 I_2 \pi/l \left[-\omega_{12}^2 A_{12} - \ddot{A}_{12}\right]}{D_2 \left(\pi/l\right)^2 + \Gamma_0 - \omega_{12}^2 I_2}$$

$$+ \gamma \cos \omega_m t \left[\left(\frac{1}{2} \Gamma_{11} \frac{2\pi}{l} + \frac{1}{2} \Gamma_{11} \mu_{21} \right) A_{21} + \left(\frac{1}{2} \Gamma_{11} \frac{2\pi}{l} + \frac{1}{2} \Gamma_{11} \mu_{22} \right) A_{22} \right] = 0$$

(16.30b)

We express $\ddot{A}_{12} + \omega_{12}^2 A_{12}$ from (16.30b) and substitute this formula to (16.30a)

$$\left[m + I_1 \left(\frac{\pi}{l} \right)^2 \right] \left[\ddot{A}_{11} + \omega_{11}^2 A_{11} \right] + \left[m + I_1 \left(\frac{\pi}{l} \right)^2 \right] \left\{ \frac{\mu_{11}}{\mu_{12}} \left(-\ddot{A}_{11} - \omega_{11}^2 A_{11} \right) \right.$$

$$+ \frac{\gamma \cos \omega_m t}{\mu_{12}} \left[\left(\frac{1}{2} \Gamma_{11} \frac{2\pi}{l} + \frac{1}{2} \Gamma_{11} \mu_{21} \right) A_{21} + \left(\frac{1}{2} \Gamma_{11} \frac{2\pi}{l} + \frac{1}{2} \Gamma_{11} \mu_{22} \right) A_{22} \right] \right\}$$

$$+ \gamma \cos \omega_m t \left[\left(\frac{1}{2} D_{11} \frac{4\pi^4}{l^4} + \frac{1}{2} \Gamma_{11} \frac{2\pi^2}{l^2} + \frac{1}{2} \Gamma_{11} \frac{\pi}{l} \mu_{21} \right) A_{21} \right.$$

$$\left. + \left(\frac{1}{2} D_{11} \frac{4\pi^4}{l^4} + \frac{1}{2} \Gamma_{11} \frac{2\pi^2}{l^2} + \frac{1}{2} \Gamma_{11} \frac{\pi}{l} \mu_{22} \right) A_{22} \right] = \frac{2q}{l} \cos \omega_q t$$

(16.31)

The system (16.28b), (16.28d) is processed analogously.

As is seen, the differential equation (16.31) with respect to A_{11} contains amplitudes A_{21} and A_{22} due to the presence of a parametric stiffness modulation. Thus, we've got coupling of three modes. Now let us consider the case of our main concern, which is

$$\omega_q = \omega_{11} + \varepsilon \sigma_q \qquad (16.32a)$$

$$\omega_m = \omega_{22} - \omega_{11} + \varepsilon \sigma_m, \quad \varepsilon \to 0 \qquad (16.32b)$$

The condition (16.32a) defines the resonant excitation, the condition (16.32b) formulates a special type of the parametric stiffness modulation. Thus, we expect resulting motions of the system to occur with a strong participation of the directly excited mode A_{11} and also with the strong

16.9. A Modal Formulation of the Control Problem

participation of a 'complimentary' mode A_{22}. We explore a possibility of vibration control when γ tends to zero, $\gamma = \varepsilon \bar{\gamma}$, $\varepsilon \to 0$. In this case, if eigenfrequencies ω_{12}, ω_{21} are not linked to frequencies ω_{11}, ω_{22}, ω_q, ω_m by any simple multiplicity relations, then the non-negligibly small modal interaction exists only between the modes defined by functions A_{11}, A_{22}, whereas participation of the 'non-resonant' modes A_{12} and A_{21} vanishes.

This discussion is aimed at further reduction of the original model (which so far has four degrees of freedom) to a kind of 'manifold' model with only two degrees of freedom. Dynamics of this reduced model is governed by the following set of ordinary differential equations

$$\left[m + I_1\left(\frac{\pi}{l}\right)^2\right]\left[1 - \frac{\mu_{11}}{\mu_{12}}\right]\left[\ddot{A}_{11} + \omega_{11}^2 A_{11}\right]$$

$$+ \gamma \cos\omega_m t \left\{ \frac{1}{\mu_{12}} \left[m + I_1\left(\frac{\pi}{l}\right)^2\right]\left(\frac{1}{2}\Gamma_{11}\frac{2\pi}{l} + \frac{1}{2}\Gamma_{11}\mu_{22}\right) \right. \quad (16.33a)$$

$$\left. + \left(\frac{1}{2}D_{11}\frac{4\pi^4}{l^4} + \frac{1}{2}\Gamma_{11}\frac{2\pi^2}{l^2} + \frac{1}{2}\Gamma_{11}\frac{\pi}{l}\mu_{22}\right)\right\} A_{22} = \frac{2q}{l}\cos\omega_q t$$

$$\left[m + I_1\left(\frac{2\pi}{l}\right)^2\right]\left[1 - \frac{\mu_{22}}{\mu_{21}}\right]\left[\ddot{A}_{22} + \omega_{22}^2 A_{22}\right]$$

$$+ \gamma \cos\omega_m t \left\{ \frac{1}{\mu_{21}} \left[m + I_1\left(\frac{2\pi}{l}\right)^2\right]\left(\frac{1}{2}\Gamma_{11}\frac{2\pi}{l} + \frac{1}{2}\Gamma_{11}\mu_{11}\right) \right. \quad (16.33b)$$

$$\left. + \left(\frac{1}{2}D_{11}\frac{4\pi^4}{l^4} + \frac{1}{2}\Gamma_{11}\frac{2\pi^2}{l^2} + \frac{1}{2}\Gamma_{11}\frac{\pi}{l}\mu_{11}\right)\right\} A_{11} = 0$$

It does not present any difficulty to apply the method of direct pertition of motions or the method of multiple scales to analyse modal interaction in this system provided that conditions (16.32a,b) hold true. However, one should not expect any new phenomena to be detected in the course of such an analysis besides those already explored in [14].

16.10 Analysis of Vibration Control for a Model Two-Degrees of Freedom Mechanical System

In this Section, our goal is to formulate the model system which exhibits qualitatively the same phenomena as those discussed for a sandwich plate model and to analyse its vibrations in the resonant excitation conditions. Now we address dynamics of such a mechanical system shown in Figure 16.8. It consists of two masses M_0, m linked to some foundations by the springs of stiffness K_0, K_m and connected by a spring of stiffness K_Δ.

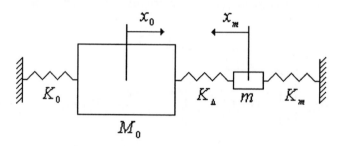

Figure 16.8. The model two-degree-of-freedom system.

An energy dissipation is taken for account by damping coefficients D_0 and D_m The equations of motions of such a system driven by a force $\overline{Q}$ are obvious

$$M_0\ddot{x}_0 = \overline{Q} - D_0\dot{x}_0 - K_0 x_0 - K_\Delta(x_0 + x_m) \qquad (16.34a)$$

$$m\ddot{x}_m = -K_m x_m - D_m \dot{x}_m - K_\Delta(x_0 + x_m) \qquad (16.34b)$$

Comparison of equations (16.33a,b) and (16.34a,b) suggests that by letting $K_\Delta = 0$ we obtain two linear uncoupled oscillators exactly as the two modes A_{11}, A_{22} become uncoupled when $\gamma = 0$. We assume

$$K_0 = K - K\gamma\cos\omega_f t \qquad (16.35a)$$

16.10. Analysis of Vibration Control

$$K_m = K - K\gamma\cos\omega_f t \quad (16.35b)$$

$$K_\Delta = K\gamma\cos\omega_f t, \quad \gamma \ll 1 \quad (16.35c)$$

The springs K_0 and K_m have modulated stiffness, whereas a spring between two masses changes its stiffness periodically, i.e., it exhibits a negative reactive force as well as a positive one. We do not discuss here a practical implementation of such a device in details, but just note that similar devices are often introduced in a control theory.

If a harmonic driving force $\overline{Q} = Q\cos\omega_q t$ is applied at the mass M_0, then equations (16.34a,b) become

$$M_0\ddot{x}_0 + D_0\dot{x}_0 + Kx_0 + Kx_m\gamma\cos\omega_f t = Q\cos\omega_q t \quad (16.36a)$$

$$m\ddot{x}_m + D_m\dot{x}_m + Kx_m + Kx_0\gamma\cos\omega_f t = 0 \quad (16.36b)$$

One may easily recognise that this set of equations is qualitatively the same as the set of equations (16.33).

We consider a resonant excitation of this system with the resonant parametric modulation by a use of the multiple scales method and by a use of the method of direct partition of motions.

16.10.1 Analysis by the method of multiple scales

Since the method of multiple scales is more involved mathematically than the method of direct partition of motions, we begin our analysis with its use. All technical details of this method may be readily found in [19,10]. We adopt two independent time-scales as $T_0 = t$ and $T_1 = \varepsilon t$, so that two-terms asymptotic expansions are formulated as

$$\begin{aligned} x_0(t) &= x_{00}(T_0, T_1) + \varepsilon x_{01}(T_0, T_1) \\ x_m(t) &= x_{m0}(T_0, T_1) + \varepsilon x_{m1}(T_0, T_1) \end{aligned} \quad (16.37)$$

Respectively, the derivatives in respect to time are split as

$$\frac{dx_0}{dt} = \frac{\partial x_{00}}{\partial T_0} + \varepsilon\frac{\partial x_{01}}{\partial T_0} + \varepsilon\frac{\partial x_{00}}{\partial T_1}, \quad \frac{dx_m}{dt} = \frac{\partial x_{m0}}{\partial T_0} + \varepsilon\frac{\partial x_{m1}}{\partial T_0} + \varepsilon\frac{\partial x_{m0}}{\partial T_1}$$

$$\frac{dx_0}{dt} = \frac{\partial^2 x_{00}}{\partial T_0^2} + \varepsilon \frac{\partial^2 x_{01}}{\partial T_0^2} + 2\varepsilon \frac{\partial^2 x_{00}}{\partial T_0 \partial T_1}, \quad \frac{dx_m}{dt} = \frac{\partial^2 x_{m0}}{\partial T_0^2} + \varepsilon \frac{\partial^2 x_{m1}}{\partial T_0^2} + 2\varepsilon \frac{\partial^2 x_{m0}}{\partial T_0 \partial T_1}$$

Asymptotic ranging of terms in equations (16.36a,b) is relevant to so-called weak resonant excitation of a weakly damped system [19, 10]

$$M\ddot{x}_0 + \varepsilon D_0 \dot{x}_0 + K x_0 + \varepsilon K x_m \gamma \cos\omega_f t = \varepsilon Q \cos\omega_q t$$
$$m\ddot{x}_m + \varepsilon D_m \dot{x}_m + K x_m + \varepsilon K x_0 \gamma \cos\omega_f t = 0$$
(16.38)

Substitution of these expansions to equations (16.38) gives up to terms to order ε

$$M_0 \left(\frac{\partial^2 x_{00}}{\partial T_0^2} + \varepsilon \frac{\partial^2 x_{01}}{\partial T_0^2} + 2\varepsilon \frac{\partial^2 x_{00}}{\partial T_0 \partial T_1} \right) + K(x_{00} + \varepsilon x_{01})$$

$$+ \varepsilon \gamma K \frac{1}{2} \left(e^{i\omega_f t} + e^{-i\omega_f t} \right) x_{m0} + \varepsilon D_0 \frac{\partial x_{00}}{\partial T_0} = \varepsilon Q \frac{1}{2} \left(e^{i\omega_q t} + e^{-i\omega_q t} \right)$$

$$m \left(\frac{\partial^2 x_{m0}}{\partial T_0^2} + \varepsilon \frac{\partial^2 x_{m1}}{\partial T_0^2} + 2\varepsilon \frac{\partial^2 x_{m0}}{\partial T_0 \partial T_1} \right) + K(x_{m0} + \varepsilon x_{m1})$$

$$+ \varepsilon \gamma K \frac{1}{2} \left(e^{i\omega_f t} + e^{-i\omega_f t} \right) x_{00} + \varepsilon D_m \frac{\partial x_{m0}}{\partial T_0} =$$

In particular, the system of equations to order ε^0 is trivial

$$M_0 \frac{\partial^2 x_{00}}{\partial T_0^2} + K x_{00} = 0 \tag{16.39a}$$

$$m \frac{\partial^2 x_{m0}}{\partial T_0^2} + K x_{m0} = 0 \tag{16.39b}$$

It describes free undamped vibrations of uncoupled oscillators and its solution is

$$x_{00}(T_0, T_1) = A(T_1)\exp(i\omega_0 T_0) + \overline{A}(T_1)\exp(-i\omega_0 T_0), \tag{16.40a}$$

16.10. Analysis of Vibration Control

$$x_{m0}(T_0,T_1) = B(T_1)\exp(i\omega_m T_0) + \overline{B}(T_1)\exp(-i\omega_m T_0), \quad (16.40b)$$

here $\omega_0 = \sqrt{K/M_0}$, $\omega_m = \sqrt{K/m}$ and the second terms in (16.40a,b) are complex conjugates of the first ones.

The system of equations to order ε^1 is inhomogeneous

$$M_0 \frac{\partial^2 x_{01}}{\partial T_0^2} + Kx_{01} = \frac{1}{2}Q\left[\exp(i\omega_q t) + \exp(-i\omega_q t)\right] -$$

$$2M_0\left[i\omega_0 \frac{dA}{dT_1}\exp(i\omega_0 T_0) - i\omega_0 \frac{d\overline{A}}{dT_1}\exp(-i\omega_0 T_0)\right] -$$

$$\frac{1}{2}\gamma K\left[\exp(i\omega_f t) + \exp(-i\omega_f t)\right]\left[B\exp(i\omega_m T_0) + \overline{B}\exp(-i\omega_m T_0)\right] -$$

$$D_0\left[i\omega_0 A\exp(i\omega_0 T_0) - i\omega_0 \overline{A}\exp(-i\omega_0 T_0)\right]$$

$$(16.41a)$$

$$m\frac{\partial^2 x_{m1}}{\partial T_0^2} + Kx_{m1} =$$

$$-2m\left[i\omega_m \frac{dB}{dT_1}\exp(i\omega_m T_0) - i\omega_m \frac{d\overline{B}}{dT_1}\exp(-i\omega_m T_0)\right]$$

$$-\frac{1}{2}\gamma K\left[\exp(i\omega_f t) + \exp(-i\omega_f t)\right]\left[A\exp(i\omega_0 T_0) + \overline{A}\exp(-i\omega_0 T_0)\right]$$

$$-D_m\left[i\omega_m B\exp(i\omega_m T_0) - i\omega_m \overline{B}\exp(-i\omega_m T_0)\right]$$

$$(16.41b)$$

Now let us formulate conditions of the excitation and the stiffness modulation as

$$\omega_q = \omega_0 + \varepsilon\sigma_q \quad (16.42a)$$

$$\omega_f = \omega_m - \omega_0 + \varepsilon\sigma_f \quad (16.42b)$$

Then to ensure a uniform validity of the asymptotic expansion (16.37) the secular terms should be eliminated from the right hand sides of equations (16.41). An elementary algebra gives the following two equations obtained by equating to zero all terms containing $\exp(i\omega_0 T_0)$ and $\exp(i\omega_m T_0)$ in the right hand sides of the first and the second equation, respectively

$$\frac{1}{2}Q\exp(i\sigma_q T_1) - 2i\omega_0 M_0 \frac{dA}{dT_1} - \frac{1}{2}\gamma K B \exp(-i\sigma_f T) - i\omega_0 D_0 A = 0$$

$$-2i\omega_m m \frac{dB}{dT_1} - \frac{1}{2}\gamma K A \exp(i\sigma_f T) - i\omega_0 D_m B = 0$$

As is well-known, equating to zero terms containing $\exp(-i\omega_0 T_0)$ and $\exp(-i\omega_m T_0)$ in these equations results in the same system. A solution for this system is sought as

$$A(T_1) = a(T_1)\exp[i\varphi(T_1)]$$

$$B(T_1) = b(T_1)\exp[i\psi(T_1)]$$

where $a(T_1), b(T_1), \varphi(T_1), \psi(T_1)$ are real-valued functions which determine amplitudes and phases of complex functions $A(T_1), B(T_1)$.

$$\frac{1}{2}Q\exp(i\sigma_q T_1 - i\varphi) - 2i\omega_0 M_0 \left(\frac{da}{dT_1} + ia\frac{d\varphi}{dT_1}\right)$$
$$-\frac{1}{2}\gamma K b \exp(i\psi - i\sigma_f T_1 - i\varphi) - i\omega_0 D_0 a = 0 \quad (16.43a)$$

$$-2i\omega_m m\left(\frac{db}{dT_1} + ib\frac{d\psi}{dT_1}\right)$$
$$-\frac{1}{2}\gamma K a \exp(-i\psi + i\sigma_f T_1 + i\varphi) - i\omega_1 D_m b = 0 \quad (16.43b)$$

It is more convenient to introduce a set of phases as

$$\alpha = \varphi - \sigma_q T_1, \quad \beta = \psi - \varphi - \sigma_f T_1, \quad \beta = \psi - \varphi - \sigma_f T_1$$

16.10. Analysis of Vibration Control

Then derivatives of original phases on slow time become

$$\frac{d\varphi}{dT_1} = \frac{d\alpha}{dT_1} + \sigma_q, \quad \frac{d\psi}{dT_1} = \frac{d\beta}{dT_1} + \frac{d\varphi}{dT_1} + \sigma_f$$

Separation of real and imaginary parts in equations (16.43) gives

$$\frac{1}{2}Q\cos\alpha + 2\omega_0 M_0 a\left(\frac{d\alpha}{dT_1} + \sigma_q\right) - \frac{1}{2}\gamma Kb\cos\beta = 0$$

$$-\frac{1}{2}Q\sin\alpha - 2\omega_0 M_0 \frac{da}{dT_1} - \frac{1}{2}\gamma Kb\sin\beta - \omega_0 D_0 a = 0$$

$$2m\omega_m b\left(\frac{d\beta}{dT_1} + \frac{d\alpha}{dT_1} + \sigma_q + \sigma_f\right) - \frac{1}{2}\gamma Ka\cos\beta = 0$$

$$-2m\omega_m \frac{db}{dT_1} + \frac{1}{2}\gamma Ka\sin\beta - \omega_m D_m b = 0$$

These equations should be explicitly formulated for the derivatives on slow time as

$$\frac{d\beta}{dT_1} = -\sigma_f + \frac{\gamma Ka\cos\beta}{2m\omega_m b} + \frac{Q\cos\alpha}{4\omega_0 M_0 a} - \frac{\gamma Kb\cos\beta}{4\omega_0 M_0 a} \quad (16.44a)$$

$$\frac{d\alpha}{dT_1} = -\sigma_q - \frac{Q\cos\alpha}{4\omega_0 M_0 a} + \frac{\gamma Kb\cos\beta}{4\omega_0 M_0 a} \quad (16.44b)$$

$$\frac{db}{dT_1} = \frac{\gamma Ka\sin\beta}{4m\omega_m} - \frac{D_m b}{2M_0} \quad (16.44c)$$

$$\frac{da}{dT_1} = -\frac{Q\sin\alpha}{4\omega_0 M_0} - \frac{\gamma Kb\sin\beta}{4\omega_0 M_0} - \frac{D_0 a}{2M_0} \quad (16.44d)$$

It is convenient to transform these equations to non-dimensional variables and to introduce the following parameters

$$\mu = \frac{m}{M_0}, \quad \delta = \frac{D_m}{D_0}, \quad \eta = \frac{D_0}{\omega_0 M_0}, \quad \hat{Q} = \frac{Q}{K} \quad \hat{\sigma}_q = \mp\frac{\sigma_q}{\omega_0} \quad \hat{\sigma}_f = \mp\frac{\sigma_f}{\omega_0}$$

Then equations (16.44) becomes (a prime denotes the derivative on slow time T_1)

$$\omega_0^{-1}\beta' = -\hat{\sigma}_f + \frac{\gamma a \cos\beta}{2b\sqrt{\mu}} + \frac{\hat{Q}\cos\alpha}{4a} - \frac{\gamma b \cos\beta}{4a} \qquad (16.45a)$$

$$\omega_0^{-1}\alpha' = -\hat{\sigma}_q - \frac{\hat{Q}\cos\alpha}{4a} + \frac{\gamma b \cos\beta}{4a} \qquad (16.45b)$$

$$\omega_0^{-1}b' = \frac{\gamma a \sin\beta}{4\sqrt{\mu}} - \frac{\eta \delta b}{2\mu} \qquad (16.45c)$$

$$\omega_0^{-1}a' = -\frac{\hat{Q}\sin\alpha}{4} - \frac{\gamma b \sin\beta}{4} - \frac{\eta a}{2} \qquad (16.45d)$$

To seek for a stationary solution of these ordinary differential equations, all the derivatives should be set to zero, $\alpha' = \beta' = a' = b' = 0$ and the resulting transcendent equations with respect to stationary amplitudes and phases α, β, a, b should be solved. This is easily done analytically, and

$$a = \frac{1}{2}\hat{Q}\left\{\left[2\hat{\sigma}_q - \frac{\gamma^2(\hat{\sigma}_q + \hat{\sigma}_f)\sqrt{\mu}}{4F^2}\right]^2 + \eta^2\left[1 + \frac{\delta\gamma^2}{4\sqrt{\mu}F^2}\right]^2\right\}^{-1/2},$$

(16.46a)

$$b = \frac{\gamma a}{2F}, \qquad (16.46b)$$

$$\tan\beta = \frac{\delta\eta}{\mu(\hat{\sigma}_q + \hat{\sigma}_f)}, \qquad (16.46c)$$

16.10. Analysis of Vibration Control

$$\tan \alpha = \frac{\eta \left(1 + \dfrac{\gamma^2 \delta}{4\sqrt{\mu} F^2}\right)}{2\hat{\sigma}_q - \dfrac{\gamma^2 \sqrt{\mu}(\hat{\sigma}_q + \hat{\sigma}_f)}{4F^2}} \qquad (16.46d)$$

In these equations, $F = \sqrt{\dfrac{\delta^2 \eta^2}{\mu} + \mu(\hat{\sigma}_q + \hat{\sigma}_f)^2}$

To judge upon stability of this solution, a Jacobean of the original system (16.45) is set up and its eigenvalues are calculated when the amplitudes and phases given by formulas (16.46) are substituted there. If all four eigenvalues have negative real parts, when this solution is stable, otherwise it is not. Besides this stability check, a direct numerical integration of the system (16.45) may be performed. Or course, in a case of stable motions, this procedure converges to the values, predicted by formulas (16.46).

It is relevant to specify the condition $\dfrac{b}{a} > 1$ to judge upon the efficiency of vibration suppression, meaning that if this condition holds true, then a desired level of control of vibrations of a protected mass is achieved. In the case of an undamped system, this condition requires the amplitude of the stiffness modulation to obey the following inequality: $\gamma > 2(\hat{\sigma}_f + \hat{\sigma}_q)\sqrt{\mu}$. Apparently, as is seen from this formula with the growth in any of detuning parameters it becomes necessary to increase an amplitude of the stiffness modulation to have the condition $\dfrac{b}{a} > 1$ held true.

In Figure 16.9, a dependence of non-dimensional amplitudes a (curve 1) and b (curve 2) on the damping parameter η is presented for $\hat{\sigma}_q = 0.001$.

Other parameters of the model system are $\hat{Q} = 0.01$, $\mu = 0.02$, $\delta = 1$, $\gamma = 0.01$, $\hat{\sigma}_f = 0.0001$. As is seen, amplitude a of the protected mass is much smaller, than the amplitude of an absorber b when damping is fairly small. Actually, as follows from formula (16.46a) in the case of perfect tuning of an external force and an absence of damping, $\hat{\sigma}_q = \eta = 0$, the amplitude of a protected mass is zero exactly. In the present case, since $\hat{\sigma}_q \neq 0$ and $\hat{\sigma}_f \neq 0$ the amplitude a has a certain finite value at $\eta = 0$. Simultaneously, the amplitude b of the absorber is large and it is close to its limit value of $b = \dfrac{\hat{Q}}{\gamma}$, which follows

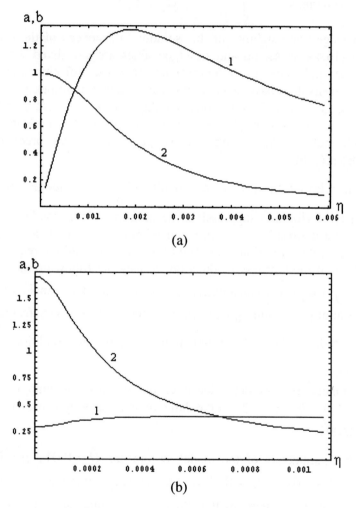

Figure 16.9. A dependence of non-dimensional amplitudes a (curve 1) and b (curve 2) on the damping parameter, (a) $\hat{\sigma}_q = 0.001$, (b) $\hat{\sigma}_q = 0.006$

16.10. Analysis of Vibration Control

from (16.46b) for $\hat{\sigma}_q = \eta = 0$. An increase in damping coefficient reduces the efficiency of vibration absorption, so that the amplitude a grows with the growth in η, whereas the amplitude b decreases. The condition $\dfrac{b}{a} = 1$ is reached at $\eta \approx 0.0007$. In Figure 16.9b, similar graphs are plotted for $\hat{\sigma}_q = 0.006$.

This is the case of an excitation which is 'softer' than is the previous one. However, it is notable that the amplitude a does not vanish at $\eta = 0$ and the condition $\dfrac{b}{a} > 1$ holds true until $\eta \approx 0.00064$, which is slightly less, than in the previous case. Comparison of results illustrated by Figure 7+2a,b leads to a rather trivial conclusion – the efficiency of a control mechanism drops down as parameters of the system deviate from their 'ideal' or 'nominal' values.

In Figure 16.10a, a dependence of amplitudes a (curve 1) and b (curve 2) on the parameter of stiffness modulation γ is presented for $\hat{Q} = 0.01$, $\mu = 0.02$, $\delta = 1$, $\hat{\sigma}_f = 0.0001$.

Other parameters of the model system are $\eta = 0.001$ and $\hat{\sigma}_q = 0.005$. The amplitude a of 'protected' mass drops down while

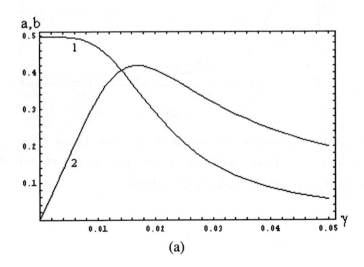

(a)

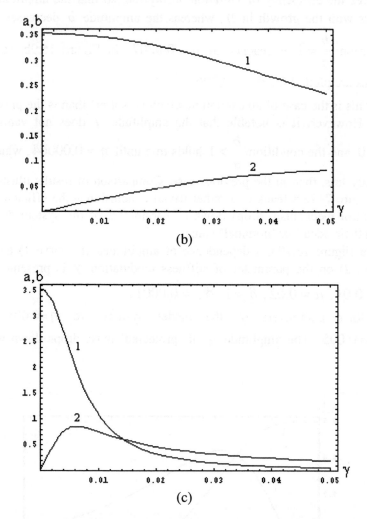

Figure 16.10. A dependence of amplitudes a (curve 1) and b (curve 2) on the parameter of stiffness modulation γ, (a) $\eta =0.001$ and $\hat{\sigma}_q = 0.005$ (b) ($\eta =0.01$ $\hat{\sigma}_q = 0.005$ (c) $\eta =0.001$ $\hat{\sigma}_q = 0.0005$

16.10. Analysis of Vibration Control

the amplitude b of an absorber grows. Condition $\dfrac{b}{a} > 1$ holds true starting from $\gamma = 0.014$. For a system which is damped more heavily ($\eta = 0.01$) efficiency of suppression of vibrations is much smaller, as seen from Figure 16.10b.

Curves in Figure 16.10c are plotted for the same set of parameters as Figure 16.10a, but the excitation is tuned closer to the resonance, $\hat{\sigma}_q = 0.0005$. Although curves qualitatively are similar to those in Figure 16.10a, it is clear that efficiency of vibration control is getting higher as excitation tends to be resonant.

It is remarkable that although as before the condition $\dfrac{b}{a} > 1$ holds true starting from $\gamma = 0.014$, the controlled amplitude decreases approximately in 5 times, $\dfrac{a|_{\gamma=0}}{a|_{\gamma=0.014}} \approx 5$ for $\hat{\sigma}_q = 0.0005$ whereas for $\hat{\sigma}_q = 0.005$ this decrease is only $\dfrac{a|_{\gamma=0}}{a|_{\gamma=0.014}} \approx 1.25$. Thus, the efficiency of a control mechanism drops down as parameters of the system deviate from their 'ideal' values.

In a particular case, when $\hat{\sigma}_q = 0$ (the perfect external tuning) and $\eta = 0$ (the undamped system), the formulas for the controlled amplitude and for the amplitude of an absorber are remarkably simple:

$$a = \frac{2\hat{\sigma}_f \sqrt{\mu}}{\gamma^2} \hat{Q} \qquad (16.47a)$$

$$b = \frac{\hat{Q}}{\gamma} \qquad (16.47b)$$

They should be compared with the predictions given by a use of the method of direct partition of motions in the next subsection.

16.10.2 Analysis by the method of direct partition of motions

Now we apply the method of direct partition of motions, see [1] for details, and for simplicity consider an undamped system

$$M_0 \ddot{x}_0 + K x_0 + K x_m \gamma \cos\omega_f t = Q \cos\omega_q t \quad (16.48a)$$

$$m \ddot{x}_m + K x_m + K x_0 \gamma \cos\omega_f t = 0 \quad (16.48b)$$

Motion of an attachment is sought as $x_m = \tilde{x}_m(t)\cos\omega_f t$, and substitution of this formula to (16.48b) gives $(K - m\omega_f^2)\tilde{x}_m = -\gamma K x_0$, so that equation (16.48a) is transformed as

$$M_0 \ddot{x}_0 + K\left[1 - \frac{\gamma^2 \cos^2 \omega_f t}{1 - (\omega_f/\omega_m)^2}\right] x_0 = Q \cos\omega_q t \quad (16.49)$$

We assume that $x_0 = \hat{x}_0 \cos\omega_q t$, $\tilde{x}_m = \hat{x}_m \cos\omega_q t$ and $\omega_q \ll \omega_f$, so that as suggested by the method of direct partition of motions averaging over a period gives $<\cos^2\omega_f t> = 1/2$. Then equation (16.49) is reduced to a simple form

$$\left[1 - \left(\frac{\omega_q}{\omega_0}\right)^2 - \frac{1}{2}\frac{\gamma^2}{1 - \left(\frac{\omega_f}{\omega_m}\right)^2}\right]\hat{x}_0 = \hat{Q} \equiv \frac{Q}{K} \quad (16.50)$$

Now let us rewrite this equation in terms introduced in the course of analysis by the multiple scales method (we note that $\mu = \frac{m}{M_0} \equiv \left(\frac{\omega_0}{\omega_m}\right)^2$)

$$\left[1 - (1+\hat{\sigma}_q)^2 - \frac{1}{2}\frac{\gamma^2}{1 - (1+\hat{\sigma}_f \sqrt{\mu})^2}\right]\hat{x}_0 = \hat{Q} \quad (16.51)$$

16.10. Analysis of Vibration Control

If we assume that vibrations are excited precisely at the frequency $\omega_q = \omega_0$ and retain only the first term in expansion (since we have already adopt $\mu \equiv \sqrt{\dfrac{\omega_0}{\omega_m}} \ll 1$) we get $\left(1+\hat{\sigma}_f\sqrt{\mu}\right)^2 \approx 1+2\hat{\sigma}_f\sqrt{\mu}$.

Then

$$\hat{x}_0 = \frac{4\hat{\sigma}_f\sqrt{\mu}}{\gamma^2}\hat{Q}, \qquad (16.52a)$$

$$\hat{x}_m = \frac{2\hat{Q}}{\gamma} \qquad (16.52b)$$

which are twice larger than the predictions given by formulas (16.47). However, these results indeed match each other perfectly, since a full actual response found by the multiple scales method is two times larger, than (16.47), see its full formulation (16.40) which includes both $A(T_1)$, $B(T_1)$ and their complex conjugates.

To confirm the excellent agreement between predictions given by the method of multiple scales and the method of direct partition of motions, consider vibrations of a model system with the following set of parameters $\hat{Q} = 0.01$, $\mu = 0.02$, $\delta = 1$, $\gamma = 0.01$, $\eta = 0$. In Figure 16.11a, a dependence of the amplitude $\hat{x}_0 \equiv 2a$ of protected mass is plotted versus $\hat{\sigma}_f$ for $\hat{\sigma}_q = 0$ and the two curves practically merge each other.

In this case, the amplitude of absorber is given by the both theories simply as $\hat{x}_m \equiv 2b = 2\hat{Q}/\gamma$. As is seen from formulas (16.46a) and (16.51), a certain combination of two detuning parameters $\hat{\sigma}_f$ and $\hat{\sigma}_q$ may produce a resonant behaviour of the system at a frequency, shifted from an original resonance. This aspect for a sandwich beam has been studied in [14]. In Figure 16.11b,c a dependence of amplitudes $\hat{x}_0 \equiv 2a$ and $\hat{x}_m \equiv 2b$ on $\hat{\sigma}_f$ is presented for $\hat{\sigma}_q = 0.001$.

Both theories give results hardly distinguished from each other. Thus, we conclude that both asymptotic methods demonstrate a possibility of vibration control by the parametric stiffness modulations.

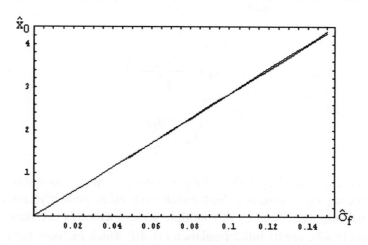

Figure 16.11a. A dependence of the amplitude $\hat{x}_0 \equiv 2a$ on $\hat{\sigma}_f$ for $\hat{\sigma}_q = 0$.

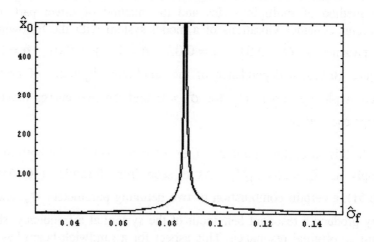

Figure 16.11b. A dependence of the amplitude $\hat{x}_0 \equiv 2a$ on $\hat{\sigma}_f$ for $\hat{\sigma}_q = 0.001$.

16.10. Analysis of Vibration Control

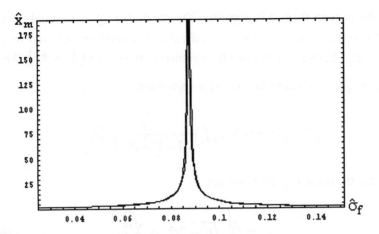

Figure 16.11c. A dependence of the amplitude $\hat{x}_m \equiv 2b$ on $\hat{\sigma}_f$ for $\hat{\sigma}_q = 0.001$.

16.10.3 'Standard' dynamical absorber

To conclude this section, we consider a classical dynamic absorber of vibrations of an one degree of freedom mechanical system. The protected system is modeled by a mass and a spring. Its eigenfrequency is $\omega_0 = \sqrt{K/M_0}$. This system is excited by a periodic force $Q\cos\omega_q t$. To avoid its resonant vibrations which occur at the frequency $\omega_q \approx \omega_0$, another mass m is attached to the protected system by a spring of the stiffness χK. Then equations of forced motions of this composed two-degree-of-freedom system become

$$M_0\ddot{x}_0 + K(1+\chi)x_0 - \chi K x_m = Q\cos\omega_q t \qquad (16.53)$$
$$m\ddot{x}_m + \chi K x_m - \chi K x_0 = 0$$

This elementary linear problem has a solution

$$\left[1 - \left(\frac{\omega_q}{\omega_0}\right)^2 + \chi - \frac{\chi}{1-\left(\frac{\omega_q}{\omega_1}\right)^2}\right] x_0 = \hat{Q} \qquad (16.54)$$

16. The Active Control of Vibrations of Composite Beams

As is seen from this equation, the parameter χ is uniquely defined by the excitation conditions, specifically, it should be selected to make $\omega_1 = \omega_0$, because it is already assumed that $\omega_q = \omega_0(1+\hat{\sigma}_q)$. Then it is $\chi = \sqrt{\mu}$ and formula (16.54) is re-written

$$\left[1+\sqrt{\mu}-(1+\hat{\sigma}_q)^2 + \frac{\sqrt{\mu}}{1-(1+\hat{\sigma}_q)^2}\right]x_0 = \hat{Q}$$

Taking the limit $\hat{\sigma}_q \ll 1$ we get

$$x_0 = \hat{Q}\left[\sqrt{\mu} - 2\hat{\sigma}_q + \frac{\sqrt{\mu}}{2\hat{\sigma}_q}\right]^{-1} \tag{16.55}$$

Similar transformation of formula (16.52a) gives

$$x_0 = \hat{Q}\left[-2\hat{\sigma}_q + \frac{\gamma^2}{4\hat{\sigma}_f \sqrt{\mu}}\right]^{-1} \tag{16.56}$$

Comparison of these formulas suggests that to reach the same amplitude the following condition should hold

$$\sqrt{\mu} - \frac{\sqrt{\mu}}{2\hat{\sigma}_q} = \frac{\gamma^2}{4\hat{\sigma}_f \sqrt{\mu}}$$

or asymptotically ($\hat{\sigma}_q \ll 1$)

$$\hat{\sigma}_f \approx -\frac{\gamma^2}{4\mu}\hat{\sigma}_q.$$

This formula shows that, since $\gamma \ll 1$, to get the same level of control, the frequency of parametric modulation ω_f should be closer to the eigenfrequency ω_m, than a driving frequency ω_q to the eigenfrequency ω_0.

16.11 Conclusions

An investigation is done into an active control of vibrations of composite beams by means of the parametric stiffness modulation. The model of sandwich beam composition is considered. Hamilton principle is used to derive equations of motions. The suggested way for an active control is based on the interaction between vibrations at the different modes. Specifically, possibilities of an energy channeling from a comparatively low-frequency symmetric mode of vibrations to a high-frequency skew-symmetric mode is explored. The mechanism of modal coupling is related to the concept of vibrational rheology that displays a link between 'slow' motions and 'hidden' fast modulation. The theoretical description of this phenomenon of modal coupling is given by a use of the method of direct partition of motions.

The following results are obtained

parametric modulation of stiffness at the resonant frequency of dominantly shear vibrations can completely eliminate the resonant behavior of a beam due to a significant increase in the first eigenfrequency of a beam

this effect is achievable by fairly small amplitude of the stiffness modulation, associated with small configuration changes in components of a core ply. These geometry changes are likely to be feasible by means of the standard piezo-electric acuators or acuators made of shape memory alloy

possibilities of control are limited by a necessity to perform stiffness modulations at the rather narrow frequency band. Although no special investigation into this subject has been performed, it is anticipated, that detuning from resonant fast modulation should not exceed 1% to obtain the effect of parametric control

the influence of an acoustic medium is shown to be not critical for the suggested way of a parametric control. Due to an existence of two frequency spectra, relevant to dominantly flexural (low-frequency) and dominantly shear (high-frequency) vibrations, modulation frequency is rather insensitive to a presence of an acoustic medium, whereas a controlled frequency is effected by it. Thus, modulations at the same 'fast' control frequency are capable to shift resonant frequency both with and without fluid loading the role of internal damping is shown to be controversial. If it is large, then amplitudes of forced response diminishes, but an efficiency of stiffness modulation drops because an imaginary part of eigenfrequency grows and it is impossible to match a purely real modulation frequency with it. If it is very small, then control effect shows up and shifts the resonant frequency, but amplitudes at the 'new' shifted resonant frequency are large. It is shown that there is an 'optimal' internal damping that does not yet destroy parametric modulation, but already reduces amplitudes of resonant vibrations.

dynamics of an elementary two-degree-of-freedom model mechanical system is considered to illustrate the mechanism of modal interaction, which is involved in the suggested way of an active control of vibrations of sandwich plates. It is shown that the method of direct partition of motions and the method of multiple scales give identical results in the case, when the controlled resonant frequency is much lower than the frequency of the stiffness modulation.

References

1. Blekhman I.I., *Vibrational Mechanics* (Fizmatgiz, Moscow, 1994) (in Russian).
2. Crighton D.G. et al., *Modern Methods in Analytical Acoustics* (Springer, Berlin, 1992).
3. Culshaw B., *Smart Structures and Materials* (Artech House, Boston, 1996)
4. Fidlin A., On the Separation of Motions in Systems with a Large Fast Excitation of General Form., *Eur. J. Mech A/Solids*, **18** (1999), p. 527–538.
5. Fuller C.R., Elliot S.J. and Nelson P.A., *Active Control of Vibration* (Academic Press, London, 1995).
6. Grigoliuk E.I., Chulkov P.P., *A theory of multilayered plates* (Nauka, Moscow, 1975) (in Russian).
7. Krylov V. and Sorokin S. V., Dynamics of Elastic Beams with Controlled Distributed Stiffness Parameters, *Smart Materials and Structures*, **6** (1997), p.573–582.
8. Lurie K.A., Effective Properties of Smart Elastic Laminates and the Screening Phenomenon, *Int. J. Solids Structures*, **34** (1997), p.1633–1643.
9. Lurie K. A., The problem of Effective Parameters of a Mixture of Two Isotropic Dielectrics Distributed in Space-Time and the Conservation Law for Wave Impedance in One-Dimensional Wave Propagation, *Proc. Roy. Soc. London, Ser. A*, **454** (1998), p.1767–1779.
10. Nayfeh A.H. and Mook D.T., *Non-linear Oscillations* (John Wiley, New York, 1979).
11. Skvortsov V. R., Symmetrically Inhomogeneous Through Thickness Plate as a Sandwich Plate Having a Soft Core. *Transactions of Academy of Sciencies of U.S.S.R. Mechanics of a Rigid Body*.**1** (1993), p. 162-168 (in Russian).
12. Sorokin S. V., Vibrations of and Sound Radiation from Sandwich Plates in Heavy Fluid Loading Conditions. *Journal of Composite Structures* **48** (2000), p. 219–230.
13. Zenkert D., *An Introduction to Sandwich Construction* (Chameleon Press, London, 1995).
14. Sorokin S. V., Ershova O.A., Grishina S. V. The Active Control of Vibrations of Composite Beams by Parametric Stiffness Modulation, *European Journal of Mechanics. A/Solids*, **19** (2000), p. 873-890.
15. Sorokin S. V., Grishina S. V., Ershova O.A. Analysis and Control of Vibrations of Honeycomb Plates by Parametric Stiffness Modulations, *Smart Materials and Structures*, **10**(5), p. 1031-1045
16. Blekhman I.I. *Vibrational Mechanics – Nonlinear Dynamic Effects, General Approach, Applications* (Singapore: World Scientific, 2000).

17. Eriksen J.L., Kinderlehler D., Kohn R.V. and Lions J.L. (eds) *Homogenization and effective moduli of materials and media,* (Springer-Verlag, Berlin, 1986).
18. Pugatchev S.I. *Piezo-Ceramic Actuators* (Sudostroenie, Leningrad, 1986) (in Russian)
19. Nayfeh A.H. *Perturbation Methods* (Wiley-Interscience, New York, 1973).
20. Fidlin A. On Asymptotic Properties of Systems with Strong and Very Strong High-Frequency Excitation, *Journal of Sound and Vibration,* **235**(2), p. 219–233.
21. Morse S.A. (ed) *Control Using Logic-Based Switching* (Springer Verlag, Berlin, 1996).
22. Junger M.G. and Feit D. *Sound, Structures and their Interaction* (Cambridge Massachusetts: MIT Press, 1992).
23. Dasgupta A. and Alghamdi A., A Transient Response of an Adaptive Beam with Embedded Piezoelectric Microactuators, *Proc. SPIE* **2190** (1994) p. 153–164.

Part V

Vibrational Hydrodynamics and Hydraulics

Chapter 17

Vibrational Hydrodynamics and Hydraulics

I. I. Blekhman

*Institute for Problems of Mechanical Engineering,
Russian Academy of Sciences
and
Mekhanobr-Tekhnika Corp.
22 Liniya 3, V.O., 199106, St. Petersburg, Russia*

17.1 Reinolds' Equation as an Equation of Vibrational Mechanic

It was shown in the book [1] that the classical equations of the turbulent flow of the viscous fluid, obtained by Reinolds, can be regarded as the main equations of vibrational mechanics. The turbulent stresses play in these equations the role of vibrational forces.

These equation were obtained by Reinolds by means of averaging Navier-Stokes' equations. Unlike the approach of vibrational mechanics, Reinolds did not consider the complementary equations of fast motions. Due to that, there appeared the so-called *problem of closure* – finding the turbulent stresses by using the experimental data and certain assumptions. Solving the corresponding equations of fast motions certainly presents essential difficulties since one is to find the self-oscillatory, regular or chaotic, solutions of nonlinear differential equations in partial derivatives. Along with that this solution is more simple than that of the initial equations since the equations of fast motions can be solved approximately, assuming that the slow components and the slow time are constant. It seems that this circumstance gives certain hopes for the success in using the approach of vibrational mechanics in the theory of turbulence.

It should be noted that this approach was in fact successfully used in hydrodynamics when solving problems on *vibrational convection* (see, say, [2, 3, 4]).

17.2 The Analog of Bernoulli's Equation for the Flows, Subjected to Vibration

Under the action of vibration on the flow of ideal incompressible fluid, let the velocity u and the pressure p be presented in the form

$$u = U + u', \quad p = P + p' \qquad (17.1)$$

where U and P, u' and p' are the "slow" and "fast" components respectively, the mean for the period of averaging values of the fast components being equal to zero, i. e. $\langle u' \rangle = 0$, $\langle p' \rangle = 0$.
Let then the slow motion be stationary, i.e. $\partial U / \partial t = 0$ and $\partial P / \partial t = 0$ (then the motion as a whole is, certainly, not stationary). In this case the use of the approach of vibrational mechanics leads to the equation

$$\frac{1}{2}\rho(U_2^2 - U_1^2) + P_2 - P_1 + P_{2v} - P_{1v} = 0 \qquad (17.2)$$

where ρ is the density of the fluid, U_1 and U_2 are the corresponding values of slow velocities, $\Delta P = P_2 - P_1$ is the "ordinary" pressure difference and $\Delta P_v = P_{2v} - P_{1v}$ is the additional one caused by the vibration and calculated on the basis of the solution of the equations of fast motions.

Equation (17.2) can be called *equation of vibrational hydraulics*. It can be regarded as an analogue of the classical equation of Bernoulli. The additional pressure difference ΔP_v can be both positive and negative and therefore can cause both the increase and decrease of the speed of flow.

For a number of problems of the theory of vibrational pumps, equations of the type of (17.2) were obtained in section 10.6 of the book [1].

References

1. Blekhman I.I. *Vibrational Mechanics – Nonlinear Dynamic Effects, General Approach, Applications.* Singapore: World Scientific (2000).
2. Zenkovskya S. M. and S. M. Simonenko, On the effect of high frequency vibrations on the appearance of convection, *AN SSSR, Fluid and Gas Mechanics*, 5 (1966) (in Russian).
3. Babushkin I.A., Zavarkyn M.P., Zorin S.V., and Putilin G.F., Controling the convective stability by vibrational fields, *Proceedings of the 2nd All Union Congress on Nonliner Oscillations of Mechanical Systems*, Gorky, Part 1 (1990) 22 (in Russian)
4. Gershuni G.Z., Zhukhovitsky E.M., and Nepomnyaschy A.A., *Stability of convective flows* (Nauka, Moscow, 1989), p.318 (in Russian).

Chapter 18

On the Vibro-Jet Effect and on the Phenomena of Vibrational Injection of the Gas into Fluid

I. I. Blekhman, L.I. Blekhman, L.A. Vaisberg, V.B. Vasilkov, K.S. Yakimova

Mekhanobr-Tekhnika Corp.
and
*Institute for Problems of Mechanical Engineering,
Russian Academy of Sciences
22 Liniya 3, V.O., 199106, St. Petersburg, Russia*

18.1 On Phenomenon under Consideration

This section considers two peculiar nonlinear phenomena. The first one, called *vibro-jet effect*, was known long ago [1, vol.4]. It consists in the fact that when a plate with conic holes vibrates in liquid, slow flows of the fluid appear in the direction of the narrowing of the holes (Fig. 18.1, a)

The vibro-jet effect is successfully used in a number of technical devices [1, vol. 4; 2, 3]. Along with this, there are data that this phenomenon was the cause of some aviation catastrophes: due to vibration the fuel stopped coming from tanks (Fig. 18.1, b), i.e. there was an effect of *vibrational closure of the holes*. In this case the pressure, facilitating the discharge of the fuel, is balanced by the counter-pressure, appearing due to vibration.

The second phenomenon – the *vibrational injection of gas into fluid* has been discovered recently [4]. It consists in the sucking of the air or another gas into a vessel with the fluid, vibrating in this gas, through the holes in its lower part (Fig. 18.2,a).

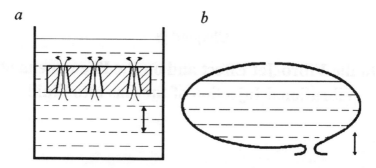

Figure 18.1. Vibro-jet effect: *a* – the generation of slow flows of the fluid through the conic holes in a vibrating plate, *b* – the effect of closing the holes in the vibrating tank with the fluid

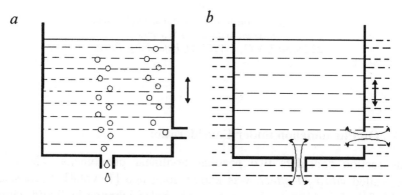

Figure 18.2. Phenomenon of vibrational injection: *a* – the injection of gas into a fluid, *b* – the injection of fluid into a fluid

Though injection will also take place in case the vessel vibrates in the fluid, i.e. we may also speak here about a phenomenon of the *vibrational injection of the fluid into the fluid* (Fig. 18.2,b). In this case the phenomenon is related to the vibro-jet effect, the only difference being the fact that the holes in the vessel need not necessarily be conic and the fluids inside and outside the vessel are not supposed, generally speaking, to be the same.

The theory of vibro-jet effect was considered in [2] and in the book [3]. Here we propose a general consideration, comprising, as special cases, the theory of both phenomena.

18.2 Common Expression for the Gas or Fluid Discharge through a Hole in Vibrating Vessel

The general scheme of the system under consideration is shown in Fig. 18.3. It is assumed that the vessel contains a certain fluid 1, and outside there is another fluid or gas 2. The space inside and outside the vessel will be designated by us by the same figures. The vessel can be both open and closed. The vessel performs vertical oscillations according to the law

$$y = -A\sin \omega t \qquad (18.1)$$

where A is the oscillation amplitude and ω is the oscillation frequency. The hole is assumed to be at the bottom of the vessel.

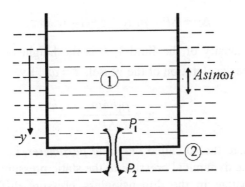

Figure 18.3. General scheme of the system "a vessel with a hole, containing the fluid, and vibrating in the air or in the fluid."

Volume discharge of the fluid or gas at their stationary outflow through a hole in a thin wall of an immobile vessel is determined by the known formula of hydraulics (see, say, [5]):

$$Q = \mu F \sqrt{2\Delta P/\rho} \qquad (18.2)$$

where ΔP is the pressure difference at the inlet and outlet of the hole, ρ is the density of the medium, flowing out, F is the area of the hole, μ is the dimensionless parameter (so-called a *coefficient of the discharge through the hole*).

In the ideal case $\mu = 1$, and at the free discharge of the liquid into the atmosphere from the vessel with the liquid column h, we have $\Delta P = \rho g h$, and formula (18.2) corresponds to the known formula of Torricelli: $v = Q/F = \sqrt{2gh}$ (g is the free fall acceleration).

We will assume as a supposition requiring an experimental testing (see below), that formula (18.2) is also valid for the instantaneous discharge at the pressure difference $|\Delta p|$ that varies in time. In this case, the coefficient μ is assumed to be dependent on both the direction of the outflow, which can be conditioned by the shape of the channel of the hole and the properties of the liquid flowing out.

Designating the static pressure difference at the hole by $\Delta P = P_1 - P_2 > 0$, we will have the expression for the pressure difference of the vessel under vibration

$$\Delta p = \Delta P - \rho_1 h \ddot{y} = \Delta P (1 - w \sin \tau) \qquad (18.3)$$

Here ρ_1 is the density of the fluid 1, h is the height of the fluid in the vessel, $\ddot{y}$ is the acceleration of the vessel, $\tau = \omega t$. By

$$w = \frac{\rho_1 h A \omega^2}{\Delta P} \qquad (18.4)$$

we will designate the so called *overload parameter*, i.e. the ratio of the amplitude of the dynamical pressure to the static pressure difference. The plot of the change in the dimensionless pressure difference $\Delta p / \Delta P$ is presented in Fig. 18.4.

Let us consider first of all the case $w > 1$. As one can see, in this case in the interval of the dimensionless time $\delta_0 < \tau < \pi - \delta_0$ where

$$\delta_0 = \arcsin 1/w, \qquad (18.5)$$

the value Δp is negative, and according to what was said, the fluid or gas 2 flows in into the vessel from the space 2, while in the interval $\pi - \delta_0 < \tau < 2\pi + \delta_0$ the value Δp is positive, and the fluid 1 flows out from the vessel into the space 2. Periods of flowing in and flowing out are respectively:

$$T_- = (\pi - 2\delta_0)/\omega, \quad T_+ = (\pi + 2\delta_0)/\omega. \qquad (18.6)$$

18.2. Common Expression for the Discharge

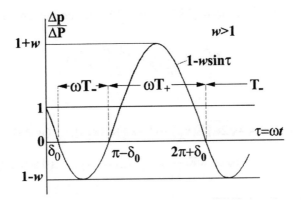

Figure 18.4. Dependence of the pressure difference on time.

The average for the period $T = 2\pi/\omega = T_+ + T_-$ the volume discharge per second of the medium flowing out and flowing into are (in accordance with the choice of the direction of the axis y, the volumes, flowing out of the vessel, are considered to be positive while those, flowing into the vessel, are considered to be negative):

$$Q_+^T = \frac{\omega}{2\pi}\mu_+ F \sqrt{\frac{2\Delta P}{\rho_+}} \int_{(\pi-\delta_0)/\omega}^{(2\pi+\delta_0)/\omega} \sqrt{1-w\sin\omega t}\, dt,$$

$$Q_-^T = -\frac{\omega}{2\pi}\mu_- F \sqrt{\frac{2\Delta P}{\rho_-}} \int_{\delta_0/\omega}^{(\pi-\delta_0)/\omega} \sqrt{w\sin\omega t - 1}\, dt \qquad (18.7)$$

$$(w \geq 1)$$

where μ_+ and μ_- denote the coefficients of the discharge in the corresponding directions, $\rho_+ = \rho_1$ and $\rho_- = \rho_2$ are the corresponding densities of the fluids.

At $0 \leq w \leq 1$ at any moment of time we have $\Delta p > 0$ and fluid 1 keeps on leaking from the vessel, so $T_- = 0$, $T_+ = T = 2\pi/\omega$. In this case

$$Q_+^T = \frac{\omega}{2\pi}\mu_+ F \sqrt{\frac{2\Delta P}{\rho_+}} \int_0^{2\pi/\omega} \sqrt{1-w\sin\omega t}\, dt, \quad Q_-^T = 0 \qquad (18.8)$$

$$(0 \leq w \leq 1)$$

18. Vibro-jet Effect. Phenomena of the Vibrational Injection

As a result the expressions (18.7) and (18.8) can be presented as

$$Q_\pm^T = \pm \frac{1}{2\pi} \mu_\pm F \sqrt{\frac{2\Delta P}{\rho_\pm}} J_\pm(w) \qquad (18.9)$$

where

$$J_\pm(w) = 4\sqrt{2w}[E(k_\pm) - (1 - k_\pm^2)K(k_\pm)], \quad (w \geq 1)$$
$$J_+(w) = 4\sqrt{1+w}\,E(\frac{1}{k_+}); \quad J_-(w) = 0, \quad (0 \leq w \leq 1) \qquad (18.10)$$

where $k_\pm = \sqrt{(w \pm 1)/2w}$ and **K** and **E** are the complete elliptic integrals of the first and second time respectively.

The plots of functions $J_\pm(w)$ are presented in Fig. 18.5.

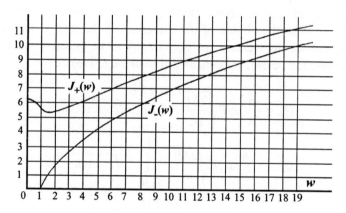

Figure 18.5. Plots of the functions $J_\pm(w)$

While making practical calculations one can use the following approximate formula:

$$J_\pm(w) = \frac{4}{3}\sqrt{w \pm 1}\left(\frac{\pi}{2} \pm \arcsin\frac{1}{w}\right), \quad (w \geq 1) \qquad (18.11)$$

obtained by the approximation of the integrands in (18.7) by the parabolas. At comparison with the experimental data, the values of the

18.2. Common Expression for the Discharge

discharge, averaged for the corresponding time of the outflow, are also of interest

$$Q_+^{T*} = \frac{T}{T_+} Q_+^T, \quad Q_-^{T*} = \frac{T}{T_-} Q_-^T \qquad (18.12)$$

It is not difficult to show that relations (18.7) correspond to the equation of vibrational hydraulics (17.2). Actually, taking into account that $Q = FU_2$, $\rho = \rho_1$, $|\Delta P| = P_1 - P_2$, we will rewrite relation (18.2) for the case of the stationery outflow in the form:

$$\frac{1}{2}\rho_1 U_2^2 - \mu^2 \Delta P = 0 \qquad (18.13)$$

This equation differs from the Bernulli equation only by the coefficient μ^2, introduced in hydraulics in order to take into account the loss of the pressure and some other factors.

If the vessel is vibrating, then, in accordance with the first of the formulas of (18.7), (taking into account that $Q_+^T = FU_2$ and $\rho_+ = \rho_1$), we have for the liquid outflow

$$\frac{1}{2}\rho_1 U_2^2 - \frac{1}{4\pi^2}\mu_+^2(P_1 - P_2)J_+^2(w) = 0. \qquad (18.14)$$

On the other hand, according to the equation of vibrational hydraulics (17.2), in this case ($U_1 = 0$ and taking into account the coefficient μ^2) the following relation should be satisfied

$$\frac{1}{2}\rho_1 U_2^2 + \mu^2(P_2 - P_1 + P_{2V} - P_{1V}) = 0 \qquad (18.15)$$

Comparing equalities (18.14) and (18.15) we come to the following expression for the vibrational pressure difference:

$$\Delta P_V = P_{2V} - P_{1V} = \left[\frac{1}{4\pi^2}\frac{\mu_+^2}{\mu^2}J_+^2(w) - 1\right](P_1 - P_2) \qquad (18.16)$$

Expressions obtained are used below for the consideration of the phenomena that are being discussed.

18.3 On the Theory of Vibro-jet Effect

In the case of a vibro-jet effect the fluids inside and outside the vessel are the same, so that $\rho_+ = \rho_- = \rho$, $w > 1$, and for the total average discharge of Q^T we obtain the expression

$$Q^T = Q_+^T + Q_-^T = \frac{F\sqrt{2\Delta P/\rho}}{2\pi}[\mu_+ J_+(w) - \mu_- J_-(w)] \quad (18.17)$$

If the following inequality is satisfied

$$\mu_+ / \mu_- < J_-(w)/J_+(w), \quad (18.18)$$

then in spite of the positive static pressure difference $\Delta P = P_1 - P_2 > 0$, the flowing of the fluid into the vessel takes place (for the period more fluid flowing in than flowing out), and when

$$\mu_+ / \mu_- = J_-(w)/J_+(w), \quad (18.19)$$

then the *effect of vibro-closing the vessel*, described above, takes place.

At $w > 5$ when using formulas (18.11), within the error of about 5%, the condition of the appearance of the inverse flow (18.18) can be presented as

$$\frac{\mu_+}{\mu_-} < 1 - \frac{2.27}{w} \quad (18.20)$$

and the condition of the vibrational closing as

$$\frac{\mu_+}{\mu_-} = 1 - \frac{2.27}{w} \quad (18.21)$$

Thus at $w = 5$ the effects under consideration take place when the coefficients μ differ approximately by 45%, and at $w = 10$ when they differ by 23%.

18.4 On the Theory of the Vibrational Injection of Gas into the Fluid

In this case there is fluid inside the vessel and there is gas outside. Believing that the vessel is open and that the pressure on the surface of the fluid and at the outlet from the hole is the same, we will have

$$\Delta P = \rho_1 g h \quad (18.22)$$

where g is the accelaration of the free fall. Then formulas (18.7) for the average for the period $T = 2\pi/\omega$ gas and liquid discharge $Q_+^T = Q_f^T$ and $Q_-^T = Q_g^T$ will be presented accordingly in the form

$$Q_+^T = Q_f^T = \frac{1}{2\pi}\mu_f F \sqrt{2gh} J_+(w)$$

$$Q_-^T = Q_g^T = -\frac{1}{2\pi}\mu_g F \sqrt{2gh\frac{\rho_f}{\rho_g}} J_-(w) \quad \text{at } w \geq 1 \quad (18.23)$$

$$Q_-^T = Q_g^T = 0 \quad \text{at } 0 \leq w \leq 1$$

Here, unlike (18.7) the values referring to the fluid are marked with the symbol "f", while those referring to the gas are marked with the symbol "g". In view of (18.22) expression (18.4) for the *overload parameter* takes the form

$$w = \frac{A\omega^2}{g} \quad (18.24)$$

It is natural to expect that the process of vibro-injection will take place only in the case if the volume of the gas, entering the vessel in the time T_- during every vibration period, will be sufficient to form bubbles with a diameter d, equal to that of the hole. This condition is illustrated in Fig. 18.6.

Fig. 18.6,*a* shows the sequential phases of the process in the case when the volume of the gas that enters the vessel is sufficient to form bubbles, while Fig. 18.6,*b* shows the case when this volume is insufficient and the vibro-injection does not take place.

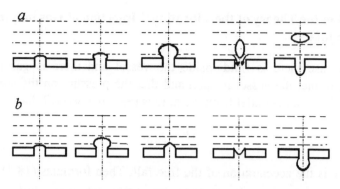

Figure 18.6. The sequential phases of the process of the suction of gas through the hole: a - the case when the volume of the gas that enters the vessel is sufficient to form bubbles, b - the case when this volume is insufficient and the vibro-injection does not take place.

The condition of forming bubbles leads to the inequality

$$\frac{2\pi}{\omega} Q^T = \frac{\mu_g F}{\omega}\sqrt{2gh\frac{\rho_f}{\rho_g} J_-(w)} > \frac{\pi d^3}{6} \qquad (18.25)$$

from which the following condition of the phenomenon of vibro-injection is obtained

$$h > h_*, \qquad h_* = \frac{2}{9}\frac{\rho_g}{\rho_f}\frac{d^2\omega^2}{\mu_g^2 J_-^2(w) g} \qquad (18.26)$$

In other words, the process takes place if the height of the column of fluid in the vessel exceeds a certain critical value h_*. Though since the ratio of the densities of gas and of fluid is very small, when vibration is intensive enough ($w > 3$) the value h_* is quite small; it can, however, be essential when vibration is not intensive ($w < 1.5$ and J_- is relatively small; see Fig. 18.5).

18.5 Results of the Experiments

Results of the investigation of the vibro-jet effect are stated in work [2]. Here we give but several results of the experiments on studying the

18.5. Results of the Experiments

vibrational injection. These experiments were made on a special vibrational stand elaborated at the "Mekhanobr" Institute several years ago. This stand after a number of modifications still remains the best and most universal means for the laboratory experiments of vibrational processes. On the counter of that stand an open glass cylindrical vessel was fixed, it was 300 mm high with the inner diameter 58 mm. In the center of the bottom of the vessel there was a hole with a diameter allowing a change. A certain level of water was kept in the vessel. The vessel was endowed with vertical vibration whose frequency and amplitude might be changed within certain limits.

The process of vibrational injection was observed in the stroboscopic light. That helped to establish that air was sucked into the vessel in the form of bubbles, one bubble being formed during each oscillation period. Figure 18.7 shows a photo of a vibrating vessel with the air being intensively sucked into it.

Figure 18.7. Vibrational injection of gas into fluid.

In the process of the experiment they measured the average volume discharge of water Q_f^T, flowing out of the hole. Figure 18.8 (a solid line) shows a plot of the dependence of that discharge $Q_{f\,\exp}^T$ on the overload parameter $w = A\omega^2 / g$ with the fixed values of the oscillation amplitude $A = 2.5$ mm, of the height of the column of water in the vessel $h = 200$ mm and the diameter of the hole $d = 2.6$ mm. As one can see, with the increase of w from $w = 1$, the discharge first diminishes and then begin to grow. Such type of dependence as well as of the dependence $J_+(w)$ (Fig. 18.5) is explained by the counteraction of two factors. On the one hand when w grows, and according to (18.8) the time interval T_+, in which the water flows out of the vessel, decreases, though it is always not less than the vibration half-period π / ω. On the other hand, when w increases, the pressure difference at the hole increases too. As a result, with the growth of w the effect of the second factor begins to dominate. It is remarkable that the corresponding minimum for the dependence $J_+(w)$ is shifted to the left with respect to the minimum of the dependence $Q^T_{f\exp}(w)$. We will not dwell here on the explanation of this regularity.

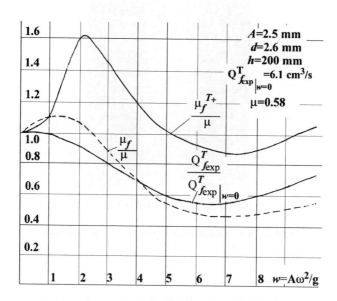

Figure 18.8. Dependence of the water discharge through the holes and the coefficients μ_f and $\mu_f^{T_+}$ on the overload parameter w.

18.5. Results of the Experiments

For the discharge of the air $Q_g^T(w)$ only the second factor is valid. The dependence $J_-(w)$ is monotonous. A sufficiently intensive sucking-in of the air into the vessel, observed in the conditions of the experiment, begins with the value $w \approx 2.5$, while according to the above-presented theory it must begin at w somewhat more than 1. This can be explained by the effect of the surface tension when bubbles are being formed.

To compare the results of the theory proposed with the experimental data it seems interesting to compare the values of the discharge coefficient μ_f with its value μ for the case of the stationary flow of the fluid through the hole. Here we can distinguish the value μ_f, found by formula (18.23) in accordance with the discharge, measured experimentally (we will arbitrarily call it "the seeming value"), and the value μ_f^{T*}, found in accordance with the average discharges, calculated by formula (18.12) in view of the actual time of the flow out. Plots of those dependences are given in Fig. 18.8. (in dotted and solid lines respectively). As one can see, the values μ_f^{T*} at $w > 4$ do not differ much from the value μ, which, it seems, speaks in favor of the elements of the theory of the phenomenon stated above.

References

1. *Vibration in technology*, Handbook in 6 volumes, **4** (Mashinostroenie, Moscow, 1978-1981), p. 351 (in Russian).
2. Blekhman I.I., Vaisberg L.A., and Korovnikov A.N., Analysis of hydrodynamics of vibrational screen with a sieve oscillating in aqueous medium, *Interdepartmental Proceedings "Investigations of the processes, machines and apparatus for classifying the material according to size"* (Mekhanobr, Leningrad, 1988) 35-46 (in Russian).
3. Blekhman I.I. *Vibrational Mechanics – Nonlinear Dynamic Effects, General Approach, Applications.* Singapore: World Scientific (2000) 509.
4. Blekhman I.I., Blekhman L.I., Vaisberg L.A., Vasilkov V.B., Yakimova K.S. Phenomenon of vibrational injection of gas into liquid. In collection *"Scientific discoveries"*, Moscow, Russian Academy of Natural Sciences (2002) 60-61(in Russian).
5. Chugaev R.R. *Hydraulics.* (Energoizdat, Leningrad. Otdelenie 1982) 672 (in Russian).

Part VI

Some Mathematical Supplement and Generalization

Part IV

Chapter 19

On Asymptotic Analysis of Systems with Fast Excitation

A. Fidlin

LuK GmbH &Co, Industriestr.3, D-77815 Germany

19.1 Introduction. Classification of Systems with Fast Excitation. Weakly Excited Systems

The effect of high frequency excitation on the low frequency motions of dynamic systems is considered. In section 19.1 it is suggested to differ between weak, strong and very strong high frequency excitations. Several approaches and difficulties connected with the analysis of these systems are discussed. Systems with strong excitation are examined in the most general form in section 19.2. Systems with very strong excitation are considered in section 19.3 for the case important in mechanics: fast oscillating inertia matrix. Two mathematical examples of systems with strong excitation, demonstrating the principal difference between systems with the velocity independent excitation and systems with the excitation terms containing velocities are analysed in section 19.4. As an example, the responses of a one degree of freedom system to strong and very strong, high frequency external and parametric excitation are compared in section 19.5.

Systems with high frequency excitation have recently attracted great attention. The conventional technical approach to systems with high frequency excitation, i.e. the excitation, where the frequency significantly exceeds the essential natural frequencies of the system, is that this excitation is almost unessential for the low frequency motions of the system, because of its strong filtering properties. However, it is well known, that this point of view is not always correct. If the high frequency excitation is strong enough, it is able to significantly change the properties of the system with respect to slow motions. Numerous examples of such systems can be found in the works of Bogoliubov and Mitropolskii, Nayfeh, Blekhman, Thomsen and others [1, 2, 3, 4, 5, 6, 7, 8]. The main approach for the useful analysis of these systems is the separation of motions.

The separation of motions is one of the main ideas for asymptotic analysis of oscillating systems with small or big parameters. It is connected with the fact, that solutions of many types of dynamic systems can be represented as a superposition of fast oscillations and slow evolution. These slow motions are the main reason for interest in most applications. Powerful asymptotic methods, such as the averaging method of Bogoliubov and Mitropolskii [1], are in fact nothing but a practical realisation of this idea. The same could be said about the direct separation of motions method, which originated in the works of Kapitsa [10] and was most generally formulated by Blekhman [3, 4]. The method of multiple scales developed in the works of Nayfeh [2, 9], is similar in substance to the averaging method.

First of all, in this section we are going to consider systems of mechanical type, i.e. systems described by second order differential equations in the following form:

$$\ddot{x} = \omega^{\alpha} \Phi(x, \dot{x}, t, \omega t) \qquad (19.1)$$

Here x is an n-dimensional vector of the generalized coordinates and $\dot{x}$ is a vector of the generalized velocities, $\omega \gg 1$ is a big parameter. We take Φ to be an n-dimensional vector of forces, which depends 2π-periodically on the fast time $\tau = \omega t$.

Depending on the magnitude of the integer parameter α we shall distinguish between *systems with weak* ($\alpha = 0$), strong ($\alpha = 1$) and very strong ($\alpha = 2$) *excitation*. Weak excitation is trivial, but it illustrates in the best way, how asymptotic methods are used for such problems. Let us use, for example, the averaging technique and rewrite our system as a system of 2n first order equations:

$$\dot{x} = y; \quad \dot{y} = \Phi(x, y, t, \tau) \qquad (19.2)$$

19.2. Systems with Strong Excitation

Converting to the fast time τ as an independent variable, we get

$$x' = \varepsilon y; \quad y' = \varepsilon \, \Phi(x,y,t,\tau); \quad t' = \varepsilon, \quad \varepsilon = \omega^{-1} \ll 1 \quad (19.3)$$

If function Φ is smooth enough, we get the standard form system for averaging and can apply it directly. The equation of the first approximation takes the form:

$$\ddot{\xi} = \langle \Phi(\xi,\dot{\xi},t,\tau) \rangle \quad (19.4)$$

Here $\langle \Phi \rangle$ is the average of Φ with respect to the fast time τ, and $(\xi,\dot{\xi})$ are asymptotically close to the solution of the original system $(x,\dot{x})$ for the time interval $\tau = O(\omega)$ or $t = O(1)$.

The obtained solution is a superposition of small fast oscillations and slow evolution of the system for both the generalized co-ordinates and the generalized velocities. This case is well known, so more complicated case of the systems with strong excitation should be considered.

19.2 Systems with Strong Excitation. General Analysis

Particular *systems with strong excitation* are usual in various applications. A lot of them are studied in detail in the works of Blekhman [3, 4]. Here we consider the case when the motion of the system is described by the equation of the type:

$$\ddot{x} = \Phi_0(x,\dot{x},t,\tau) + \omega \Phi_1(x,\dot{x},t,\tau) \quad (19.5)$$

with the initial conditions

$$x\big|_{t=0} = x_0, \quad \dot{x}\big|_{t=0} = \dot{x}_0 \quad (19.6)$$

Equation (19.5) differs from those considered in Part I and in book [4] by the fact that all the forces acting on the system are supposed to be depended on τ and consisted of two parts which differ by an order with respect to ω. In writing equation (19.5) the dependence of the function Φ_1 on $\dot{x}$ is quite essential. The main results of this section are connected with the presence of this dependence. Such a system deserves special consideration.

Usually these systems are used if we are interested in analysing motions of a machine supposing, that the inertia of its housing is significantly larger then the inertia of its moving parts. If the mass or inertia of the mechanism's moving parts is not small (for example the type of modern mechanisms often found in crank gears or vane pumps), equations of motion containing fast oscillating inertia coefficients will be obtained. These equations contain big, fast oscillating terms depending not only on the generalized coordinates but also on the system's generalized velocities.

Another example of systems with strong excitation depending on the first derivative of the unknown function, appears if we investigate vibrations or wave propagation in inhomogenious media. For example, the longitudinal waves in a rod with periodic or quasi-periodic structure. In this case the typical equations with strong excitation of general form appear naturally with respect to the spatial coordinates. Equations with the slow modulated, high frequency excitation, which we are going to analyze in section 5 of this chapter, are first of all typical for this group of applications.

Firstly, a general mathematical approach to the systems (19.1) is given.

Theorem 1

Consider system (19.1) with

$\Phi_0 : R^{2n+2} \to R^n$, $\Phi_0 \in C^1(R^{2n+2})$;

$\Phi_1 : R^{2n+2} \to R^n$, $\Phi_1 \in C^2(R^{2n+2})$,

$x, x_0 \in D_x \subset R^n$; $\dot{x}, v_o \in D_v \subset R^n$, $\omega \gg 1$

Suppose
1. *All the functions are 2π-periodic and $\langle \ \rangle$ means the averaging with respect to τ.*
2. *The general 2π - periodic with respect to τ bounded solution of the system of n first order differential equations*

$$\partial u/\partial \tau = \Phi_1(X,u,t,\tau), \ \langle u(X,t,\tau)\rangle = \dot{X} \Rightarrow u = U(X, \dot{X}, t, \tau), \quad (19.7)$$

is known for $X \in D_X, \dot{X} \in D_V$ and $(u + \dot{X}) \in D_V$.

19.2. Systems with Strong Excitation

Consider together with (19.5) a system of ordinary differential equation, which do not contain the fast time τ.

$$M(X,\dot{X},t)\ddot{X} = V(X,\dot{X},t),$$
$$X\big|_{t=0} = x_0, \quad \dot{X}\big|_{t=0} = v_0 - \partial\Psi/\partial\tau\big|_{t=0,\tau=0}, \quad (19.8)$$

with

$$M(X,\dot{X},t) = \langle W_*^T(\partial U/\partial \dot{X})\rangle; \quad V(X,\dot{X},t) = \langle W_*^T \Phi_0(X,U,t,\tau)\rangle$$
$$+\langle W_*^T \{(\partial\Phi_1/\partial x)\Psi - (\partial U/\partial X)\dot{X} - \partial U/\partial t\}\rangle, \quad (19.9)$$

$$\Psi = \int_0^\tau (U - \dot{X})d\tau$$

where W_* is the fundamental matrix of solutions for system

$$\partial W_*/\partial \tau = -(\partial\Phi_1/\partial \dot{x})^T W_*. \quad (19.10)$$

Suppose that its solutions belong to the interior subset of $D_X \otimes D_V$.

Then solutions of (19.5) and (19.8) are asymptotically close to each other for the time interval $t = O(1)$.

The proof of the Theorem 1 based on the generalized averaging method can be found in [11]. Here it is illustrated by the formal multiple scales technique. In order to apply it, we have to convert from a system of ordinary differential equations (19.5) to a system with partial derivatives and two independent variables t and τ:

$$\frac{\partial^2\varphi}{\partial t^2} + 2\omega\frac{\partial^2\varphi}{\partial t \partial \tau} + \omega^2\frac{\partial^2\varphi}{\partial \tau^2}$$
$$= \Phi_0\left(\varphi, \frac{\partial\varphi}{\partial t} + \omega\frac{\partial\varphi}{\partial \tau}, t, \tau\right) + \omega\Phi_1\left(\varphi, \frac{\partial\varphi}{\partial t} + \omega\frac{\partial\varphi}{\partial \tau}, t, \tau\right). \quad (19.11)$$

The relationship between (19.5) and (19.7) is given by a condition, that if $\varphi(t,\tau)$ is a solution of equation (19.11), then this solution taken along the straight line $\tau = \omega t$, i.e. $x = \varphi(t, \omega t)$ is a solution of equation (19.5). In other words, system (19.11) is more general than equation (19.5), so we are free in choice of boundary conditions for this system. The only restriction is that the straight line $\tau = \omega t$ should be in the inner part of the considered area. We require $\varphi(t,\tau)$ to be a 2π-periodic function of τ and try to find $\varphi(t,\tau)$ as a formal asymptotic expansion in terms of the small parameter $\varepsilon = 1/\omega$:

$$\varphi(t,\tau) = \psi_0(t,\tau) + \varepsilon\psi_1(t,\tau) + \varepsilon^2\psi_2(t,\tau) + \dots \quad (19.12)$$

Substituting this expansion into (19.11) and balancing the terms with equal orders of ε we obtain:

$$\varepsilon^{-2}: \quad \frac{\partial^2 \psi_0}{\partial \tau^2} = 0, \quad (19.13)$$

$$\varepsilon^{-1}: \quad \frac{\partial^2 \psi_1}{\partial \tau^2} + 2\frac{\partial^2 \psi_0}{\partial t \partial \tau} = \Phi_1\left(\psi_0, \frac{\partial \psi_1}{\partial \tau} + \frac{\partial \psi_0}{\partial t} + \omega\frac{\partial \psi_0}{\partial \tau}, t, \tau\right) \quad (19.14)$$

$$\varepsilon^0: \quad \frac{\partial^2 \psi_2}{\partial \tau^2} + 2\frac{\partial^2 \psi_1}{\partial t \partial \tau} + \frac{\partial^2 \psi_0}{\partial t^2} = \Phi_0 + \frac{\partial \Phi_1}{\partial x}\psi_1 + \frac{\partial \Phi_1}{\partial \dot{x}}\left(\frac{\partial \psi_1}{\partial t} + \frac{\partial \psi_2}{\partial \tau}\right). \quad (19.15)$$

The last step must be justified, because the second argument of all the functions on the right hand sides of the equations is given as:

$$\partial\psi_1/\partial\tau + \partial\psi_0/\partial t + \omega\partial\psi_0/\partial\tau \quad (19.16)$$

This expression can take values of the magnitude order of the big parameter ω. So it can create in our equations terms of any order, depending on the kind how Φ_1 depends from $\dot{x}$. If we require Φ_1, as usual, to be a bounded function in the vicinity of the solution of the averaged system, we reduce the problem to *a posteriory* check of our assumptions about the magnitude order of $\dot{x}$ in the vicinity of the found solution.

However, in this case the problem is insignificant. The general solution of equation (19.13) has the following form:

$$\psi_0(t,\tau) = X(t) + A(t)\tau \quad (19.17)$$

According to the periodicity of ψ_0, we get $A(t) = 0$. Hence, $\psi_0 = X(t)$. This depends only on the slow time t and the large terms in

19.2. Systems with Strong Excitation

the equations (19.14) and (19.15) disappear automatically. However, just this problem prevents the general analysis of systems with very strong excitation.

The objective of the following analysis is to discover differential equations for the still unknown function $X(t)$, which do not contain the fast time τ.

Equation (19.14) after substituting in it the solution of equation (19.13) takes the form:

$$\partial^2 \psi_1 / \partial \tau^2 = \Phi_1 \left(X, \dot{X} + \partial \psi_1 / \partial \tau, t, \tau \right) \tag{19.18}$$

It is natural to call this equation "The Equation of Fast Motions". It is a differential equation with only partial derivatives with respect to τ. So we can take $X, \dot{X}$ and t to be constant parameters during solving (19.18). Referring to the second condition of the Theorem (*cf.* (19.7), (19.9)) we can rewrite the solution of equation (19.18) as follows:

$$\psi_1 = \Psi(t, \tau) + X_1(t) \tag{19.19}$$

The new unknown function $X_1(t)$ is a small slow correction to the main slow part of the solution $X(t)$. System (19.7) is significantly simpler than the original system (19.5), because its order is twice lower and we can take all the functions of slow time t to be constant. In other words we are not interested in the slow evolution of the system here, but only in its high-frequency oscillations.

Let us move on now to the equation of second approximation (19.15) as follows:

$$\frac{\partial v}{\partial \tau} = \frac{\partial \Phi_1}{\partial \dot{x}} v + \Phi_0 + \frac{\partial \Phi_1}{\partial x} \psi_1 - \frac{\partial^2 \psi_1}{\partial t \partial \tau} - \ddot{X}; \quad v = \frac{\partial \psi_2}{\partial \tau} + \frac{\partial \psi_1}{\partial t}. \tag{19.20}$$

This is a system of n first order linear inhomogeneous equations with periodic coefficients. As it is known from the general theory of linear systems with periodic coefficients, for the existence of periodic solutions to (19.20) the projections of the inhomogeneous parts of the equations on the solutions of the system conjugated to the homogeneous one must vanish. (A reference to the classical work of Poincarè [13] seems to be in order here and to underscore the close relationship of Poincarè's method to the method of multiple scales, which is in this case a procedure to find periodic with respect to τ solutions of (19.11) with a not isolated unperturbed solution (19.17)).

$$\left\langle W_*^T \left\{ \Phi_0 + (\partial \Phi_1/\partial x)\psi_1 - \partial^2 \psi_1/\partial t \partial \tau - \ddot{X} \right\} \right\rangle = 0. \tag{19.21}$$

W_* is defined in (19.10). In order to get the final form of (19.21) let us notice that

$$\left\langle W_*^T \partial \Phi_1/\partial x \right\rangle = 0. \tag{19.22}$$

Due to this identity, one can show, that (19.21) does not contain $X_1(t)$, and reduce this equation as follows:

$$\left\langle W_*^T \right\rangle \ddot{X} = \left\langle W_*^T \left\{ \Phi_0 + (\partial \Phi_1/\partial x)\Psi - \partial^2 \Psi/\partial t \partial \tau \right\} \right\rangle \tag{19.23}$$

Lastly, function Ψ depends on t both direct and indirect through functions $X(t)$ and $\dot{X}(t)$. Under $\partial/\partial t$ we understand here the full partial derivative with respect to t. Taking this into consideration and using the partial partial derivatives we will obtain the final explicit form of equation (19.23) given in (19.8), (19.9).

Equations (19.9) do not contain fast time and determine slow evolution of the solutions to the original system (19.5). That's why they could be called "Equations of Slow Motions" or "Main Equation of Vibrational Mechanics" (see chapter 2). Function $V(X, \dot{X}, t)$ is natural to call "vibration force" and Matrix $M(X, \dot{X}, t)$ can be interpreted as a matrix of the averaged system's efficient mass with respect to slow motions. This matrix depends on the solution of the equations of fast motions, i.e. on the amplitude of fast excitation. So one can say that the action of high-frequency oscillation, with respect to slow motions, leads not only to the appearance of vibrational force but also to the change of the effective mass of the system. This circumstance is mentioned in [4]. Along with that, multiplying equation (19.9) on the left by the matrix $M_0 \left\langle W_*^T (\partial U/\partial \dot{X}) \right\rangle^{-1}$ one can always transfer this equation to the form

$$M_0 \ddot{X} = V_*$$

where $M_0 > 0$ is a certain constant, i.e. one can speak only about the appearance of the vibrational force like it is interpreted, as a rule, in [4]. The vibrational force, introduced in such a way, in the case when

$\langle W_*^T \rangle \ne$ const is of a formal character. Due to this, the values $M(X,\dot{X},t)$ and $V(X,\dot{X},T)$ defined in accordance with (19.9) can be called physical effective mass and vibrational force respectively.

If function Φ_1 does not depend on $\dot{x}$, system (19.23) goes over into equation

$$\ddot{X} = \langle \Phi_0 + (\partial \Phi_1/\partial x)\Psi \rangle. \tag{19.24}$$

19.3 Systems with Very Strong Excitation in a Special Case of Fast Oscillating Inertial Coefficients

Now we are going to consider systems with very strong excitation. Unfortunately it is not possible to perform this analysis for an arbitrary function Φ_1. That's why we are forced to restrict the analysis to a special form:

$$\ddot{x} = \omega^2 \Phi_0(t,\tau) + \omega \Phi_1(t,\tau)\dot{x} + \Phi_2(t,\tau)\dot{x} + \Phi_3(x,t,\tau); x(0) = x_0;$$
$$\dot{x}(0) = v_0; \quad \tau = \omega t \tag{19.25}$$

Such a system appears naturally in mechanics, if the corresponding kinetic energy contains inertial coefficients depending on the fast time: $T = \frac{1}{2} \dot{x}^T J(t,\tau)\dot{x}$ and the external excitation is very strong.

The analysis of this system is from the technical point of view very similar to the case of the strong excitation. It would be natural to expect strong response, i.e. solutions with $x \gg O(1)$, for systems with very strong excitation. But under some special conditions solutions $x = O(1)$ (weak response) are also possible. This case is considered in the next theorem.

Theorem 2

Consider system (19.25) with

$F_1, \Phi_1 : R^2 \to R^{n \otimes n}; \; F_2 : R^{n+2} \to R^n; \; \Phi_0 : R^2 \to R^n;$
$F_2 \in C^1(R^{n+2}), \; F_1, \Phi_0, \Phi_1 \in C^2(R^2),$
$x, x_0 \in D_X \subset R^n; \; \dot{x}, v_0 \in D_V \subset R^n \; \omega \gg 1.$

19. On Asymptotic Analysis of Systems with Fast Excitation

Suppose

1. All the functions are 2π - periodic and $\langle\ \rangle$ means the averaging with respect to τ.

2. The bounded, 2π - periodic with respect to τ solution to the system of n first order linear differential equations is known for $u \in D_V$:

$$\partial u/\partial \tau = \Phi_1(t,\tau)u + \Phi_2(t,\tau),$$
$$\langle u(t,\tau)\rangle = 0 \Rightarrow \langle W_*^T \Phi_2 \rangle = 0, \quad \text{with } \partial W_*/\partial \tau = -\Phi_1^T W_*. \quad (19.26)$$

3. The bounded, 2π - periodic solution to the system of n first order linear differential equations is known for $(\dot X_0 + u) \in D_V$:

$$\partial v/\partial \tau = \Phi_1 v + \Phi_2 u - \partial u/\partial t,$$
$$\langle v(t,\tau)\rangle = \dot X_0 \Rightarrow \langle W_*^T(\Phi_2 u - \partial u/\partial t)\rangle = 0. \quad (19.27)$$

Consider together with (19.25) a system of ordinary differential equation, which do not contain the fast time τ:

$$M(X,\dot X,t)\ddot X = V(X,\dot X,t), X|_{t=0} = x_0 - \Psi|_{t=0,\tau=0}$$
$$\dot X|_{t=0} = v_0 - v - \omega u|_{t=0,\tau=0}, \quad (19.28)$$

with

$$M(X,\dot X,t) = \langle W_*^T \rangle;$$
$$V(X,\dot X,t) = \langle W_*^T \Phi_2(t,\tau)\rangle \dot X + \langle W_*^T \Phi_0(X+\Psi,t,\tau)\rangle + \langle W_*^T(\Phi_2 v - \partial v/\partial t)\rangle,$$
$$\Psi = \int_0^\tau (u - \dot X_0)d\tau$$

(19.29)

Suppose that its solutions belong to the interior subset of $D_X \otimes D_V$.

Then solutions of (19.25) and (19.28) are asymptotically close to each for the time interval $t = O(1)$.

The proof of this theorem is very similar to the previous one and can be obtained using both the multiple scales and the modified averaging method.

19.4 Two Mathematical Examples of Systems with Strong Excitation

Now we are going to consider two mathematical examples of systems with strong excitation, illustrating the main qualitative difference between the systems with strong terms depending on the velocities (19.5) and systems where these terms depend only on the coordinates or the corresponding equations of slow motions (19.8), (19.9) and (19.24).

Firstly, following to [11] we consider a system with one degree of freedom. Secondly an example of a system with two degrees of freedom is analyzed. It is necessary to emphasize that these examples are pure mathematical. Further mechanical examples are discussed in the next section.

As the first example we are going to analyze the following equation:

$$\ddot{x} + \beta \dot{x} + x = a\omega \dot{x}\cos(\omega t) \qquad (19.30)$$

Let us assume that a and β both are magnitude order 1 and $\omega \gg 1$ is the big parameter. Then according to our designations $\Phi_0 = -\beta \dot{x} - x$, $\omega \Phi_1 = a\omega \dot{x}\cos\tau$. The equation of fast motions takes form:

$$\partial u / \partial \tau = au\cos\tau. \qquad (19.31)$$

Its periodic solution according to (19.7) is

$$u = \dot{X} \, e^{a\sin\tau} / I_0(a). \qquad (19.32)$$

Here $I_0(a) = \dfrac{1}{2\pi}\int_0^{2\pi} e^{a\sin\vartheta} d\vartheta$ is the modified Bessel's function of 0-order. The linear homogeneous system (19.10) has the form $\partial W / \partial \tau = -aW_* \cos\tau$ and its periodic solution is $W_* = e^{-a\sin\tau}$, $W_* > 0$. Substituting these expressions into (19.9) and averaging the corresponding term one obtains the following equation of slow motions:

$$\ddot{X} + \beta \dot{X} + I_0^2(a)X = 0. \tag{19.33}$$

Equation (19.33) is a very simple linear differential equation with constant coefficients. Its solution, together with the fast component (19.32) gives us the approximate solution to (19.30). A comparison of the analytic solution with a numeric one can be found in Fig. 19.1 and Fig. 19.2 for two different values of big parameter (ω = 50 – Fig. 19.1; ω = 10 – Fig. 19.2). In both cases the calculations were done for $\beta = 0$, $a = 1 \Rightarrow I_0(1) \approx 1,266$.

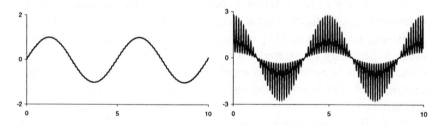

Figure 19.1. Comparison between analytic (——) and numeric (——) solutions.

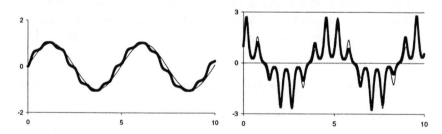

Figure 19.2. Comparison between analytic (——) and numeric (——) solutions for smaller values of the big parameter.

In the first case there is no visible difference between two predictions. For smaller values of ω the trajectories diverge already at early times.

Let us notice some features of this solution.

- It is typical for systems with strong excitation that the solution is a superposition of slow component and fast oscillations. These are small respectively to the generalized co-ordinate x, but their derivatives are not small respectively to the generalized velocities $\dot{x}$.
- Big fast oscillations of the damping coefficient change the frequency of slow free oscillations of the system. This frequency can be controlled by the excitation amplitude. This result is similar to that reported by Nayfeh and Nayfeh [12] for another particular system. This effect is typical for systems with strong excitation, but in this particular case it is caused by the changed effective mass of the system. This thesis is even more clearly illustrated by the next example.

As the second mathematical example we are going to consider a system with two degrees of freedom.

$$\begin{aligned}\ddot{x}_1 &= a\omega \dot{x}_2 \cos\omega t + F_1(x_1, x_2) \\ \ddot{x}_2 &= -a\omega \dot{x}_1 \cos\omega t + F_2(x_1, x_2)\end{aligned} \quad (19.34)$$

Let us assume that a is magnitude order 1 and $\omega \gg 1$ is the big parameter. F_1 and F_2 are arbitrary bounded functions. Then according to our designations

$$\Phi_1 = a\cos\tau Q\dot{x}; \quad \Phi_0 = \begin{bmatrix} F_1 \\ F_2 \end{bmatrix}; \quad Q = \begin{bmatrix} 0 & 1 \\ -1 & 0 \end{bmatrix}; \quad x = \begin{bmatrix} x_1 \\ x_2 \end{bmatrix}. \quad (19.35)$$

The equation of fast motions takes form:

$$\partial u/\partial \tau = aQu\cos\tau. \quad (19.36)$$

Its periodic solution according to (19.7) is

$$u = W_* \dot{X}/J_0(a). \quad (19.37)$$

Here $J_0(a) = \dfrac{1}{2\pi}\displaystyle\int_0^{2\pi}\cos(a\sin\tau)d\tau$ is the Bessel's function of 0-order.

W_* is the fundamental matrix to the linear homogeneous system (19.10) taking in this particular case the form $\partial W_*/\partial \tau = -QW_* a\cos\tau$:

$$W_* = \begin{bmatrix} \cos(a\sin\tau) & \sin(a\sin\tau) \\ -\sin(a\sin\tau) & \cos(a\sin\tau) \end{bmatrix}. \tag{19.38}$$

Substituting these expressions into (19.9) and averaging the corresponding terms one obtains the following equation of slow motions:

$$\ddot{X} = J_0^2(a)F(X). \tag{19.39}$$

This system is extremely interesting. It is common for systems with fast excitation, which does not depend on the generalized velocities, that there appear additional slow vibration forces. These terms are usually added to the existing slow forces. Here we have got a totally different result. The existing slow forces at least in the considered approximation are *multiplied by the factor*, which depends on the amplitude of fast excitation. And even more. Due to the oscillating character of the Bessel's function J_0 it has an infinite number of zeros. It means there is a countable number of excitation amplitudes corresponding to these zeros, for which the equation of slow motions takes the form $\ddot{X} = 0$ for arbitrary functions F, *i.e.* the system does not react to an arbitrary slow effect. This unusual property is in some sense similar to the quantization effect in physics. It is interesting to notice that this effect still remains for the higher order approximations and the orbit calculated numerically for the first zero of the Bessel's function is shown in Fig. 19.3.

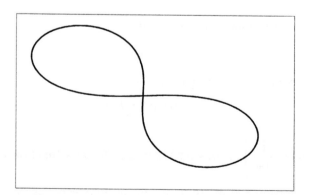

Figure 19.3. Stationary orbit to system (19.34) in the co-ordinates plane (x_1, x_2) for $a= 2.40482$ and $\omega =300$.

It must be emphasized that system (19.34) does not exist in mechanics and the considered example has pure mathematical character. Nevertheless it demonstrates clearly the principal difference between systems containing fast excitation terms depending on the velocities and those, which do not contain these terms. The obtained result is quite unexpected and it would be interesting to analyze if there are physical systems described by these or similar equations.

19.5 Response of a One Degree of Freedom System to Strong and Very Strong, High Frequency External and Parametric Excitation.

Now we are going to consider a more physical example, which has simple mechanical origins. The equations with strong excitation appear naturally, as it was mentioned above, in systems with oscillating inertia coefficients and in continuous systems with periodic structure. Let us take firstly the simplest example of such an equation in case of a strong external excitation:

$$(\dot{x}/(1+c(t)\cos\omega t))^{\cdot} + x + \alpha x^3 = \omega f(t)\sin(\omega t + \theta). \quad (19.40)$$

We can rewrite this equation as follows:

$$\ddot{x} = -(x + \alpha x^3)(1 + c\cos\tau) + \dot{x}\dot{c}\cos\tau(1 + c\cos\tau)^{-1} \\ + \omega f \sin(\tau + \theta)(1 + c\cos\tau) - \omega \dot{x} c \sin\tau(1 + c\cos\tau)^{-1}. \quad (19.41)$$

Referring back to the general form (19.5) we can sign

$$\Phi_0 = -(x + \alpha x^3)(1 + c\cos\tau) + \dot{x}\dot{c}\cos\tau(1 + c\cos\tau)^{-1}; \\ \Phi_1 = f\sin(\tau + \theta)(1 + c\cos\tau) - \dot{x}c\sin\tau(1 + c\cos\tau)^{-1}. \quad (19.42)$$

The corresponding equation of fast motions has the form:

$$\partial u/\partial \tau = -uc\sin\tau(1 + c\cos\tau)^{-1} + f\sin(\tau + \theta)(1 + c\cos\tau). \quad (19.43)$$

Its solution, which fulfils the condition $\langle u \rangle = \dot{X}$, is

$$u = \left(\dot{X} + \tfrac{1}{2} fc\cos\theta - f\cos(\tau+\theta)\right)(1+c\cos\tau). \qquad (19.44)$$

Solution of the system conjugated to the homogeneous part of the equation of second approximation

$$\partial W_* / \partial \tau = cW_* \sin\tau (1+c\cos\tau)^{-1} \qquad (19.45)$$

is also simple to find:

$$W_* = (1+\cos\tau)^{-1}. \qquad (19.46)$$

Averaging the corresponding terms one obtains:

$$M = 1; V = -X - \alpha X^3 - \tfrac{1}{2}(fc)^\bullet \cos\theta.$$

Finally, we receive the equation of slow motions:

$$\ddot{X} + X + \alpha X^3 = -\tfrac{1}{2}(fc)^\bullet \cos\theta. \qquad (19.47)$$

This equation has several interesting peculiarities. Firstly, in this case under the suppositions made about the orders of the values in this approximation, neither the effective mass, nor the natural frequency is transformed. Instead of it we have another interesting phenomenon – transformation of the slow excitation's character. There are both external and parametric high-frequency excitations in the original system. If the high frequency excitations are slowly modulated through the functions $f(t)$ and $c(t)$, it means, that we do not only deal with the high frequency but also with the low frequency parametric and external excitations of the original system. In the equation of slow motions the parametric excitation disappeared. The external excitation is transformed in an unexpected way. It got to be proportional to the first derivative of the product of the slow variable amplitudes of both external and parametric high-frequency excitations.

The properties of the averaged system are illustrated through the following figures, obtained by numeric simulation of the full equation (19.40). Figure 19.4 shows the solution for the parameter set: $\alpha = 0$; $\omega = 100$; $\theta = 0$.

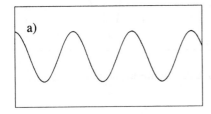
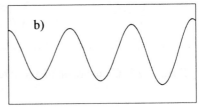

Figure 19.4. System with strong excitation, displacement;
a) modulation, which do not excite slow resonance:
$$f = 1 + 0.5\sin t; \quad c = \tfrac{1}{3}(1 + 0.5\sin t)^{-1};$$
b) modulation exciting slow resonance:
$$f = 1; \quad c = \tfrac{1}{3}(1 + 0.5\sin t)^{-1}.$$

This figure illustrates the typical character of the solutions of systems with strong excitation, which is a superposition of slow motion and fast oscillations. The amplitude of the fast velocity oscillations is comparable with the amplitude of its slow evolution.

In case a) we have $(fc)^{\bullet} = 0$. As it can be seen, in this case we have stationary oscillations of the averaged system. Figure 19.4, b) shows the results of the simulation for the case $(fc)^{\bullet} \neq 0$. In this case, according to the prediction, we have a typical picture of the external non-parametric resonance with the linearly increasing amplitude of slow oscillations.

In these cases there is no visible difference between analytic and numeric predictions. Figure 19.5 shows the comparison of analytic and numeric solutions for $\omega = 5$. We can see, that, although in this case the small parameter is not small enough, the asymptotic solution still gives the qualitative character of the system's movement. However, the quantitative differences are significant.

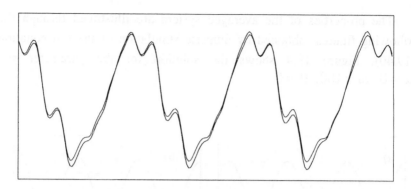

Figure 19.5. Comparison of asymptotic and numeric solutions. Travels.

We are going now to consider the same system but with very strong excitation.

$$(\dot{x}/(1+c(t)\cos\tau))\dot{} + x + \alpha x^3 = \omega^2 f(t)\sin(\tau+\theta). \qquad (19.48)$$

This equation can be rewritten as follows:

$$\ddot{x} = \omega^2 f \sin(\tau+\theta)(1+c\cos\tau) - \omega\dot{x}c\sin\tau(1+c\cos\tau)^{-1} \\ + \dot{x}\dot{c}\cos\tau(1+c\cos\tau)^{-1} - (x+\alpha x^3)(1+c\cos\tau). \qquad (19.49)$$

Referring back to the general form (19.25) we can sign:

$$\Phi_0 = f\sin(\tau+\theta)(1+c\cos\tau); \quad \Phi_1 = -c\sin\tau(1+c\cos\tau)^{-1}; \\ \Phi_2 = \dot{c}\cos\tau(1+c\cos\tau)^{-1}; \quad \Phi_3 = -(x+\alpha x^3)(1+c\cos\tau). \qquad (19.50)$$

The corresponding equation of fast motion (19.26) has the following form:

$$\partial u/\partial \tau = f\sin(\tau+\theta)(1+c\cos\tau) - c\sin\tau(1+c\cos\tau)^{-1}u. \qquad (19.51)$$

This equation does not differ from the already solved equation (19.43). The only difference is that now its average has to vanish, i.e. $\langle u \rangle = 0$. The corresponding solution is:

19.5. Response of a One Degree of Freedom System

$$u = \left(\tfrac{1}{2} fc\cos\theta - f\cos(\tau+\theta)\right)(1+c\cos\tau)$$
$$\Psi = -f\sin(\tau+\theta) - \tfrac{1}{4} fc\sin(2\tau+\theta) + \tfrac{1}{2} fc^2 \sin\tau\cos\theta. \quad (19.52)$$

Equation (19.27) can be rewritten as follows:

$$\partial v/\partial\tau = -vc\sin\tau(1+c\cos\tau)^{-1} - (1+c\cos\tau)\frac{\partial\left(u(1+c\cos\tau)^{-1}\right)}{\partial t}. \quad (19.53)$$

This equation does not contain X_0. Meaning, this equation can not be used to determine the slow part of the solution. However, it gives us a restriction, which is able to be fulfilled by functions $f(t)$ and $c(t)$. This restriction is necessary for the solutions of this type to exist. This condition is not difficult to find if we, as usual, require the periodicity of the solutions of (19.53):

$$(fc)^{\cdot}\cos\theta = 0. \quad (19.54)$$

Assuming, that this condition is fulfilled, we can find the function v satisfying the periodicity condition $\langle v \rangle = \dot{X}_0$:

$$v = \left(\dot{X}_0 - \tfrac{1}{2}\dot{f}c\sin\theta + \dot{f}\sin(\tau+\theta)\right)(1+c\cos\tau). \quad (19.55)$$

Finally, averaging the corresponding terms according to (19.29) we obtain the equation of slow motions:

$$\ddot{X}_0 + \left\{1 + \tfrac{3}{2}\alpha f^2\left(1 + \tfrac{1}{16}c^2 + \tfrac{1}{4}c^4\cos^2\theta - c^2\cos^2\theta\right)\right\}X_0 + \alpha X_0^3$$
$$= \tfrac{1}{2}(\dot{f}c)^{\cdot}\sin\theta + \tfrac{3}{16}\alpha f^3 c\sin\theta\left(1 - \tfrac{1}{4}c^4\cos^2\theta\right). \quad (19.56)$$

The main properties of this system are very similar to those of the system with strong excitation. The frequency of the free oscillations of the averaged system depends on the amplitude of the high-frequency excitation. If both external and parametric excitations are slowly modulated, we can find both the parametric and external slow excitations of the averaged system. However, they are significantly changed. For the slow parametric excitation to exist, the system has to be nonlinear ($\alpha \neq 0$) and it is therefore necessary to have a modulated external excitation ($f \neq 0$).

If we take the simplest linear situation, we obtain an equation, which is very similar to the situation of strong excitation:

$$\ddot{X}_0 + X_0 = \tfrac{1}{2}(\dot{f}\, c)^{\bullet} \sin\theta. \qquad (19.57)$$

The only difference is that the external slow excitation depends on the second derivative of the slow modulation. Another point is, that in this case the excitation is proportional to the sinus of the phase difference between the external and the parametric high-frequency excitations – not to the cosine as in the previous case.

It should be noticed, that a particular example of this system, without parametric excitation, was considered by Nayfeh and Nayfeh [12]. Equation (19.57) conforms completely to their results.

Figure 19.6 represents the direct numeric simulation results with regard to the full equations (19.48). The calculation was carried out using the following parameters:

$$\alpha = 0;\ \ \theta = 0;\ \ \omega = 100;\ \ f = 1 + 0.5\sin t;\ \ c = \big(3(1+0.5\sin t)\big)^{-1}.$$

At the beginning of this analysis we have assumed, that the necessary condition for the existence of such solutions $(f\, c)^{\bullet} \cos\theta = 0$ is fulfilled. However, the equation can be analysed in the event of this condition being unfulfilled. A solution with the amplitude order 1 does not exist in this case.

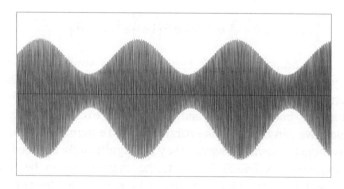

Figure 19.6. Weak response in a system with very strong excitation. Displacement.

19.5. Response of a One Degree of Freedom System

However, there are solutions with bigger amplitudes. In order to demonstrate this, the simplest linear situation will be taken and the scale of the variable x changed:

$$x = \omega z \qquad (19.58)$$

For the new variable z an equation will be obtained, as follows:

$$\ddot{z} = -z(1+c\cos\tau) + \frac{\dot{z}c\cos\tau}{1+c\cos\tau} + \omega f \sin(\tau+\theta)(1+c\cos\tau) - \omega\frac{\dot{z}c\sin\tau}{1+c\cos\tau}. \qquad (19.59)$$

This is only the linear variant of the previously analyzed equation with strong excitation (19.41). Its solutions are known.

Hence, in systems with very strong excitation we should distinguish between two types of solutions. These will be referred to as "weak response" and "strong response". The strong response with a slow amplitude, significantly bigger than 1, can exist under more general conditions than the weak response. If some additional restrictions are placed on the character of the excitation's modulation, the weak response appears side by side with the strong response. Its amplitude has the magnitude order of 1.

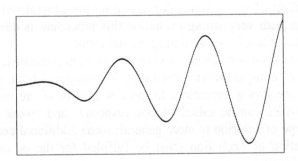

Figure 19.7. Strong response in a system with very strong excitation.

Figure 19.7 represents the direct numeric simulation results with regard to the full equations (19.49). The calculation was carried out using the following parameters:

$$\alpha = 0; \ \theta = 0; \ \omega = 100; \ f = 1; \ c = (3(1+0.5\sin t))^{-1}.$$

In this situation the weak response does not exist. The reaction of the system is typical for the external resonance, as predicted by the equation of slow motions (19.47). The amplitude of the oscillations has from the very beginning the magnitude order ω.

19.6 Conclusions

In this chapter the dynamics of systems are considered. These are described through systems of second order differential equations. These systems are subjected to a strong (magnitude order of big parameter ω) or very strong (magnitude order of ω^2) high-frequency excitation. Equations of this type occur naturally if we analyze, for example, dynamics of a machine housing, which inertia depends significantly on the position of the moving parts of the internal mechanism. Another, perhaps even more important example of systems with strong excitation depending on the first derivative of the unknown function, appears if we investigate vibrations or wave propagation in the inhomogenious media. For example, the longitudinal waves in a rod with a periodic or quasi-periodic structure.

A general separation of motions procedure is developed for systems with strong excitation. Two mathematical examples illustrate the principal effect of strong terms depending on the generalized velocities. For systems with very strong excitation this procedure is developed for the important case of fast oscillating inertial terms.

An example of a mechanical system with both parametric excitation and external (non-parametric) excitation is considered both with strong and with very strong excitation. In the last case there are two types of solutions, which can be called "strong response" and "weak response". The first type of solution is more general; some additional conditions in the modulation of excitation must be fulfilled for the of second type solutions to exist.

References

1. Bogoliubov N. N. and Mitropolskii Yu. A.: *Asymptotic methods in the theory of nonlinear oscillations*. Gordon and Breach, New York, 1961.
2. Nayfeh A.H.: *Perturbation Methods*. New-York: Wiley-Interscience, 1973.
3. Blekhman I.I.: *Synchronisation in Science and Technology*. New York: ASME Press, 1988.
4. Blekhman I. I.: *Vibrational mechanics. Nonlinear dynamic effects, general approach, Applications*. World Scientific, Singapore – New Jersey – London – Hong Kong, 2000.
5. Thomsen J. J.: Vibration suppression by using self-arranging mass: effects of adding restoring force. *J. Sound Vibr.*, 197 (1996), pp. 403 – 425.
6. Thomsen J. J. and Tcherniak D.: Slow effects of fast harmonic excitation for elastic structures. *Nonlinear Dynamics*, 17 (1998), pp. 227 – 246.
7. Thomsen J. J.: Using fast vibrations to quench friction induced oscillations. *J. Sound Vibr.*, 228 (1999) 5, pp. 1079 – 1102.
8. Bishop S. R. and Sudor D. J.: Inverted dynamics of a tilted parametric pendulum. *Europ. J. Mech. A/Solids*, 18 (1999), pp. 517 – 526.
9. Nayfeh A. H. and Mook D. T.: *Nonlinear Oscillations*. Wiley, New York, 1979.
10. Kapitsa P. L.: Dynamic stability of a pendulum with oscillating suspension point. *J. Exper. Theor. Physics*, 21 (1951) 5, pp. 588 – 597 (in Russian).
11. Fidlin A. On the separation of motions in systems with the big fast excitation of general form. *Europ. J. Mech. A/Solids*, 18 (1999), pp. 527 – 539.
12. Nayfeh A. H. and Nayfeh S. A.: The response of nonlinear systems to modulated high frequency input. *Nonlinear Dynamics*, 7 (1995), pp. 310 – 315.
13. Poincaré H. *Les méthodes nouvelles de mèchanique céleste* (3 vol.), Dover – New York, 1957.
14. Fidlin A. On asymptotic properties of systems with strong and very strong high-frequency excitation. *J. Sound Vibr.*, 235 (2000) 2, pp. 219 – 233.

Chapter 20

On the Averaging of Discontinuous Systems

A. Fidlin

LuK GmbH &Co, Industriestr.3, D-77815 Germany

20.1 Introduction. Types of Discontinuities in Oscillating Systems. Short Review of the Investigations

Discontinuous systems are usual in the modeling of mechanical systems. In any case it is some kind of idealization, which can be more or less useful for the analysis of systems behaviour. There are different sources of discontinuities in mechanical systems. In any case the discontinuities are connected with very quick changes of system's properties. The processes of the main interest should be much slower then the characteristic time of these changes, if the discontinuous idealization may be applied. Sometimes the discontinuity can be caused through an explosion like firings in combustion engines, sometimes it is a result of the discontinuous control: such discontinuous control laws are typical for optimal control with constraints [55]. Many interesting ideas connected with the averaging of discontinuous systems were inspired by the optimal control problems [35, 36, 52, 53, 54].

There are two main natural sources of discontinuities in mechanics. The first source is dry friction. The second one are impacts due to unilateral constraints.

The discontinuity of dry friction is connected with the ability of nature to resist differently against the beginning of motion from equilibrium (sticking) and against the existing motion (slipping). The ability of a friction contact to resist against an applied external force without any motion displays that in such a situation friction can be interpreted as a constraint. Many mechanical interfaces are characterized by a form of dry friction where the force-velocity curve has negative slope at low velocities. Initially, friction decreases as the contacting object starts to move, whereas at higher velocities the friction force increases again; in particular this characterizes surfaces with boundary lubrication. The initial negative slope corresponds to negative damping, and may thus cause oscillations that grow in amplitude, until a balance of dissipated and induced energy is attained, as pointed out already by Lord Rayleigh [61].

The characteristic change in friction coefficient with velocity has been explained quite convincingly by Tolstoi [70], who considers the normal separation distance between the friction surfaces as a key to the specific shape of the friction curve. However it is currently unclear to which extent the friction-velocity relationship, obtained experimentally during a quasi-static change of velocity, can be used to describe friction forces in dynamics, for example during the stationary oscillations [73].

There are two main phenomena connected with the dry friction in oscillating systems. The first one are stick-slip vibrations well known in many kinds of engineering systems and everyday life, e.g. as sounds form a bowed violin, squeaking chalks and shoes, creaking doors, squealing tramways, chattering machine tools, and grating brakes.

Numerous works has been devoted to the study of friction-induced oscillations. For ease of setup and interpretation an idealized physical system consisting of a mass sliding on a moving belt has been considered very often. Panovko and Gubanova [44], for such a system with a friction characteristic having minimum coefficient at velocity v_m, show that self-excited oscillations occur only when the belt velocity is lower than v_m. Tondl [71], Nayfeh and Mook [40], and Mitropolskii and Nguyen [38] describe self-excited oscillations of the mass-on-belt system, presenting approximate expressions for the vibration amplitudes for the case where there is no sticking between mass and belt. Ibrahim [25, 26] and McMillan [34] presents and discusses the basic mechanics of friction and friction models, and provides reviews on relevant literature. A very readable historical review on dry friction and stick-slip phenomena is given by Guran et al. [23], and a large survey on friction literature until 1992 by Armstrong-Hélouvry et al. [2].

Much research seems concentrated in determining the onset of stick-vibrations in order to avoid these totally. However, stick-slip vibrations

20.1. Introduction. Short Review of the Investigation

might be acceptable in applications, provided their amplitudes are sufficiently small. Therefore, simple expressions providing immediate insight into the influence of parameters on vibration amplitudes are believed to be useful. There are, however, very few works providing expressions for stationary stick-slip amplitudes. Armstrong-Hélouvry [3] performed a perturbation analysis for a system with Stribeck friction and frictional lag, predicting the onset of stick-slip for a robot arm. Elmer [14] discusses stick-slip and pure-slip oscillations of the mass-on-belt system with no damping and different kinds of friction functions, provides analytical expressions for the transfer between stick-slip and pure-slip oscillations, and sketch typical local and global bifurcation scenarios. Thomsen [68] and Thomsen and Fidlin [69] sets up approximate expressions for stick-slip oscillations of a friction slider, which are accurate for not very small differences in static and kinetic friction.

The second phenomenon connected with dry friction is the vibration induced displacement and transportation. The intensive research in this area was started 1964 due to the book by Blekhman and Dzhanelidze [10]. This book has shown one of the basic mathematical methods for the analysis of discontinuous systems – to split the considered motion into several time intervals. In each of these intervals the system's behaviour is continuous and all the discontinuities are replaced by the switching conditions between different time intervals and subsystems. Many industrial applications for this phenomenon, like vibrating conveyers and screens followed this book (see for example [11]).

The recent interest to such systems was inspired by the development of electronic devices combined with piezoelectric ceramics and ultrasonic motors as excitation sources for the mechanical part of system [72]. Several works by the Danish school should be mentioned here [29, 37, 66, 67], where the Direct Separation of Motions [12] was used for systems with friction. But in special cases, where the first approximation is not sufficient, special methods were developed and successfully used by Fidlin and Thomsen [18, 19].

The second and even more difficult source of discontinuities in mechanical systems is impacts in systems with unilateral constraints. Solid bodies impact theory is very complex and not really completed yet [13, 21, 27, 28]. But in many important cases the short duration of impact enables to avoid the impact process analysis and to replace it through some kinematic relationships. The corresponding stereo-mechanical impact theory, originated by Newton and Huygens, uses the so-called "Newton's impact law" together with the impulse conservation law the in order to find the velocities of two bodies after their central collision. The corresponding restitution coefficient

$$R = \left| \frac{v_{2+} - v_{1+}}{v_{2-} - v_{1-}} \right|, \quad 0 \leq R \leq 1 \tag{20.1}$$

expresses the energy dissipation during impact.

Significant generalizations of this theory for non-central collisions, collisions with constraints and collisions with friction can be found in [1, 8, 28, 39, 51, 75]. This approach allows the uniform description of different impacts, including the absolutely elastic ($R=1$) and absolutely inelastic ($R=0$) impacts. This simplicity and uniformity makes the stereo-mechanical approach very popular in applications. A lot of works devoted to dynamics of vibrating machines with impacts are based on this approach. The works by Babitskii [5, 8], Ivanov [28], Kobrinskii [31], Nagaev [39], Peterka [47] should be mentioned here.

Under this idealization systems with impacts can not be described by continuous ordinary differential equations even in the most simple cases. They are naturally described by sequences of differential equations between the collisions and some additional kinematic conditions, connecting one equations set before the collision with the next one after it. The jump of the velocities taking place as a result of the impact is in general not small.

Three different types of motion are possible in systems with normal collisions. The first one are motions with simple collisions. It is typical for these motions, that, firstly, a finite number of collisions takes place in each finite time interval, and, secondly, that the order of systems describing the motions between the collisions is the same for the whole analysis. The second one are motions with finite time intervals of contact (similar to sticking in systems with dry friction). These motions are typical (but not necessary) for systems with completely inelastic impacts and combined with the so-called quasi-plastic impact for systems with small restitution coefficient. The quasi-plastic impact, being an infinite sequence of impacts taking place within a finite time interval [39], is the third and the most difficult type of motions in one dimensional impact oscillators. From this point of view it is sensible to distinguish between

- constant order discontinuous systems (for example slipping systems with dry friction or systems with instantaneous impacts)

- variable order discontinuous systems (for example sticking systems with dry friction or systems with finite time intervals of contact).

Quasi-plastic impact is a mixture of both variants.

It should be also distinguished between periodic, transient and chaotic motions in discontinuous systems from the dynamical point of view. The recent interest to chaotic motions in discontinuous and impact oscillators [3, 9, 16, 20, 24, 26, 30, 48, 49, 50, 51, 56, 57, 58, 59, 60, 64, 65, 74] is mainly motivated by the unwanted vibrations, like gear-rattling [30, 50] or braces squeal. For machines using vibrations like vibration crushers, screens or grinding machines the simplest periodic regimes are of main interest. All these machines are processing bulk materials, for which the zero restitution coefficient is the most appropriate approximation for the impacts between the material layer and the machines working surface [11, 33].

Different mathematical methods are used for the analysis of these simplest regimes. The classical approach was developed for the piecewise linear systems. According to this approach the equations of motions have to be integrated between the collisions, and the kinematic impact conditions are used to switch from one interval of the solution to another one. This approach is straight forward and still the most effective one for the numeric simulations. Its analytical realization is useful only for the most simple systems [31]. Any nonlinearity makes this analytical approach inapplicable.

Another method useful for numeric realization is the method of point maps [22]. This method was also successfully used for the existence and stability analysis of periodic solutions. For example the completely nonlinear problem of self-synchronization in systems with impacts was considered using this method [76]. This method has also helped to consider systems with quasi-plastic impacts [39].

Further development of analytic methods was connected with the ideas of small parameter and perturbations. Poincare's method, the first method of small parameter from the historical point of view, is very effective for the periodic solutions analysis. It was generalized for the discontinuous systems of constant order by Kolovsky [32] and for discontinuous systems of variable order by Nagaev [39]. These methods helped to analyze existence and stability of periodic solutions in slightly nonlinear systems with impacts, but the necessity to stabilize the technologically best impact regimes made the transient analysis inevitable.

The first and very tempting idea was to apply harmonic linearization for the transient analysis of impact oscillators [5, 6]. This approach is very easy. Numerous comparisons of the corresponding approximate results for the simplest piecewise linear systems with exact and numeric solutions [5, 8], shows that results of harmonic linearization are accurate enough for systems with strong filtering properties of secondary structures. The discrepancy increases with increasing energy dissipation and outside of the very small vicinity of resonance. Another point, especially noticeable in [6], is that the harmonic linearization does not distinguish between different impact regimes, i.e. the variety of impact oscillations vanishes completely, but, as a rule, when we apply it, we would like to make sure that the result is accurate enough especially with respect to vibro-shock systems.

The first successful attempt to apply averaging for the impact oscillator transient analysis was in [7]. The further development was performed by Zhuravlev [77, 78]. He suggested to apply discontinuous variable transformations in order to regularize impact oscillators, i.e. to eliminate the strong impact discontinuities for the given sequence of collisions within each oscillations period. This approach was used for the analysis of impact resonance damper with almost absolutely elastic collisions [79]. Later it was generalized for the case of non-symmetric limitationes [28] and for absolutely inelastic impacts [17, 46].

A special interest is devoted to interactions between impact oscillators and energy source of limited power. The main peculiarity of these systems is the principal strong nonlinear resonance. This problem is deeply connected with the general problem of nonlinear resonance and accuracy of asymptotic methods at very long time intervals $t = O(1/\varepsilon^2)$. Fundamental mathematical results concerning this problem can be found in [4, 15, 41, 42, 43, 45, 62, 63]. Combined with inelastic impact oscillator this problem was considered in [46]. Main advantage of the approach based on the averaging technique is that it allows not only to consider the existence and stability areas of stationary resonance solutions in the parameter space, but also to investigate their attraction areas in the phase space. The last point is inevitable both for the problem of passing through resonance and for the stabilization of an impact machine in the resonance.

The objective of this chapter is to show how averaging can be used to find first or higher order approximations for discontinuous systems, described by ordinary differential equations. These systems cannot be transformed to the "standard form" or their right-hand sides do not fulfil some conditions necessary to apply the continuous averaging. So it is suggested to use forms which are natural for the discussed class of problems and to prove averaging theorems for these forms. Three theorems validating the use of averaging for constant order discontinuous systems are formulated in section 20.2. Two theorems concerning averaging in variable order discontinuous systems are formulated in section 20.3. Problems of the variable order discontinuous averaging in the vicinity of a strong non-linear resonance is discussed in section 20.4. The described mathematical methods, based on the generalized averaging method can be useful for the analysis of oscillations in mechanical systems with dry friction and collisions. Section 20.5 is devoted to analysis of a quasi-elastic system with impacts (a classical impact oscillator). A simple example demonstrating the discontinuous variables transformation technique is considered here. The same system excited by an inertial source of limited power illustrates in section 20.6, how the discussed methods can be applied in case of the non-linear resonance. Impact oscillator with inelastic impacts (variable order system) is considered in section 20.7.

20.2 Averaging of Constant Order Discontinuous Systems

Two theorems are formulated below. The first one deals with systems whose right hand sides are small, but fulfil the Lipschitz-condition only in some integral sense. The second one considers systems whose right-hand sides are not small for short time intervals. Only the periodic case is analyzed.

Theorem 1:

Consider initial value problems

$$\dot{x} = \varepsilon Z(x,t) E(g(t) + f(x)), \quad x(0) = x_0, \qquad (20.2)$$

where E(t) is a one-step function and

$$\dot{\xi} = \varepsilon \varsigma(\xi), \quad \xi(0) = x_0$$
$$\varsigma(\xi) = \langle Z(\xi,t) E(g(t) + f(\xi)) \rangle \qquad (20.3)$$

with

$$Z: R^{n+1} \to R^n; \quad f: R^n \to W_f \subset R^1; \quad g \; R^1 \to W_g \subset R^1 \; ; \; W_f \subset W_g,$$
$$x, \xi, x_0 \in D \subset R^n, t \in [0, \infty), \varepsilon \in (0, \varepsilon_0].$$

Suppose
1. *Z and f are measurable functions of t for constant x and ε.*
2. *All the functions are T-periodic with respect to t.* $\langle \; \rangle$ *means the average.*
3. *Z is a bounded Lipschitz-continuous function in x on D, i.e.*

$$\|Z(x,t)\| \le M_Z; \quad \|Z(x_1,t) - Z(x_2,t)\| \le \lambda_Z \|x_1 - x_2\| \quad (20.4)$$

4. *f is a bounded differentiable function in x on D with Lipschitz-constant λ_f, i.e.*

$$\|f(x)\| \le M_f; \quad \|f(x_1) - f(x_2)\| \le \lambda_f \|x_1 - x_2\|. \quad (20.5)$$

5. $g(t) \in C^{(1)}[0, \infty)$, *an equation*

$$g(t) = const \in W_f. \quad (20.6)$$

has m solutions $t_{i0}, i = 1, \ldots, m$ for $t \in [0, T)$ and

$$|g'(t_{i0})| \ge G > 0. \quad (20.7)$$

(The last condition is illustrated in Fig. 20.1)
6. *All the constants does not depend on ε, and ξ belongs to the interior subset of D on the time scale 1/ε.*
Then

$$\|x(t) - \xi(t)\| \le c_1 \varepsilon e^{c_2 \varepsilon t}. \quad (20.8)$$

This Theorem validates the first order averaging of constant order discontinuous systems. The absence of order changes (like sticking in systems with friction or finite time intervals of contact in systems with collisions) is guaranteed by the condition (20.7).

Now we can formulate the second theorem validating the higher order averaging in the constant order discontinuous systems.

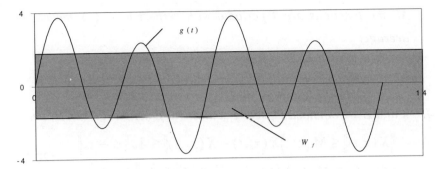

Figure 20.1. The requirements on $g(t)$.

Theorem 2:
Consider initial value problems of the form

$$\dot{x} = \varepsilon X(x,t) + Z(x,t)[E(g(t) + \varepsilon f(x,t)) - E(g(t))], \quad x(0) = x_0, \quad (20.9)$$

where $E(t)$ is a one-step function

$$E(\dot{x}) = \begin{cases} 1 & \text{if } \dot{x} > 0 \\ 1/2 & \text{if } \dot{x} = 0 \\ 0 & \text{if } \dot{x} < 0 \end{cases} \quad E(\dot{x}) + E(-\dot{x}) = 1, \quad E(\dot{x}) - E(-\dot{x}) = \operatorname{sgn}(\dot{x})$$

and

$$\dot{\xi} = \varepsilon \Xi(\xi) + \varepsilon \varsigma(\xi), \quad \xi(0) = x_0$$
$$\Xi(\xi) = \langle X(\xi,t) \rangle; \quad \varepsilon \varsigma(\xi) = \langle Z(\xi,t)[E(g(t) + \varepsilon f(\xi,t)) - E(g(t))] \rangle \quad (20.10)$$

with

$$X, Z : R^{n+1} \to R^n; \quad f : R^{n+1} \to R; \quad g : R \to R,$$
$$x, \xi, x_0 \in D \subset R^n, t \in [0, \infty), \varepsilon \in (0, \varepsilon_0].$$

Suppose
1. *X, Z and f are measurable functions of t for constant x and ε.*
2. *f in addition is a piecewise differentiable function with respect to t, and its derivative is bounded:*

$$\sup \left\| \frac{\partial f}{\partial t} \right\| \leq F_t. \quad (20.11)$$

3. All the functions are T-periodic with respect to t. $\langle \ \rangle$ means the average:

$$\langle f(t)\rangle = \frac{1}{T}\int_0^T f(t)dt$$

4. X and Z are bounded Lipschitz-continuous functions in x on D, i.e.

$$\|X(x,t)\| \le M_X; \quad \|X(x_1,t) - X(x_2,t)\| \le \lambda_X \|x_1 - x_2\|$$
$$\|Z(x,t)\| \le M_Z; \quad \|Z(x_1,t) - Z(x_2,t)\| \le \lambda_Z \|x_1 - x_2\|$$
(20.12)

5. f is a bounded Lipschitz-continuous function in x on D, i.e.

$$\|f(x,t)\| \le M_f; \quad \|f(x_1,t) - f(x_2,t)\| \le \lambda_f \|x_1 - x_2\|. \quad (20.13)$$

6. $g(t) \in C^{(1)}[0,\infty)$, an equation

$$g(t) = 0 \quad (20.14)$$

has m solutions $t_{i0}, i = 1,\ldots,m$ for $t \in [0,T)$ and

$$|g'(t_{i0})| \ge G > 0. \quad (20.15)$$

7. All constants do not depend on ε, and ξ belongs to the interior subset of D on the time scale 1/ε.

Then the solutions of (20.9) and (20.10) are asymptotically close to each other, i.e. the error one makes on using the averaged system instead of the original one is small for the asymptotically long time interval:

$$\|x(t) - \xi(t)\| \le c_1 \varepsilon e^{c_2 \varepsilon t}. \quad (20.16)$$

The proofs of these theorems can be found in [19]. Statements similar to Theorem 1 were formulated and proved in [54, 55, 56] based on the ideas of differential inclusions.

For the future analysis of impact oscillators we'll also need the following statement.

Theorem 3:

Consider a system

$$\frac{dx}{d\varphi} = \varepsilon X(\varphi,x) \text{ if } \varphi \ne \pi n; \quad \Delta x \equiv x_+ - x_- = \varepsilon f(x_-) \text{ if } \varphi = \pi n \quad (20.17)$$

with $x \in D \subset R^n$. φ is the independent variable, function X satisfies the requirements of the Theorem 2, x_- and x_+ are the values of the phase variables before and after the passage of φ through the value πn.

Suppose additionally f(x) is bounded and Lipschitz-continuous in D. The averaged system

$$\frac{d\xi}{d\varphi} = \varepsilon \Xi(\xi) + \frac{\varepsilon}{\pi} f(\xi), \quad \Xi(\xi) = \langle X(\varphi, \xi) \rangle_\varphi. \tag{20.18}$$

Under these conditions solutions of the initial value problems (20.17) and (20.18) are asymptotically close to each other, i.e.

$$\|x(\varphi) - \xi(\varphi)\| \leq \varepsilon O(1), \quad \text{for } \varphi = O(1/\varepsilon). \tag{20.19}$$

Proof of this theorem does not differ from the proof of the standard averaging (see [79]).

A collision oscillator illustrating the use of these theorems is considered in section 20.5. Numerous examples of friction oscillators can be found in [18, 19, 38, 40, 69, 79].

20.3 Averaging of Variable Order Discontinuous Systems

Now we are going to formulate the averaging theorems for the variable order discontinuous systems, i.e. systems described by differential equations and finite switch conditions (for example between sticking and slipping in friction systems or between time intervals of separate and time intervals of coupled motion in inelastic collision oscillators).

Theorem 4:

Consider a system described over certain times by differential equations, and at other time intervals by differential and finite relations of the following form:

$$\begin{aligned} \dot{x} &= \varepsilon X(x, yM(t), t, \varepsilon), & \dot{y}M(t) &= \varepsilon Y(x, yM(t), t, \varepsilon) M(t) \\ x(0) &= x_0; & y(2\pi n) &= G(x(2\pi n), \varepsilon), \quad n = 0, 1, 2, \ldots \end{aligned} \tag{20.20}$$

Here M(t) is a 2π-periodic piecewise-constant function:

$$M(t) = \begin{cases} 1, & 0 \leq t < \pi \\ 0, & \pi \leq t < 2\pi \end{cases}, \quad M(t + 2\pi) = M(t), \tag{20.21}$$

together with averaged equations

$$\dot{\xi} = \varepsilon \Xi(\xi,\eta,\varepsilon) \equiv \varepsilon \langle X(\xi,\eta M(t),t,\varepsilon) \rangle; \quad \eta = G(\xi,\varepsilon); \quad \xi(0) = x_0,$$
(20.22)

with

$$X: R^{k+l+2} \to R^k; \quad Y: R^{k+l+2} \to R^l; \quad G: R^{k+1} \to D_y \subset R^l;$$

$$x, x_0, \xi \in D_x \subset R^k; \quad y, \eta \in D_y \subset R^l; \quad t \in [0,\infty); \quad \varepsilon \in (0,\varepsilon_0]$$

Notice that vector function y(t) is a solution of an infinite sequence of systems of differential equation, each of which acts together with the continuous equations for x in the time intervals $2\pi n \le t < (2n+1)\pi$. In the time intervals $(2n+1)\pi \le t < (2n+2)\pi$ there are only equations for x. At the end of this interval new initial conditions for y are imposed. Suppose

1. *X and Y are measurable functions of t for constant x, y and ε.*
2. *All the functions are 2π-periodic with respect to t. $\langle \ \rangle$ means the average:*

$$\langle f(t) \rangle = \frac{1}{2\pi} \int_0^{2\pi} f(t) dt$$

3. *X and Y are bounded Lipschitz-continuous functions in x and y on $D_x \cup D_y$, i.e.*

$$\|X(x, yM(t), t, \varepsilon)\| \le M_X; \quad \|Y(x, yM(t), t, \varepsilon)\| \le M_Y;$$
$$\|X(x_1, y_1 M(t), t, \varepsilon) - X(x_2, y_2 M(t), t, \varepsilon)\| \le \lambda_{XX}\|x_1 - x_2\| + \lambda_{XY}\|y_1 - y_2\|M(t).$$
$$\|Y(x_1, y_1 M(t), t, \varepsilon) - Y(x_2, y_2 M(t), t, \varepsilon)\| \le \lambda_{YX}\|x_1 - x_2\| + \lambda_{YY}\|y_1 - y_2\|M(t)$$
(20.23)

4. *G is a bounded function, satisfying Lipschitz-condition in x on D_X together with its first partial derivatives with respect to x, i.e.*

$$\|G(x,t)\| \le M_G; \quad \|G(x_1,t) - G(x_2,t)\| \le \lambda_G \|x_1 - x_2\|;$$

$$\left\| \frac{\partial G(x,t)}{\partial x} \bigg|_{x_1} - \frac{\partial G(x,t)}{\partial x} \bigg|_{x_2} \right\| \le \lambda_{G1} \|x_1 - x_2\|$$
(20.24)

5. All constants do not depend on ε, and ξ belongs to the interior subset of D_X on the time scale $1/\varepsilon$.

Then the solutions of (20.20) and (20.22) are asymptotically close to each other, i.e. the error one makes on using the averaged system instead of the original one is small for the asymptotically long time interval:

$$\|x-\xi\| \le c_1 \varepsilon^2 t e^{c_2 \varepsilon t}; \quad \|yM(t)-\eta M(t)\| \le c_3 \varepsilon^2 t e^{c_4 \varepsilon t} M(t). \quad (20.25)$$

Remark. The above discussion remains valid if the function M consists of a finite number of isolated impulses of arbitrary fixed length in the interval $0 \le t < 2\pi$.

Systems like (20.20) do not appear directly in the analysis of discontinuous oscillators. Mach more important are systems with slowly variable length of the impulses in the function M. This case can be reduced to the previous one. It is considered in the next theorem.

Theorem 5:

Consider following problem:

$$\dot{x} = \varepsilon X(x, yM_1(z,\alpha), \varphi, zM_1(z,\alpha), \varepsilon); \quad x(0) = x_0$$
$$\dot{y}M_1(z,\alpha) = \varepsilon Y(x, yM_1(z,\alpha), \varphi, zM_1(z,\alpha), \varepsilon) M_1(z,\alpha)$$
$$y(t_n) = G(x(t_n)) + \varepsilon G_1(x(t_n), \varphi(t_n), \varepsilon)$$
$$\dot{z}M_1(z,\alpha) = [1 + \varepsilon Z(x, yM_1(z,\alpha), \varphi, zM_1(z,\alpha), \varepsilon)] M_1(z,\alpha); \quad z(t_n) = 2\pi n$$
$$\dot{\varphi} = 1 + \varepsilon \Phi(x, yM_1(z,\alpha), \varphi, zM_1(z,\alpha), \varepsilon); \quad \varphi(t_n) = 2\pi n + \varepsilon F(x(t_n), \varphi(t_n), \varepsilon)$$
$$(20.26)$$

Here x and y are, as before, vector functions of arbitrary finite dimensions, z and φ are scalar phases, the requirements on the functions X, Y and G are the same as in the Theorem 3, F has in addition bounded partial derivatives with respect to its first two arguments.

The function $M_1(z,\alpha)$ is defined as follows (see Fig. 20.2):

$$M_1(z,\alpha) = \begin{cases} 1, & 0 \le z < \alpha \\ 0, & \alpha \le z < 2\pi \end{cases}, \quad M(z+2\pi,\alpha) = M(z), \quad (20.27)$$

the switchover point α may depend on slow variables: $\alpha = \alpha(x, y, z-\varphi)$, but it should remain in the interval $\alpha \in [0, 2\pi[$. Otherwise function M_1 would not be defined correctly.

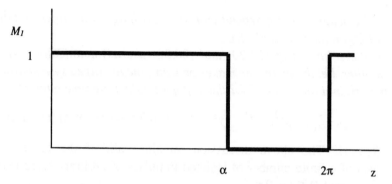

Figure 20.2. On the definition of the function M_1.

An averaged system

$$\dot{\xi} = \varepsilon \langle X(\xi, G(\xi,\varepsilon)M_1(\psi,\alpha), \psi, \psi M_1(\psi,\alpha), \varepsilon) \rangle_\varphi ; \quad \dot{\psi} = 1$$
$$\eta M_1(\psi,\alpha) = G(\xi,\varepsilon)M_1(\psi,\alpha); \quad \varsigma M_1(\psi,\alpha) = \psi M_1(\psi,\alpha)$$
(20.28)

is considered together with the system (20.26).

Then the solutions of (20.26) and (20.28) are asymptotically close to each other, i.e. the error one makes on using the averaged system instead of the original one is small for the asymptotically long time interval:

$$\|x-\xi\| = O(\varepsilon), \quad \|y-\eta\|M_1 = O(\varepsilon), \quad \|(z-\varphi)-(\varsigma-\psi)\|M_1 = O(\varepsilon).$$
(20.29)

Proofs of these theorems can be found in [17].

20.4 On the Averaging of Discontinuous Systems in the Vicinity of a Strong Nonlinear Resonance

Strong nonlinear resonance problems are very important both from the theoretical and from the practical point of view. There are many mathematical works devoted to this problem [4, 15, 22, 41, 42, 43, 45, 46, 62, 63, 79]. The main result here it that the resonant motions take

20.4. Discontinuous Systems in the Vicinity of a Strong Nonlinear Resonance

place in a $\sqrt{\varepsilon}$-vicinity of a resonance surface. The main mathematical problem is to get an asymptotic procedure valid not only for the time interval $O(1/\sqrt{\varepsilon})$ - this is guaranteed by the standard averaging, but for a significantly longer time interval $O(1/\varepsilon)$. One of the ways to solve this problem is the so-called hierarchical averaging [45] giving less accurate solutions valid for longer times.

All the considered theorems can be combined with the hierarchical averaging procedure for the analysis of strong nonlinear resonance problems. Here such a combination of Theorem 4 is taken as an example [17, 46]. All other cases can be analyzed in a similar way.

It is very important for the practical design of vibration devices and machines to take nonlinear properties of the drive into account. Often it is possible to transform the equations of motion in this vicinity to the so-called hierarchical form (*cf.* [45, 46] and section 20.6 of this chapter)

$$\dot{x}_+ = \varepsilon X_+(x_+, yM(t), \alpha, t, \varepsilon);$$
$$\dot{y}M(t) = \varepsilon Y(x_+, yM(t), \alpha, t, \varepsilon)M(t); \, y(2\pi i) = G(x_+(2\pi i), \alpha(2\pi i), \varepsilon) \quad (20.30)$$
$$\dot{\alpha} = \sqrt{\varepsilon} A_1(x_+, yM(t), \alpha, \varepsilon) + \varepsilon A_2(x_+, yM(t), \alpha, t, \varepsilon).$$

Here the functions X_+, A_1 and A_2 are 2π-periodic in φ and t, bounded and satisfy together with their first partial derivatives Lipschitz-conditions with respect to the arguments x, $yM(t)$. All other conditions are the same as in the Theorem 3. Here x_+ is a vector, variable α is considered to be the semi-slow phase, i.e. the function A_1 is separated from zero:

$$A_1 \geq A_0 > 0.$$

The hierarchic averaging scheme can be formulated for the system (20.30) as follows.

The first step of the hierarchical averaging is the variable order averaging with respect to the fast variable t (second approximation with respect to the small parameter $\sqrt{\varepsilon}$). It can be performed as usual by help of the almost identical variable transformation:

$$x_1 = \xi_1 + \varepsilon u(\xi_1, \eta_1, \beta_1, t, \varepsilon); \quad y_1 = \eta_1 + \varepsilon v(\xi_1, \eta_1, \beta_1, t, \varepsilon)$$
$$\alpha_1 = \beta_1 + \sqrt{\varepsilon} w_1(\xi_1, \eta_1, \beta_1, t) + \varepsilon w_2(\xi_1, \eta_1, \beta_1, t, \varepsilon) \quad .(20.31)$$

The corresponding averaged equations have the following form:

$$\dot{\xi}_1 = \varepsilon \Xi(\xi_1, G(\xi_1,\beta_1,\varepsilon), \beta_1,\varepsilon) + \varepsilon^{3/2} O(1)$$
$$\dot{\beta}_1 = \sqrt{\varepsilon} B_1(\xi_1, G(\xi_1,\beta_1,\varepsilon), \beta_1) + \varepsilon B_2(\xi_1, G(\xi_1,\beta_1,\varepsilon), \beta_1,\varepsilon) + \varepsilon^{3/2} O(1)$$
$$\eta_1 = G(\xi_1,\beta_1,\varepsilon) + \varepsilon O(1)$$

(20.32)

with

$$\Xi(\xi_1,\eta_1,\beta_1,\varepsilon) = \langle X_+(\xi_1,\eta_1 M(t),\beta_1,t,\varepsilon)\rangle_t,$$
$$\Psi(\xi_1,\eta_1,\beta_1,\varepsilon) = \langle Y(\xi_1,\eta_1 M(t),\beta_1,t,\varepsilon)\rangle_t,$$
$$B_1(\xi_1,\eta_1,\beta_1) = \langle A_1(\xi_1,\eta_1 M(t),\beta_1,t)\rangle_t; B_2(\xi_1,\eta_1,\beta_1,\varepsilon) = \left\langle A_2 + \frac{\partial A_1}{\partial \beta_1} w_1 \right\rangle_t$$
$$w_1 = \int_0^t [A_1(\xi_1,\eta_1 M(\tau),\beta_1,\tau) - B_1(\xi_1,\eta_1,\beta_1)]d\tau.$$

(20.33)

It is necessary to notice, that B_1 is a separated from zero 2π-periodic function of β_1 like A_1. It means that β_1 is a semi-slow phase. The second step of the hierarchic scheme is the averaging with respect to this phase, performed as usual by means of the following almost identical variables transformation:

$$\xi_1 = \rho_2 + \sqrt{\varepsilon} p(\rho_2,\gamma_2,\varepsilon), \quad \beta_1 = \gamma_2 + \sqrt{\varepsilon} q(\rho_2,\gamma_2,\varepsilon). \quad (20.34)$$

Here p and q are bounded 2π-periodic functions of γ_2 with zero average. The corresponding equations are:

$$\dot{\rho}_2 = \varepsilon P(\rho_2,\varepsilon) + \varepsilon^{3/2} O(1); \dot{\gamma}_2 = \sqrt{\varepsilon}\Gamma_1(\rho_2,\gamma_2,\varepsilon) + \varepsilon \Gamma_2(\rho_2,\varepsilon) + \varepsilon^{3/2} O(1),$$
$$\eta_1 = G(\rho_2 + \sqrt{\varepsilon} p(\rho_2,\gamma_2,\varepsilon), \gamma_2 + \sqrt{\varepsilon} q(\rho_2,\gamma_2,\varepsilon), \varepsilon) + \varepsilon O(1)$$

(20.35)

20.4. Discontinuous Systems in the Vicinity of a Strong Nonlinear Resonance

with

$$P(\rho_2,\varepsilon)=\left\langle \frac{\Xi(\rho_2,G(\rho_2,\gamma_2,\varepsilon),\gamma_2,\varepsilon)}{B_1(\rho_2,G(\rho_2,\gamma_2,\varepsilon),\gamma_2)}\right\rangle_{\gamma_2};$$

$$p(\rho_2,\gamma_2,\varepsilon)=\int_{\gamma_0}^{\gamma_2}\frac{\Xi(\rho_2,G(\rho_2,\gamma,\varepsilon),\gamma,\varepsilon)-P(\rho_2,\varepsilon)}{B_1(\rho_2,G(\rho_2,\gamma,\varepsilon),\gamma)}d\gamma$$

$$\Gamma_1(\rho_2,\gamma_2,\varepsilon)=B_1(\rho_2,G(\rho_2,\gamma_2,\varepsilon),\gamma_2)$$

$$\Gamma_2(\rho_2,\varepsilon)=\left\langle \frac{\frac{\partial \Gamma_1}{\partial \rho_2}p+B_2(\rho_2,G(\rho_2,\gamma_2,\varepsilon),\gamma_2,\varepsilon)}{B_1{}^2(\rho_2,G(\rho_2,\gamma_2,\varepsilon),\gamma_2)}\right\rangle_{\gamma_2}\left\langle B_1^{-2}(\rho_2,G(\rho_2,\gamma_2,\varepsilon),\gamma_2)\right\rangle_{\gamma_2}^{-1}.$$

The system (20.35) is absolutely equivalent to the original system (20.30). Now we are going to consider the hierarchically averaged system

$$\dot\rho=\varepsilon P(\rho,\varepsilon); \quad \dot\gamma=\sqrt{\varepsilon}\Gamma_1(\rho,\gamma,\varepsilon)+\varepsilon\Gamma_2(\rho,\varepsilon); \quad \eta_1=G(\rho,\gamma_2,\varepsilon)$$
(20.36)

(all the terms $O(1)$ are omitted here), together with the system (20.35). The following theorem can be formulated:

Theorem 6:

Under the formulated conditions the following estimates between the solutions of the systems (20.36) and (20.30) are valid for the time interval $O(1/\varepsilon)$:

$$\|x_+ - \rho\|\le C_{\rho 1}\sqrt{\varepsilon}; \quad \|yM(t)-\eta M(t)\|\le C_{\eta 1}\sqrt{\varepsilon} \quad \|\alpha-\gamma\|\le C_{\gamma 1}\sqrt{\varepsilon}. \quad ;$$
(20.37)

The proof of this theorem is a combination of the standard variable order discontinuous averaging [17] and the hierarchical averaging for continuous systems [45].

Remark. All the considered variable order systems are degenerated with respect to the variables y. As a result the corresponding variables are replaced by finite relationships in the first approximation of the averaging.

20.5 Constant Order Discontinuous Averaging in Systems with Quasi-Elastic Collisions.

Systems with collisions represent the most difficult case of discontinuous systems. First of all systems with collisions can not be described by systems of differential equations valid for all times. The continuous differential equations are valid between the collisions and state variables after each collision can be calculated as functions of state variables before the collision. For the collision oscillator is typical that the co-ordinates of the colliding objects stay continuous, but the jump of the velocities during the collision is not small (has the same magnitude order as the velocity itself). All these circumstances make the analysis of impact oscillators even more complex.

None of the described mathematical methods can be applied directly to collision oscillators. The corresponding equations of motion including kinematic impact conditions must be firstly transformed to one of the forms for which the validity of averaging is proved. The basic idea for such transformations is discontinuous variable changes. This idea suggested by Zhuravlev [77–79] and developed for different applications by Fidlin [17, 46] and Ivanov [28] enables to eliminate the system's main discontinuity either in general, or at least for the given collisions sequence (in case of systems with many colliding pares). Combined with the discontinuous averaging methods this idea leads to efficient technique for asymptotic analysis of impact oscillators.

The classical impact oscillator with *quasi-elastic collisions* is considered here as the first example. The aim is to demonstrate the main ideas and methods using a simple and physically clear object. Then the described methods are applied for a more complex and realistic system including such important effects as kinematic couplings and excitation source of limited power.

Figure 20.3 shows the physical example system: a mass described by its co-ordinate $s(t)$ attached to a damped linear spring and excited by a harmonic force. The mass can collide with an quasi-elastic stop at a distance Δ from its static equilibrium point. Only oscillations with collisions are considered.

The equation of motion and corresponding kinematic impact condition is, in non-dimensional form:

$$\ddot{s} + \beta \dot{s} + s = \varepsilon \sin \omega t, \text{ if } s < \Delta; \; s_+ = s_-; \; \dot{s}_+ = -R\dot{s}_-, \text{ if } s = \Delta, \quad (20.38)$$

20.5. Systems with Quasi-Elastic Collisions

Figure 20.3. Classical impact resonator.

where index "+" denotes state variables after collision and index "-" denotes the state variables before collision.

The stop at $s = \Delta$ guarantees that the variable s can be never bigger then Δ. This circumstance prompts to apply the following discontinuous variables transformation (cf. [28]) from the old variables $(s, \dot{s})$ to the new variables (z, v) :

$$s = \Delta - |z|; \quad \dot{s} = -v(\operatorname{sgn} z - e\operatorname{sgn} v), \quad e = \frac{1-R}{1+R}. \tag{20.39}$$

This variables transformation fulfils automatically the impact conditions. For example, the transition from $z<0$ to $z>0$ corresponds to $\operatorname{sgn} v = 1$. In this case according to (20.39):

$$\dot{s}_- = \frac{2v}{1+R}; \quad \dot{s}_+ = -\frac{2Rv}{1+R}. \tag{20.40}$$

The transition from $z>0$ to $z<0$ corresponds to $\operatorname{sgn} v = -1$. In this case according to (20.39):

$$\dot{s}_- = -\frac{2v}{1+R}; \quad \dot{s}_+ = \frac{2Rv}{1+R}. \tag{20.41}$$

Substituting (20.39) in (20.38) one obtains equations of motions with eliminated impacts, i.e. the new equations are valid for all times:

$$\dot{z} = v(1 - e\operatorname{sgn}(zv)); \quad \dot{v} = -\beta v + \frac{\Delta \operatorname{sgn} z - z - \varepsilon \sin \omega t \operatorname{sgn} z}{1 - e\operatorname{sgn}(zv)}. \tag{20.42}$$

Now the small parameters are introduced. We take β, ε, Δ and e to be the small parameters of the same order. These hypothesis expresses that both external excitation and energy dissipation due to external damping and collisions are small, i.e. the oscillator is quasi-elastic and its oscillations form is close to a harmonic one.

Omitting the terms $O(\varepsilon^2)$, the equations (20.42) can be rewritten as follows:
$$\dot{z} = v - ev\,\mathrm{sgn}(zv)$$
$$\dot{v} = -z - ez\,\mathrm{sgn}(zv) - \beta v + \Delta\,\mathrm{sgn}\,z - \varepsilon \sin\omega t\,\mathrm{sgn}\,z \qquad (20.43)$$

System (20.43) is quasi-linear, so it can be brought to the standard form by the Van-der-Pol transformation:
$$z = A\sin\varphi;\quad v = A\cos\varphi;\quad \omega t = \psi. \qquad (20.44)$$

Substituting (20.44) into (20.43) one obtains:
$$\dot{A} = -eA|\sin 2\varphi| - \beta A\cos^2\varphi + \Delta\cos\varphi\,\mathrm{sgnsin}\varphi - \varepsilon \sin\psi\cos\varphi\,\mathrm{sgnsin}\varphi$$
$$\dot{\varphi} = 1 - e\cos 2\varphi\,\mathrm{sgnsin}2\varphi + \beta\cos\varphi\sin\varphi - \frac{\Delta}{A}|\sin\varphi| + \frac{\varepsilon}{A}\sin\psi|\sin\varphi| \qquad (20.45)$$
$$\dot{\psi} = \omega$$

The resonance of the system (20.45) corresponds to $\omega = 2$. Considering the system in the vicinity of the resonance we introduce the small frequency difference of the same magnitude order as the introduced small parameters and a new slow variable θ, expressing the phase difference between the system's oscillations and the external excitation:
$$\omega = 2 + \delta,\quad \delta \ll 1,\quad \psi = 2\varphi + \theta. \qquad (20.46)$$

Finally we obtain the system in the standard form for the constant order discontinuous averaging (Theorem 1):
$$\dot{A} = -eA|\sin 2\varphi| - \beta A\cos^2\varphi + \Delta\cos\varphi\,\mathrm{sgnsin}\varphi - \varepsilon \sin(2\varphi + \theta)\cos\varphi\,\mathrm{sgnsin}\varphi$$
$$\dot{\theta} = \delta + 2e\cos 2\varphi\,\mathrm{sgnsin}2\varphi - \beta\sin 2\varphi + \frac{2\Delta}{A}|\sin\varphi| - \frac{2\varepsilon}{A}\sin(2\varphi + \theta)|\sin\varphi|$$
$$\dot{\varphi} = 1 - e\cos 2\varphi\,\mathrm{sgnsin}2\varphi + \frac{1}{2}\beta\sin 2\varphi - \frac{\Delta}{A}|\sin\varphi| + \frac{\varepsilon}{A}\sin(2\varphi + \theta)|\sin\varphi|$$
$$(20.47)$$

Averaging (20.47) the following equations to the first order approximation can be obtained:
$$\dot{A} = -\left(\frac{2e}{\pi} + \frac{\beta}{2}\right)A - \frac{4\varepsilon}{3\pi}\cos\theta;\quad \dot{\theta} = \delta + \frac{4\Delta}{\pi A} + \frac{4\varepsilon}{3\pi A}\sin\theta. \qquad (20.48)$$

20.5. Systems with Quasi-Elastic Collisions

The first effect, which can be noticed from (20.48) is that the influence of the impacts non-elasticity ($e \neq 0$) can be reduced to an additional linear damping by introducing the effective damping coefficient

$$\beta_e = \frac{1}{2}\beta + \frac{2e}{\pi}. \qquad (20.49)$$

The stationary solution to (20.48) can be obtained by setting its right-hand sides to zero. There are two stationary solutions with different amplitudes:

$$A_{1,2} = \frac{4}{\pi} \cdot \frac{-\Delta\delta \pm \sqrt{\varepsilon^2(\beta_e^2 + \delta^2)/9 - \beta_e^2\Delta^2}}{\beta_e^2 + \delta^2}. \qquad (20.50)$$

From (20.50) follows that some the parameters should fulfil an inequality in order the stationary resonant solution to exist:

$$\Delta^2 < \frac{\varepsilon^2}{9}\left(1 + \frac{\delta^2}{\beta_e^2}\right), \qquad (20.51)$$

i.e. the gap should not be to big.

Considering the stability of the stationary solutions (20.50) one obtains the following conditions:

$$\beta_e > 0, \quad A_1 > -\frac{4\Delta\delta}{\pi(\beta_e^2 + \delta^2)}. \qquad (20.52)$$

Comparing (20.52) and (20.50) one can see that only the solution with plus in front of the square root is stable.

Returning back to the original variables according to (20.39) and (20.44) none obtains the approximate stationary solution in form:

$$s = \Delta - |A_1 \sin t|; \qquad \dot{s} = -A_1 \cos t (\operatorname{sgn}\sin t - e \operatorname{sgn}\cos t)$$

$$A_1 = \frac{4}{\pi(\beta_e^2 + \delta^2)}\left(\sqrt{\frac{\varepsilon^2(\beta_e^2 + \delta^2)}{9} - \beta_e^2\Delta^2} - \Delta\delta\right). \qquad (20.53)$$

Figures 20.4 and 20.5 illustrate the accuracy of the approximate solution (20.53) compared with the numeric simulations performed with a commercial program ITI-SIM for the following values of parameters:

$\beta = 0.01, \quad \varepsilon = 0.01, \quad R = 0{,}95 \Rightarrow e \approx 0.0256 \quad$ (Fig. 20.4)

$\beta = 0.01, \quad \varepsilon = 0.01, \quad R = 0{,}6 \Rightarrow e \approx 0.25 \quad\quad$ (Fig. 20.5)

The accuracy seems to be quite acceptable even in the second case although the collisions in this case are not really quasi-elastic.

Figure 20.4 illustrates influence of the gap Δ on the form of the resonance curve. Interesting is that the maximal amplitude does not depend on Δ, but this constant value corresponds to different excitation frequency.

Now we are moving on to a more complex and more realistic problem, containing two additional phenomena: kinematic couplings in colliding bodies and excitation source of limited power.

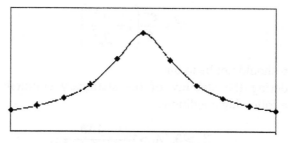

Figure 20.4. Comparison between analytic and numerical solutions. R=0.95. – Theory ♦ Numeric

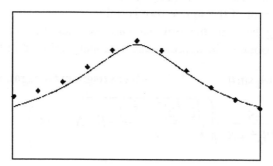

Figure 20.5. Comparison between analytic and numerical solutions. R=0.6. – Theory ♦ Numeric

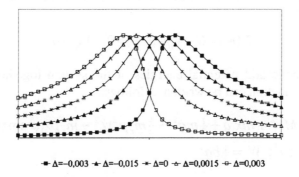

-■- Δ=-0,003 -▲- Δ=-0,015 -*- Δ=0 -△- Δ=0,0015 -▫- Δ=0,003

Figure 20.6. Resonance curves for different gap values.

20.6 Resonant Motions in a Quasi-Elastic Colliding Oscillator Excited by an Inertial Source of Limited Power

20.6.1 The problem statement.

A colliding oscillator analyzed in the previous subsection driven by an *inertial source of limited power* is investigated. Figure 20.7 shows the considered system consisting of a frame of mass M_1 in whose interior a striker of mass m is attached elastically. The striker can collide quasi-elastic with the frame at a relative distance Δ_f from the static equilibrium point. An inertial vibrator characterized by the mass m_r and radius ρ is placed on the frame and set in rotation by an induction motor. The

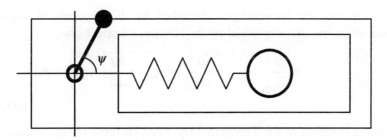

Figure 20.7. Impact resonator with inertial excitation.

analysis below is confined to linear approximation of the torque at the motor shaft:

$$P(\dot{\psi}) = U - V_R \dot{\psi}, \quad U > 0, \quad V_R > 0. \tag{20.54}$$

The kinetic and potential energies of the system together with the dissipative function can be written as follows:

$$T = \tfrac{1}{2} M \dot{x}_1^2 + m \dot{x}_1 \dot{q} + \tfrac{1}{2} m \dot{q}^2 + \tfrac{1}{2} m_r \rho^2 \dot{\psi}^2 - m_r \rho \dot{x}_1 \dot{\psi} \sin\psi$$
$$\Pi = \tfrac{1}{2} c q^2; \quad W = \tfrac{1}{2} b \dot{q}^2 \tag{20.55}$$

Here x_1 is the co-ordinate of the frame, $x_2 = x_1 + q$ is the mass of the striker, c and b are the stiffness and the damping of the spring correspondingly, $M = M_1 + m + m_r$ is the total mass of the system.

Now the equations of motion between the collisions and the kinematic collision conditions must be formulated. For the time intervals between the collisions one obtains:

$$\begin{cases} M \ddot{x}_1 + m \ddot{q} - m_r \rho \left(\ddot{\psi} \sin\psi + \dot{\psi}^2 \cos\psi \right) = 0 \\ m \ddot{x}_1 + m \ddot{q} + b \dot{q} + c q = 0 \\ m_r \rho^2 \ddot{\psi} - m_r \rho \ddot{x}_1 \sin\psi = U - V_R \dot{\psi} \end{cases} \quad \text{if } q < \Delta_f. \tag{20.56}$$

Now the kinematic conditions for the collisions have to be formulated. The main peculiarity of the system is that the rotor is placed on the colliding frame. It means that not only the velocities of the colliding bodies, but also the rotor's angular velocity is discontinuous at the time point of the collision. The collisions take place when $q = \Delta_f$. The collisions are conventional described by the generalized impulse conservation law accomplished by the Newton's impact law:

$$p_{1+} \equiv \left. \frac{\partial T}{\partial \dot{x}_1} \right|_+ = p_{1-} \equiv \left. \frac{\partial T}{\partial \dot{x}_1} \right|_- ; p_{\psi+} \equiv \left. \frac{\partial T}{\partial \dot{\psi}} \right|_+ = p_{\psi-} \equiv \left. \frac{\partial T}{\partial \dot{\psi}} \right|_- ; \dot{q}_+ = -R \dot{q}_-.$$
$$\tag{20.57}$$

In the notation of the Theorem 3 (cf. (20.17)) these conditions can be formulated as follows:

$$\Delta \dot{x}_1 = -\frac{m\Delta \dot{q}}{M - m_r \sin^2 \psi}; \quad \Delta \dot{\psi} = -\frac{m \sin\psi \Delta \dot{q}}{\rho(M - m_r \sin^2 \psi)},$$
$$\Delta \dot{q} = -(1+R)\dot{q}_-; \quad \text{if } q = \Delta_f \tag{20.58}$$

Equations (20.56) and (20.58) describe motions of the considering system completely.

20.6.2 Dimensionless formulation. Small parameters.

Firstly, we are going now to eliminate the variable x_1 and to introduce dimensionless parameters and variables:

$$k = \sqrt{\frac{c}{m}\frac{M}{M-m}}, \beta = \frac{bM}{km(M-m)}, \mu = \frac{m}{M-m}, \varepsilon = \frac{m_r}{M}, \delta = \frac{\Delta}{\rho},$$
$$u = \frac{U}{m\rho^2 k^2}, v = \frac{VR}{m\rho^2 k^2}, e = \frac{1-R}{1+R}, y = \frac{q}{\rho}, \tau = kt, \frac{d(\ )}{d\tau} = (\ )'. \tag{20.59}$$

Now the equations (20.56) and the impact conditions (20.58) can be rewritten as follows:

$$\begin{cases} y'' + \beta y' + y = -\varepsilon(1+\mu)(\psi'' \sin\psi + \psi'^2 \cos\psi) \\ \psi'' = u - v\psi' + \varepsilon(\psi'' \sin\psi + \psi'^2 \cos\psi) \\ \quad + \sin\psi[\beta\mu(1+\mu)y' + \mu(1+\mu)y + \varepsilon\mu(\psi'' \sin\psi + \psi'^2 \cos\psi)] \end{cases} \quad \text{if } y < \delta \tag{20.60}$$

$$\Delta y' = -\frac{2}{1+e}y'_-; \quad \Delta \psi' = -\frac{\mu \sin\psi}{(1+\mu)(1-\varepsilon \sin^2 \psi)}\Delta y', \quad \text{if } y = \delta. \tag{20.61}$$

We suppose now $\beta, \varepsilon, \mu, \delta, e, u, v$ to be small parameters of the same order and omit the terms $O(\varepsilon)$:

$$\begin{cases} y'' + \beta y' + y = -\varepsilon \psi'^2 \cos\psi \\ \psi'' = u - v\psi' + \varepsilon \psi'^2 \sin\psi \cos\psi + \mu y \sin\psi \end{cases} \quad y < \delta, \tag{20.62}$$
$$\Delta y' = -2y'_-; \quad \Delta \psi' = -\mu \cos\psi \Delta y' \quad y = \delta.$$

System (20.62) shows that both variables y and ψ are discontinuous. But the arts of the discontinuity of these variables are significantly different. The jump of the variable y is strong: $\Delta y = O(y_-)$. The discontinuity of the variable ψ is to the contrary weak: $\Delta \psi = O(\mu)$. Due to this circumstance different methods can be applied for the averaging of these discontinuities. Variable ψ can be averaged "directly" according to Theorem 3. Variable y must be regularised by means of the discontinuous variables transformation similar to (20.39).

20.6.3 Discontinuous variables transformation. Main resonance.

In order to regularize the variable y the following discontinuous variable transformation can be applied:

$$y = \delta - |z|; \quad y' = -w \operatorname{sgn} z (1 - e \operatorname{sgn}(zw)). \tag{20.63}$$

The collision in new variables take place at $z = 0$. New variables are governed by the following equations

$$\begin{aligned}
& z' = w - ew \operatorname{sgn}(wz); \\
& w' = -z - ez \operatorname{sgn}(wz) - \beta w + \delta \operatorname{sgn} z + \varepsilon \omega^2 \cos\psi \operatorname{sgn} z \\
& \psi' = \omega; \quad \omega' = u - v\omega + \varepsilon \omega^2 \sin\psi \cos\psi - \mu |z| \sin\psi; \\
& \Delta \omega = 2\mu |w| \sin\psi, \text{ when } z = 0.
\end{aligned} \tag{20.64}$$

In order to obtain a system in standard form the Van-der-Pol transformation can be applied:

$$z = A\sin\varphi; \qquad w = A\cos\varphi. \tag{20.65}$$

Substituting (20.65) into (20.64) one obtains:

$$A' = -eA|\sin 2\varphi| - \beta A\cos^2\varphi + \delta \cos\varphi \operatorname{sgn} \sin\varphi + \varepsilon \omega^2 \cos\psi \cos\varphi \operatorname{sgn} \sin\varphi$$

$$\omega' = u - v\omega + \varepsilon \omega^2 \sin\psi \cos\psi - \mu A|\sin\varphi|\sin\psi, \quad \Delta\omega = 2\mu A \sin\psi, \text{if } \varphi = \pi n$$

$$\psi' = \omega$$

$$\varphi' = 1 - \cos 2\varphi \operatorname{sgn} \sin 2\varphi + \frac{1}{2}\beta \sin 2\varphi - \frac{\delta}{A}|\sin\varphi| - \frac{\varepsilon}{A}\omega^2 \cos\psi |\sin\varphi|.$$

$$\tag{20.66}$$

20.6. Resonant Motions in a Quasi-Elastic Colliding Oscillator

The main principal resonance in the system (20.66) corresponds to the surface $\omega = 2$. We are going to consider the motions in the vicinity of the resonance surface following the hierarchical averaging technique described in section 20.4.

Two half-slow variables can be introduced in the $\sqrt{\varepsilon}$-vicinity of the resonance:

$$\theta = \psi - 2\varphi; \qquad \sigma = \frac{\omega - 2}{\sqrt{\mu}}. \qquad (20.67)$$

Substituting (20.67) into (20.66) one obtains equations in standard form for hierarchical averaging:

$$A' = -eA|\sin 2\varphi| - \beta A \cos^2 \varphi + \delta \cos \varphi \operatorname{sgn} \sin \varphi$$
$$\quad + 4\varepsilon \cos(2\varphi + \theta) \cos \varphi \operatorname{sgn} \sin \theta$$

$$\theta' = \sqrt{\mu}\sigma + 2e \cos 2\varphi \operatorname{sgn} \sin 2\varphi - \beta \sin 2\varphi + 2\delta A^{-1}|\sin \varphi|$$
$$\quad + 8\varepsilon A^{-1} \cos(2\varphi + \theta)|\sin \varphi|$$

$$\sigma' = \mu^{-1/2}\left(u - 2v + 2\varepsilon\left(1 + \sqrt{\mu}\sigma\right)\sin(4\varphi + 2\theta) - \mu A|\sin \varphi|\sin(2\varphi + \theta)\right)$$
$$\quad - v\sigma$$

$$\Delta\sigma\big|_{\varphi=\pi n} = 2\sqrt{\mu}A\sin\theta$$

$$\varphi' = 1 - e\cos 2\varphi \operatorname{sgn} \sin 2\varphi + \frac{1}{2}\beta \sin 2\varphi - \delta A^{-1}|\sin \varphi|$$
$$\quad - 4\varepsilon A^{-1} \cos(2\varphi + \theta)|\sin \varphi|. \qquad (20.68)$$

20.6.4 First step of the hierarchical averaging. Stationary resonance solution.

These equations can be averaged with respect to the fast rotating phase φ. The corresponding second approximation with respect to $\sqrt{\mu}$ can be obtained by the direct averaging of the system (20.68), taking Theorem 3 into account.

$$A_1' = -\beta_e A_1 - \frac{16}{3\pi}\varepsilon \sin\theta_1; \quad \theta_1' = \sqrt{\mu}\sigma_1 + \frac{4\delta}{\pi A_1} - \frac{16}{3\pi A_1}\varepsilon\cos\theta_1$$

$$\sigma_1' = \frac{u-2v}{\sqrt{\mu}} - v\sigma_1 + \frac{8}{3\pi}\sqrt{\mu}A_1\sin\theta_1 \qquad (20.69)$$

This system is obtained by the first step of the hierarchical averaging. The effective damping β_e is defined by (20.49). But the main interesting effect of impacts in the system (20.69) is not the increased effective damping. This effect is totally similar to that considered in the simple example above. The most interesting effect is that the last term in the last equation is significantly increased due to the discontinuous condition for σ (the forth equation in (20.68)). In other words the contribution of the discontinuous jumps in the rotor's rotation speed increases significantly the "vibration torque" [12] in the last equation. It is interesting to notice that just that effect is completely lost in [78, 79] where a similar system is considered.

In order to analyze (20.69) the first order approximation can be considered firstly:

$$A_{10}' = 0; \quad \theta_{10}' = \sqrt{\mu}\sigma_{10}; \quad \sigma_{10}' = \sqrt{\mu}u_0 + \frac{8}{3\pi}\sqrt{\mu}A_{10}\sin\theta_{10}, \quad u_0 = \frac{u-2v}{\mu}.$$

The last two equation can be easily transformed to one second order differential equation, describing the so-called "equivalent pendulum" [46]:

$$\theta_{10}'' - \frac{8}{3\pi}\mu A_{10}\sin\theta_{10} = \mu u_0. \qquad (20.70)$$

Equation (20.70) describes a mathematical pendulum with a constant applied torque. Two qualitatively different types of motions are possible in such a system. If $|u_0| > \frac{8}{3\pi}A_{10}$, the only possibility is rotational motion of the pendulum, which corresponds to the breakoff of the resonance. But if $|u_0| < \frac{8}{3\pi}A_{10}$, either rotational (also resulting in the resonance breakoff) or librational motions are possible, depending on the initial condition. In other words the necessary condition for the existence of the stationary resonance is

$$|u_0| < \frac{8}{3\pi} A_{10}. \qquad (20.71)$$

If this condition is fulfilled, the system (20.69) has two different stationary solutions with the same amplitude:

$$A_1 = \sqrt{\frac{2\varepsilon u_0}{\beta_e}}; \quad \sigma_1 = 0,$$

$$\theta_1^{(1)} = -\arcsin\left(\frac{3}{8\pi}\sqrt{\frac{u_0 \beta_e}{2\varepsilon}}\right), \quad \theta_1^{(2)} = \pi + \arcsin\left(\frac{3}{8\pi}\sqrt{\frac{u_0 \beta_e}{2\varepsilon}}\right). \qquad (20.72)$$

The necessary condition for the existence of the resonance (20.71) can be now rewritten as follows:

$$0 < u_0 < \frac{128\varepsilon}{9\pi^2 \beta_e}. \qquad (20.73)$$

It is obvious from the classical equation (20.70), that only the second solution is stable (corresponds to the stable equilibrium point of the equivalent pendulum).

20.6.5 Second step of the hierarchical averaging. Attraction area of the resonant solution.

In order to perform the second step of the hierarchical averaging according to the Theorem 6 it is convenient to rewrite the equations (20.69) as follows ($\theta_1 = \pi + \xi_1$):

$$A_1' = -\beta_e A_1 + \frac{16\pi}{3}\varepsilon \sin\xi_1; \quad \xi_1' = \sqrt{\mu}\sigma_1 + \frac{4\delta}{\pi A_1} + \frac{16\pi}{3 A_1}\varepsilon \cos\xi_1;$$

$$\sigma_1' = \sqrt{\mu}\, u_0 - v\sigma_1 - \frac{8\pi}{3}\sqrt{\mu} A_1 \sin\xi_1 \qquad (20.74)$$

Now we are going to convert from the variables ξ_1, σ_1 to new variables γ, which has the sense of the phase, and g, which is related to the potential energy $\frac{1}{2}\mu g^2 \cos^2\gamma$ of the equivalent-pendulum oscillations (cf. Fig. 20.8):

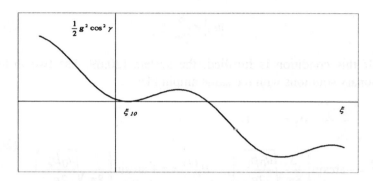

Figure 20.8. Potential energy of the equivalent pendulum.

$$g \sin \gamma = \sigma_1; \quad \frac{1}{2} g^2 \cos^2 \gamma = \frac{8\pi}{3} A_1 \left(\cos \xi_{10} - \cos \xi_1 \right) + u_0 \left(\xi_{10} - \xi_1 \right)$$

$$\text{sgn} \cos \gamma = \text{sgn} \left(\xi_{10} - \xi_1 \right); \quad \xi_{10} = \arcsin \left(\frac{3 u_0}{8 \pi A_1} \right). \tag{20.75}$$

This substitution enables to express the old variables uniquely in terms of the new ones, but it does not permit analytic solutions of the transcendental equation for ξ_1 in elementary functions. It will be assumed below that ξ_1 can be expressed in terms of the variables A_1, g, γ.

In the new variables, the equations become

$$A_1' = \frac{16\pi}{3} \varepsilon \sin \xi_1 - \beta_e A_1$$

$$g' = \frac{8\pi}{3g} \left(\cos \xi_{10} - \cos \xi_1 \right) \left(\frac{16\pi}{3} \varepsilon \sin \xi_1 - \beta_e A_1 \right) - \nu g \sin^2 \gamma$$

$$- \frac{8\pi}{3g} \left(\sin \xi_{10} - \sin \xi_1 \right) \left(\frac{16\pi}{3} \varepsilon \cos \xi_1 + \frac{4\delta}{\pi} \right)$$

$$\gamma' = \sqrt{\mu} \frac{8\pi}{3g} A_1 \frac{\left(\sin \xi_{10} - \sin \xi_1 \right)}{\cos \gamma} \tag{20.76}$$

$$- \frac{8\pi}{3g^2} \left(\cos \xi_{10} - \cos \xi_1 \right) \left(\frac{16\pi}{3} \varepsilon \sin \xi_1 - \beta_e A_1 \right) \tan \gamma$$

$$- \frac{1}{2} \nu \sin 2\gamma + \frac{8\pi}{3g^2} \left(\sin \xi_{10} - \sin \xi_1 \right) \left(\frac{16\pi}{3} \varepsilon \cos \xi_1 + \frac{4\delta}{\pi} \right) \tan \gamma.$$

20.6. Resonant Motions in a Quasi-Elastic Colliding Oscillator

It is seen from the last equation that the variable γ has the sense of the half slow phase, since the expression $(\sin\xi_{10} - \sin\xi_1)/\cos\gamma$ always remains positive, as does $(\cos\xi_{10} - \cos\xi_1)/\cos\gamma$, which can be easily demonstrated by direct expansion of the substitution equations (20.75) for ξ_1 near ξ_{10} or $g\cos\gamma \to 0$:

$$\lim_{g\cos\beta \to 0} \frac{\sin\xi_{10} - \sin\xi_1}{g\cos\gamma} = \sqrt{\frac{3\cos\xi_{10}}{8\pi A_1}}$$

$$\lim_{g\cos\beta \to 0} \frac{\cos\xi_{10} - \cos\xi_1}{g\cos\gamma} = -\sqrt{\frac{3}{8\pi A_1 \cos\xi_{10}}} \sin\xi_{10}.$$
(20.77)

There are two important singularity sources in system (20.76). The first one is connected with low energy oscillations $g \leq O(\sqrt{\mu})$. In this case the small terms in the equation for γ get bigger. This area should be excluded from the considered domain. The second singularity is caused by the separatrix of the equivalent pendulum represented by the energy level corresponding to the limitation regime:

$$\frac{1}{2}g_L^2 = \frac{16\pi}{3} A_1 \cos\xi_{10} + u_0(2\xi_{10} - \pi).$$
(20.78)

In the vicinity of the separatrix the following estimation is valid for $\gamma \approx (2n+1)\pi, n = 1, 2, \ldots$:

$$\gamma' \approx \sqrt{\mu} \left(\frac{1}{g} \sqrt{\frac{8\pi}{3} A_1 (g_L^2 - g^2) \cos\xi_{10}} + O(g_L^2 - g^2) \right).$$
(20.79)

This estimation displays that in the vicinity of the limitation curve (20.78) the velocity of γ decreases significantly and γ can not be considered as a half slow phase. This vicinity must be also excluded from the considered domain.

In other words the system (20.76) can be averaged with respect to γ in a region D_μ of the phase plane (A_1, g), which is defined by inequalities:

$$|O(\sqrt{\mu})| < g < g_L(A_1) - |O(\sqrt{\mu})|.$$
(20.80)

In this region it is possible to convert in (20.76) to the new independent variable γ and to average with respect to it. Finally the twice averaged system (the first approximation with respect to γ) is obtained:

$$A_2' = \sqrt{\mu}\, \frac{3 g_2 H_1}{8\pi A_2^2} \left(\frac{16\pi\, \varepsilon u_0}{3\mu} - \frac{\beta_e}{\mu} A_2^2 \right)$$

$$g_2' = \sqrt{\mu}\, \frac{H_3}{A_2^2} \left(\frac{16\pi\, \varepsilon u_0}{3\mu} - \frac{\beta_e}{\mu} A_2^2 - \frac{3 v g_2^2 A_2 H_2}{8\pi\mu H_3} \right).$$
(20.81)

with

$$H_1 = \frac{1}{\pi} \int_0^\pi \frac{\cos\gamma\, d\gamma}{\sin\xi_{10} - \sin\xi_1} > 0; \quad H_2 = \frac{1}{\pi} \int_0^\pi \frac{\sin^2\gamma \cos\gamma\, d\gamma}{\sin\xi_{10} - \sin\xi_1} > 0$$

$$H_3 = \frac{1}{\pi} \int_0^\pi \frac{(\cos\xi_{10} - \cos\xi_1)\cos\gamma\, d\gamma}{\sin\xi_{10} - \sin\xi_1} > 0.$$
(20.82)

Here the subscript 2 is attached to the twice-averaged variables and the prime indicates differentiation with respect to γ.

According to Theorem 6 this system is valid in D_μ with an error of $O(\sqrt{\mu})$ on a time interval $O(1/\mu)$.

Considering the behaviour of the phase paths in the (A_2, g_2) plane, it becomes possible to find the attraction area of the singular point corresponding to the stable stationary regime.

System (20.81) has the singular integral curve

$$A_2 = A_{10}; \quad g_2' = -\frac{v}{\sqrt{\mu}} \frac{3 g_2^2 H_2}{8\pi A_2},$$
(20.83)

motion along which is in the direction of the equilibrium $g_2 = 0$. This curve divides region D_μ into two parts. The integral curves of system (20.81) are given by one first order differential equation:

$$\frac{dg_2}{dA_2} = \frac{8\pi H_3}{3 H_1} \left[1 - \frac{3 v g_2^2 A_2 H_2}{8\pi H_3} \left(\frac{16\pi\, \varepsilon u_0}{3\mu} - \frac{\beta_e}{\mu} A_2^2 \right)^{-1} \right].$$
(20.84)

20.6. Resonant Motions in a Quasi-Elastic Colliding Oscillator

To determine the behaviour of the integral curves, their form near the limitation curve (20.78) should be investigated. The equation (20.78) can be rewritten as follows:

$$\frac{dg_L^2}{dA_2} = \frac{32\pi}{3}\cos\xi_{10}. \qquad (20.85)$$

The explicit form of the functions H_1, H_2, H_3 can be obtained in the vicinity of this curve, and (20.84) can be rewritten in the form:

$$\frac{d\Delta}{dA_2} = \frac{8\pi H_4}{3H_1} + \frac{H_2 \nu g_2^2 A_2}{\mu H_1}\left(\frac{16\pi}{3}\frac{\varepsilon u_0}{\mu} - \frac{\beta_e}{\mu}A_2^2\right)^{-1} + O\left(\frac{1}{\ln^2 \Delta}\right)$$

$$\Delta = \frac{1}{2}\left(g_L^2 - g_2^2\right); \quad H_1 = \frac{1}{2\pi g_2}\sqrt{\frac{8\pi A_2}{3\cos\xi_{10}}}\left(\frac{1}{\Delta}\right) + O \qquad \ln(20.86)\ (1);$$

$$H_4 = 2\cos\xi_{10}H_1 - H_3 > 0; \quad H_2 = O(1), \quad H_4 = O(1) \text{ at } \Delta \to 0.$$

Equations (20.86) shows that for $A_2 < A_{10}$ the integral curves leave the neighbourhood of the limitation curve and tend to the equilibrium point (see Fig. 20.9).

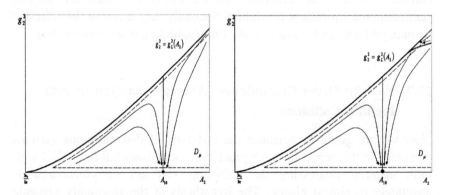

Figure 20.9. Phase plane $\left(A_2, g_2^2\right)$ for sufficient big (left) and small (right) values of the parameter ν.

The situation is much more complex for $A_2 > A_{10}$. Analyzing equation (20.86) (*cf.* [46]) it can be shown that at large v the entire region D_μ is the attraction region for the resonant regime. If to the contrary

$$v < \frac{\beta_e}{2\mu} \qquad (20.87)$$

there exists such a curve that all integral curves above it approach to the limitation curve, so that a random perturbation may terminate libration motion and initiate rotation of the equivalent pendulum. All the curves above this special integral curve reach the equilibrium point (see Fig. 20.9).

20.6.6 Concluding remarks

It should be noticed in conclusion to this section that due to the applied discontinuous hierarchical averaging it is possible to reduce the considered problem with three degrees of freedom and internal collisions to one continuous first order differential equation. This allows the complete analysis of the resonance attraction region in the phase plane. This region has the magnitude order $O(\sqrt{\mu})$ as usual in significantly nonlinear resonance problems. In the first approximation the averaged system is similar to a mathematical pendulum with an externally applied constant torque. The impacts increase firstly the vibration torque, stabilizing the equivalent pendulum. Secondly they increase the effective damping which can be added to the linear damping in the main spring.

20.7 Variable Order Discontinuous Averaging in Systems with Inelastic Collisions

Now we are going to consider the last type of discontinuous systems discussed in this work. In many practical cases connected first of all with bulk material processing the collisions can not be considered as absolutely or almost elastic. The hypothesis of the absolutely inelastic collisions is much more useful in these cases (*cf.* [11,12,33]). The main qualitative differences between systems with friction and systems with collisions described in the previous section are even stronger in case of

20.7. Variable Order Discontinuous Averaging in Systems with Inelastic Collisions

inelastic impacts. First of all systems with inelastic collisions can not be described by systems of differential equations valid for all times. The continuous differential equations are valid between the collisions and state variables after each collision can be calculated as functions of state variables before the collision. There are two different types of inelastic collisions. The first one are collisions with finite time interval of contact between the colliding bodies. In this case systems before and after the collision have different order. This type is considered in the following examples. The second type are collisions with instantaneous contact loss (instantaneous throwing in Russian literature). This type of motions is less important from the practical point of view and is not considered here. The corresponding examples can be found in [10, 11, 31].

The variable order discontinuous averaging can be applied to analyze inelastic collision oscillators. But compared to the almost elastic collision oscillator the analysis must be restricted to the given collisions sequence, for example to the given number of collisions for each oscillation period. This restriction is strong from the theoretical point of view, but it is not so critical in practice, because mainly dynamic regimes with most intensive collisions are desired for bulk material processing. The corresponding regime of collisions is usually evident, and the analysis can be concentrated on this regime.

The classical double restricted impact oscillator with inelastic collisions is considered here as the first example. The aim is to demonstrate the main ideas and methods using a simple and physically clear object. The extension of the approach a model of resonant crasher excited by an inertial source of limited power is totally similar to analysis in section 20.6. It can be found in [46]. The results are illustrated and tested by comparisons to those obtained by numerical simulation of the full equations of motion.

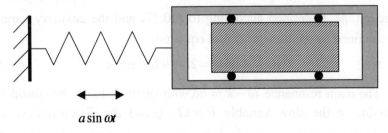

Figure 20.10. Impact oscillator with two-sided restriction.

Figure 20.10 shows the physical example system: a frame of mass M attached to a linear spring and excited kinematically according to a harmonic law. Inside of the frame there is a free moving striker of mass m. The length of the frame's interior is $2l$. The striker can collide with both sides of the frame.

The equations of motion have the form:

$$\ddot{s} = -k^2(1-\mu)s - k^2\mu(1-\mu)r + k^2 a(1-\mu)\sin\omega t$$

$$\ddot{r} = \begin{cases} k^2 s - k^2\mu r - k^2 a \sin\omega t, & |r| < l \\ 0, & |r| = l \end{cases} \qquad (20.88)$$

These equations must be supplemented by the corresponding kinematic conditions for the separation of the masses moving together:

$$r = \pm l : s = \mu r + a\sin\omega t; \quad \dot{s} = 0. \qquad (20.89)$$

Here s is the co-ordinate of the centre of mass of the system, r is the relative co-ordinate of the strike in the gap, k is the frequency of free oscillations of the frame, $\mu = m/(M+m)$ and a/l are small parameters of single order.

We now change to new phase variables with the aid of the following non-analytical transformation, which for a regime with two collisions in one oscillation period of the frame ensures exact satisfaction of the collision conditions:

$$s = A\sin\psi; \quad \dot{s} = A\cos\psi; \quad \Omega = \omega t;$$
$$\dot{r} = k\{B - A\cos\psi M_2(\varphi,\alpha)\}M_2(\varphi,\alpha)$$
$$r = l\,\mathrm{sgn}\sin\varphi - lM_2(\varphi,\alpha) \qquad (20.90)$$
$$+ \{-l + B(\varphi - [\varphi/\pi]) - A\sin\psi M_2(\varphi,\alpha)\}M_2(\varphi,\alpha)$$

Here $[z]$ is the integer part of z, function $M_2(\varphi,\alpha) = M_1(2\varphi,\alpha)\mathrm{sgn}\sin\varphi$, function M_1 is defined according to (20.27) and the auxiliary variable α is defined by the transcendental equation:

$$Ba - A\sin\psi = 2l \quad \text{when } \varphi = a \qquad (20.91)$$

The main resonance $\omega - k = k\Delta$ with small Δ has to be considered. Introducing the slow variable $\theta = \Omega - \psi$ and the dimensionless time $\tau = kt$ and omitting terms $o(\mu)$, the following system can be obtained:

20.7. Variable Order Discontinuous Averaging in Systems with Inelastic Collisions

$$\dot{A} = \tfrac{1}{2}\mu A \sin 2\psi + \mu r \cos\psi + a\sin(\psi + \theta)\cos\psi$$
$$\dot{\theta} = \Delta + \mu \sin^2 \psi + \mu r A^{-1}\sin\psi + aA^{-1}\sin(\psi + \theta)\sin\psi$$
$$\dot{B}M_2(\varphi,\alpha) = \mu A \sin\psi M_2^2(\varphi,\alpha) \qquad (20.92)$$
$$\dot{\varphi}M_2(\varphi,\alpha) = \left\{1 - \mu(\varphi - [\varphi/\pi])AB^{-1}\sin\psi\,\text{sgn}\sin\varphi\right\}M_2(\varphi,\alpha)$$
$$\dot{\psi} = 1 - \sin^2\psi - \mu r A^{-1}\sin\psi - aA^{-1}\sin(\psi + \theta)\sin\psi$$

with separation condition

$$\varphi = n\pi \Rightarrow B = A + O(\mu^2); \quad \psi = n\pi + O(\mu); \quad B\alpha - A\sin\alpha = 2l + O(\mu). \qquad (20.93)$$

System (20.92), (20.93) is in the standard form for the variable order discontinuous averaging (Theorem 5).

Averaging over the fast phase ψ and denoting the differentiation with respect to it by a prime, we arrive at the equations of the first approximation:

$$A'_1 = \tfrac{1}{2}a\sin\theta_1 - (4\mu l/\pi)\sin^4(\alpha_1/2)/(\alpha_1 - \sin\alpha_1)$$
$$\theta'_1 = \Delta + \tfrac{1}{2}\mu + (a/4l)(\alpha_1 - \sin\alpha_1)\cos\theta_1$$
$$\qquad + (\mu/\pi)\left(\sin\alpha_1 - \tfrac{1}{4}\sin 2\alpha_1 - \tfrac{1}{2}\alpha_1\right) \qquad (20.94)$$
$$B_1 = A_1$$
$$\theta_1 = 0$$
$$\alpha_1 - \sin\alpha_1 = 2l/A_1.$$

Equations (20.94) are valid for times $t = O(1/\mu)$ with the accuracy $O(\mu)$. Setting the right hand sides of (20.94) to zero we obtain a system of transcendental equations governing the stationary resonant regime.

Figure 20.11 shows the dependence of the resonant amplitude A_1 on the frequency Δ for $\mu = 0.01$ and $a/l = 0.01$. Here it is necessary to take into account the condition for the realizing the simplest collision regime: $0 < \alpha < \pi$. The existence of the considered resonance regime is ensured if

$$\Delta^2 + (2\mu/\pi)^2 < (\pi a/4l)^2. \qquad (20.95)$$

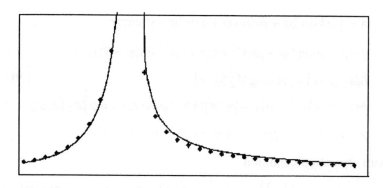

Figure 20.11. Comparison between analytic and numerical solutions.
— Theory ♦ Numeric

It is interesting to notice that the dissipative influence of inelastic impact decreases in the vicinity of the resonance at $\Delta = -\mu/2$ and the stationary amplitudes in this point are infinite.

20.8 Conclusions

Averaging is an effective method for asymptotic analysis of discontinuous systems. It can be applied to find first and higher order approximations in systems with friction and collisions. Averaging of constant order discontinuous motions (typical for slipping systems with friction and for systems with quasi-elastic collisions) is very similar to the standard averaging. Variable order discontinuous motions (typical for sticking systems with friction and for systems with inelastic collisions) are strongly degenerated and allow some additional reduction of the averaged system. Described methods are illustrated by the analysis of impact oscillators.

20.9 Concluding Remarks of the Editor of the Book

The concluding chapter differs essentially from the previous ones by the fact that it actually does not use either the method of direct separation of motions or the approach of vibrational mechanics. The author preferred to use somewhat modified classical schemes of averaging. Nevertheless,

2.8. Conclusions

the inclusion of this chapter into the book seems expedient for three reasons:

– first, the theorems, proved here, substantiate methods of averaging and extend them to the discontinuous systems (i.e. to the method of direct separation of motions as well). That is quite essential for a number of applied problems;

– secondly, the examples, considered in this chapter, are not only illustrative, they are also of applied interest;

– thirdly, on the one hand, it seemed useful to let the reader compare the difficulties, connected with the practical use of the approaches of this chapter, with the simplest version of the method of direct separation of motions when a rough solution of equations of fast motions is being used. On the other hand, it seems useful to compare the main qualitative results, obtained by means of the first approach and those, got by the second, more complicated one.

Such comparison can be made if we use as an example the problem, considered in section 20.6. To do that, we must recall the results, obtained when solving that problem by the method of direct separation of motions in its simplest version.

Following the solution, described in detail in section 6.3 of the book [12] for the case of a smooth system, it is easy to obtain in this case too the following equation of the first order (in the designations, close to those, adopted in this chapter)

$$m_r \rho^2 \dot{\Omega} = U - V_R \Omega + V(\Omega) \qquad (20.96)$$

describing the slow change in the angular velocity of the rotation of the rotor ψ:

$$\Omega = \langle \dot{\psi} \rangle \qquad (20.97)$$

The expressions for the vibrational moment $V(\Omega)$ are given in the book [12]. For both systems – the smooth one, considered there, and the system with impacts, this expression is defined by the formula

$$V(\Omega) = m_r \rho \langle \ddot{x}_1 \sin \psi \rangle \qquad (20.98)$$

where the expression for x_1 is calculated approximately as a stationary solution of the equations of motion of the oscillatory part of the system (the first two equations (1) corresponding to the rotation of the rotor with a slowly changing angular velocity $\Omega(t) = \langle \dot{\psi} \rangle$).

It is natural that the expression for $V(\Omega)$ is of a distinctly "peak" character in the vicinity of the resonances of the oscillatory part of the system. The condition of stability of the stationary solutions $\Omega = \Omega_*$ of equation (1), found from the equation

$$U - V_R \Omega_* + V(\Omega_*) = 0, \qquad (20.99)$$

is presented by the inequality

$$V'(\Omega_*) < V_R \qquad (20.100)$$

This simple consideration makes it possible to explain such most important regularities of the behavior of this type of systems as the instability of the right-hand descending part of the resonance curve of the oscillatory part of the system and the effect of Sommerfeld, connected with that instability (see section 6.3 of the book [12]).

A detailed investigation of this vibro-impact system by the approach under consideration requires a study of different regimes of motion of the bodies m and M with collisions. Then the expressions of the moment $V(\Omega)$ will prove to be different for each of such regimes. We can, however, state that the regularities, described above, in particular those of the main resonance regime, are preserved in this case too. The reader, interested in this problem, can easily perform all the necessary calculations.

As for the new regularities, found by a more complicated investigation, similar to the one, given in this chapter, there are two of them that should be mentioned.

1. There has been established a more complicated hierarchy of motions according to their tempo: the system having not only the slow or fast motions, but also the half-slow motions.

2. A hidden equivalent pendulum has been discovered in the system.

It proved possible to extend these regularities, well-known for the smooth systems, to the essentially discontinued systems of vibro-collision type.

Thus, the content of this chapter is directly connected with the problems, considered in book [12], and in this book as well, and facilitates developing a broader view of this problem.

References

1. Appell P.: *Traité de Méchanique Rationelle*, Vol. 2, Gauthier-Villars, Paris, 1953.
2. Armstrong-Hélouvry B., Dupont P. and Canudas de Wit C.: A survey of models, analysis tools and compensation methods for the control of machines with friction. *Automatica*, 30 (1994) 7, pp. 1083 – 1138.
3. Armstrong-Hélouvry B.: A perturbation analysis of stick-slip. Friction-induced vibration, chatter, squeal and chaos, *ASME* DE-Vol. 49 (1992), pp. 41 – 48.
4. Arnold V. I.: Conditions for the applicability, and estimate of the error, of an averaging method for systems which pass through states of resonance during the course of their evolution. *Soviet Math.* 6 (1965), pp. 331 – 334.
5. Babitsky V. I.: *The theory of vibro-impact systems* (in Russian). Nauka, Moscow, 1976.
6. Babitsky V. I., Borovkov B. A.: Vibroimpact motions in systems with limited excitation. *Izv. AN SSSR, Mechanika Tverdogo Tela*, 1 (1977), pp. 50 – 57 (in Russian).
7. Babitsky V. I., Kovaleva A. S., and Krupenin V. L.: Analysis of quasi-conservative vibro-impact systems using averaging method. *Izv. AN SSSR, Mechanika Tverdogo Tela*, 1 (1982), pp. 41 – 49 (in Russian).
8. Babitsky, V. I.: *Theory of Vibro-Impact Systems*. Springer-Verlag, Berlin, 1998.
9. Bishop S. R.: Impact oscillators. *Phil. Trans. R. Soc. London A* 347 (1994), pp. 347 – 351.
10. Blekhman I. I. and Dzhanelidze G. Y.: *Vibrational Displacement* (in Russian). Nauka, Moscow, 1964.
11. Blekhman I. I., Vial'tseva O. A., Vaisberg L. A. and Fidlin A. Capacity of vibrating screens with active working surfaces. In *"Analysis of processes, machines and devices for size separation"*, Leningrad, "Mekhanobr", 1988, 20 – 35. (in Russian)
12. Blekhman I. I.; *Vibrational mechanics. Nonlinear dynamic effects, general approach, Applications*. World Scientific, Singapore – New Jersey – London – Hong Kong, 2000.
13. Brogliato B.: *Nonsmooth impact mechanics*. Lecture notes in control and information sciences 220, Springer, Berlin, 1996.
14. Elmer F.-J.: Nonlinear dynamics of dry friction. *Journal of Physics A – Mathematical and General*, 30 (1997), pp. 6057 – 6063.
15. Evan-Iwanowski R. M.: *Resonance oscillations in mechanical systems*, Elsevier Publ. Co., Amsterdam, 1976.
16. Feeny B., Moon F.C.: Chaos in a forced dry friction oscillator: experiments and numerical modelling. *J. Sound. Vibr.*, 179 (1994) pp. 303 – 323.

17. Fidlin A.: On averaging in systems with a variable number of degrees of freedom. *J. Appl. Math. Mech.*, 55 (1991) 4, pp. 507 – 510.
18. Fidlin A. and Thomsen J. J.: Predicting vibration induced displacement for a resonant friction slider. *Eur. J. Mech. A/Solids*, 20 (2001), pp. 155 – 166.
19. Fidlin A.: On the asymptotic analysis of discontinuous systems. *DCAMM Report 646*, Technical University of Denmark, Lyngby, 2000 (accepted for publication in ZAMM, 2001).
20. Foale S. and Bishop S. R.: Dynamical complexities of forced impacting systems. *Phil. Trans. R. Soc. London A* 338 (1992), pp. 547 – 556.
21. Goldsmith W.: *Impact. The theory and physical behaviour of colliding solids*. Edward Arnold, London, 1960.
22. Guggenheimer J., Holmes P. J.: *Nonlinear oscillations, dynamical systems, and bifurcations of vector fields*. Springer, New York, 1983.
23. Guran A., Feenly F., Hinrichs N., Popp K.: A historical review on dry friction and stick-slip phenomena. *Appl. Mech. Rev.* 51 (1998), pp. 321 – 341.
24. Hinrichs N., Oestreich M., Popp K.: Dynamics of oscillators with impacts and friction. *Chaos, Soliton and Fractals*, 8 (1997), pp. 535 – 558.
25. Ibrahim R. A.: Friction-induced vibration, chatter, squeal and chaos: part 1 – mechanics of friction. *Friction-induced vibration, chatter, squeal and chaos, ASME* DE-Vol. 49 (1992), pp. 107 – 121.
26. Ibrahim R. A.: Friction-induced vibration, chatter, squeal and chaos: part 2 – dynamics and modelling. *Friction-induced vibration, chatter, squeal and chaos, ASME* DE-Vol. 49 (1992), pp. 107 – 121.
27. Ivanov A. P.: Visco-elastic approach to impact with friction. *Amer. Mech. Divis.*, 205 (1995), pp. 115 – 127.
28. Ivanov A. P.: *Dynamics of systems with mechanical collisions*. International Program of Education, Moscow, 1997 (in Russian).
29. Jensen J. S.: Non-trivial effects of fast harmonic excitation, *PhD thesis*, Technical University of Denmark, Lyngby, 1999.
30. Karagiannis K, Pfeiffer F.: Theoretical and experimental investigations of gear-rattling. *Nonlinear dynamics*, 2 (1991), pp. 367 – 387.
31. Kobrinskii A. E. *Dynamics of mechanisms with elastic connections and impact systems*. Iliffe, London, 1969, 363 p.
32. Kolovsky M. Z.: On the conditions for the existence of periodic solutions in systems of differential equations with discontinuous right-hand sides, containing small parameter. *Appl. Math. Mech.*, 24 (1960) 4, pp. 738 –745 (in Russian).
33. Kremer E. and Fidlin A.: One-dimensional dynamic continuum model of a free-flowing granular medium. *Soviet Physics Doklady*, Vol. 34 (1989) 12, pp. 1063-1065.

34. McMillan J.: A nonlinear friction model for self excited vibrations. *J. Sound Vibr.*, 205 (1998) 3, pp. 323 – 335.
35. Meerkov S. E.: Vibrational control Theory. *Journal of the Franklin Institute*, 303 (1977) 2, pp. 117 – 128.
36. Meerkov S. M.: Principle of vibrational control: theory and applications. *IEEE Transactions on Automatic Control*, AC-25 (1980) 4, pp. 755 – 762.
37. Miranda E. C. and Thomsen J. J. Vibration induced sliding: theory and experiment for a beam with a spring-loaded mass. *Nonlinear Dynamics*, 16 (1998) 2, pp. 167 – 186.
38. Mitropolskii Yu. A. and Nguyen V. D.: *Applied Asymptotic Methods in Nonlinear Oscillations*. Kluwer, Dordrecht, 1997.
39. Nagaev R. F.: *Mechanical processes with repeated attenuated impacts.* World Scientific Publishing, Singapore, 1999.
40. Nayfeh A. H. and Mook D. T.: *Nonlinear oscillation*. Wiley, New York, 1978.
41. Neishtadt A. I.: On the passing through resonance in a problem with two frequencies. *Soviet Physics Doclady*, 221 (1975) 2, pp. 301 – 304 (in Russian).
42. Neishtadt A. I.: On the averaging in multi-frequenced systems. *Soviet Physics Doclady*, 223 (1975) 2, pp. 314 – 317 (in Russian).
43. Neishtadt A. I.: On the averaging in multi-frequenced systems II. *Soviet Physics Doclady*, 226 (1976) 6, pp. 1295 – 1298 (in Russian).
44. Panovko Y. G. and Gubanova I. I.: *Stability and oscillations of elastic systems; paradoxes, fallacies and new concepts*. Consultants Bureau, New York, 1965.
45. Pechenev A. V.: Averaging of systems with a hierarchy of velocities of phase rotation on essentially long time intervals. *Soviet Physics Doclady*, 315 (1990) 1, pp. 28 – 31 (in Russian).
46. Pechenev A. V. and Fidlin A.: Hierarchy of resonant motions excited in a vibroimpact system with contact zones by an inertial source of limited power. *Izw. AN SSSR Mechanika Tverdogo Tela* (English Translation), 27 (1992) 4, pp. 46 – 53.
47. Peterka F.: *Introduction to Vibration of Mechanical Systems with internal Impacts*, Academia, Prague, 1981, p. 269 (in Czech)
48. Peterka F.: Dynamics of mechanical systems with impacts and dry friction, *Proc. 2nd European Nonlinear Oscillations Conference*, Prague, 1996, Vol. 1, pp. 33 – 44.
49. Peterka F.: Bifurcations and transitions phenomena in an impact oscillator. *Chaos, Solitons and Fractals*, 7 (1996) 10, pp. 1635 – 1647.
50. Pfeiffer F.: Seltsame Attraktoren in Zahnradgetrieben. *Ingenieur-Archiv* 58 (1988), pp. 113 – 125.

51. Pfeiffer F., Glocker C.: *Multibody dynamics with unilateral contacts*. Wiley, New York, 1996.
52. Plotnikov V. A.: The averaging method for differential inclusions and its application to optimal-control problems. *Differential Equations*, 15 (1979) 8, 1013 – 1018.
53. Plotnikov V. A., Zverkova T. S.: The method of averaging for the systems of standard form with discontinuous right-hand sides. *Diff. Uravn.*, 18 (1982) 6, 1091 – 1093, (in Russian).
54. Plotnikov V. A., Zverkova T. S.: Method of partial averaging in the systems of standard form with discontinuous right-hand sides. *Ukr. Math. J.*, 45 (1993) 1, 156 – 159.
55. Pontryagin L. S. et. al.: *The mathematical theory of optimal procasses*. Interscience Publ., New York, 1962.
56. Popp K., Stelter P.: Stick-slip vibrations and chaos. *Phil. Trans. R. Soc. London A* (1990), pp. 89 – 105.
57. Popp K.: Chaotische Bewegungen beim Reibschwinger mit simultaner Selbst- und Fremderregung. *Z. angew. Math. Mech.* 71 (1991), pp. 71 – 73.
58. Popp K.: Nichtlineare Schwingungen mechanischer Struckturen mit Füge- oder Kontaktstellen. *Z. angew. Math. Mech.* 74 (1994), pp. 147 – 165.
59. Popp K., Hinrichs N., Oestreich M.: Analysis of a self excited friction oscillator with external excitation. In *Dynamics with Friction/* Guran A., Pfeiffer F., Popp K. (Eds), World Scientific, Singapore, 1996, pp. 1 – 35.
60. Rand R. H. and Moon F. C.: Bifurcations and chaos in forced zero-stiffness impact oscillator. *Int. J. Nonlinear Mech.* 25 (1990) 4, 417 – 432.
61. Rayleigh J. W. S.: *The theory of Sound*, second edition, 1945 re-issue, Dover Publications, New York, 1877.
62. Sanders J. A.: On the passage through resonance. *SIAM J. Math. An.*, 10 (1979), pp. 1220 – 1243.
63. Sanders J. A. and Verhulst F.: *Averaging methods in nonlinear dynamical systems*. Springer-Verlag, New York, 1985.
64. Shaw S. W. and Holmes P. J.: A periodically forced piecewise linear oscillator. *J. Sound Vibr.* 90 (1983) 1, 129 – 155.
65. Shaw S. W.: On the dynamic response of a system with dry friction. J. Sound Vibr. 108 (1986), pp. 305 – 325.
66. Thomsen J. J.: Vibration suppression by using self-arranging mass: effects of adding restoring force. J. Sound Vibr., 197 (1996), pp. 403 – 425.
67. Thomsen J. J. :*Vibrations and Stability, Order and Chaos*. McGraw-Hill, London, 1997.
68. Thomsen J. J.: Using fast vibrations to quench friction induced oscillations. *J. Sound Vibr.*, 228 (1999) 5, pp. 1079 – 1102.
69. Thomsen J. J. and Fidlin A.: Analytical Approximations for stick-slip vibration amplitudes. *DCAMM Report No. 652*, Technical University of

Denmark, Lyngby, 2001(accepted for publication in International Journal of Nonlinear Mechanics).
70. Tolstoi D. M.: Significance of the normal degree of freedom and natural normal vibrations in contact friction. *Wear*, 10 (1967), pp. 199 – 213.
71. Tondl A.: Quenching of self excited vibrations. Elsevier, Amsterdam, 1991.
72. Ueda S. and Tomikawa Y.: *Ultrasonic Motors, Theory and Applications*. Clarendon Press, Oxford, 1993.
73. Van der Velde F. and de Baets P.: A new approach of stick-slip based on quasi-harmonic tangential oscillations. *Wear*, 216 (1998), pp. 15 – 26.
74. Vielsack P., Storz M.:Bifurcations of motion in a technical system with stick-slip and impact. *Inst. f. Wiss. Rechnen u. Math. Modellbildung*. Preprint Nr. 95/5, Uni Karlsruhe, 1995.
75. Wittenburg J.: *Dynamics of systems of rigid bodies*. Teubner, Stuttgart, 1977.
76. Zaretsky L. B.: Synchronisation of centrifugal vibrational exciters in systems with discontinuous characteristics. Eng. Jour. MTT, 1 (1968), pp. 18 – 25.
77. Zhuravlev V. F.: Method for analysing of vibroimpact systems using special functions. *Izv. AN SSSR, Mechanika Tverdogo Tela*, 2 (1976), pp. 30 – 34 (in Russian).
78. Zhuravlev V. F. and Meniailov A. I.: An investigation of a vibroimpact system with limited excitation. *Izv. AN SSSR, Mechanika Tverdogo Tela*, 2 (1978), pp. 45 – 50 (in Russian).
79. Zhuravlev V. F. and Klimov D. M.: *Applied methods in the theory of nonlinear oscillations* (in Russian). Nauka, Moscow, 1988.

Index

Activation, 256
Additional (vibrational) rigidity, 144
Antiphase solution, 226, 227
Approximate synchronization, 182
Articulated model, 120, 132
Asymptotic analysis, 333, 334
Asymptotic methods, 221
Asymptotic synchronization, 182
Averaging, 83
Averaging method, 334, 337
Averaging of discontinuous systems, 357, 370
Averaging operator, 84, 99, 100, 123

Basin of attraction, 127
Bifurcation, 96, 104, 105, 106, 108, 109, 110, 111, 118, 119, 130, 131, 132
Bookkeeping parameter, 117, 128

Cantilever pipe, 120, 132
Carried constraints, 213
Carrying bodies, 210
Carrying constraints, 213
Carrying system, 210
Center manifold, 103, 104, 119
Chaos, 74, 112
Chaotic synchronization, 180
Characteristic equation, 101, 125
Chelomei's Pendulum, 73, 74
Codimension two, 109
Coefficient of the discharge through the hole, 320
Comparison functions, 181
Conjugacy of the bifurcation points, 157
Conjugate resonances and bifurcations, 151
Conjugate resonances, 156

Constant order discontinuous systems, 360, 363
Controlled synchronization, 179
Coordinate synchronization, 184
Cubical nonlinearities, 104

Desynchronization measure, 182
Detuning parameter, 273, 276
Diagram of Ince-Strutt, 44
Differential equations, 83, 85, 100, 101, 104, 121
Direct separation of motions, 334, 359
Discontinuous systems, 357, 359, 361, 390
Discontinuous variable transformations, 362
Divergence, 97, 102, 103, 108, 109, 110, 111, 113, 115, 119, 125
Double pendulum, 74, 98, 119, 128
Dry friction, 357, 358, 359, 360, 361, 363
Duffing's differential equation, 160
Dynamic composites, 255
Dynamic instability, 97
Dynamic materials, 255, 260
Dynamic materials of the first kind or the activated dynamic materials, 256
Dynamic materials of the second kind or the kinetic dynamic materials, 256

Effect of vibro-closing the vessel, 324
Effective rigidity of the rope, 144
Eigenvalues, 102, 104, 115, 124, 125, 126
Eigenvibrational force, 19, 26
Electro-technical dynamic materials, 257
Equation of fast motions, 18, 143, 339, 343, 345, 347, 350
Equation of slow motions, 18, 143, 340, 343
Equation of vibrational hydraulics, 314
Equilibrium positions, 74
Equivalent length of the pendulum, 40
Essentially non-isochronous objects, 219
Euler operator, 29
Experimental Pendulum, 76, 77, 79, 81, 92, 95
External excitation, 341, 348, 351
Extremal synchronization, 183

Fantastic analogy, 57

Fast component, 6, 10
Fast force, 6, 10
Fast oscillating inertia coefficients, 336
Fast time, 6, 10
Fast time scale, 99
First order averaging, 364
Flexural mode, 79, 80, 96
Flexural vibrations, 76, 77, 78, 79, 82, 84, 90, 91, 93, 94, 96
Floquet-Lyapunov's theory, 43
Fluid-loaded structure, 280
Flutter, 97, 102, 103, 105, 106, 107, 109, 110, 115, 116, 119, 124, 125, 126, 132
Follower-force, 74, 110
Forced synchronization, 179
Forms, 103, 104, 105, 119
Frequency (Huygens) synchronization, 183
Frequency synchronization, 179
Friction-induced oscillations, 358

Generalized (partial) coordinate synchronization, 184
Generating equation, 41
Generative parameters, 236
Global dynamics, 110
Global Jacobian matrix, 191

Hamilton operator, 25
Hamilton variational principle, 243
Harmonic influence coefficients, 196, 205
Heavy fluid loading, 280
Hierarchical averaging, 371, 373, 383, 384, 385, 390
High frequency excitation, 333, 334
Higher order averaging, 364
Hopf bifurcations, 102, 104, 106, 108, 109, 130

Impacts, 357, 359, 360, 361, 362, 363, 377, 384, 390
Indian magic rope, 139
Induced vibrational force, 19, 26
Inelastic collisions, 390, 391
Inertia matrix of the rotors, 191

Integral criterion (extreme property) of the stability of synchronous motions, 215
Integral signs of the stability of synchronous motions, 235
Inverted rod, 75, 79, 90

Kinetization, 256
Kronecker's symbol, 43

Lagrange multipliers, 25
Lagrange-Maxwell equation, 172
Limit cycle, 106
Lyapunov exponents, 119

μ-stability, 24
Main assumption of vibrational mechanics, 14
Main equation of vibrational mechanics, 7, 18
Mathieu's equation, 43
Matrices of the inertia forces and moments, 192
Mechanics of the systems with the hidden motions, 7, 14, 19
Method of Bubnov—Galerkin, 49
Method of direct separation of motions, 74, 122, 221
Method of harmonic balance, 49
Method of multiple scales, 46, 334, 339
Modal coupling, 273
Model equations, 75, 97, 99, 101, 103, 110, 113, 114, 118, 119, 120, 122, 123, 124, 126, 127, 130, 131
Modulation of the stiffness, 260
Monodromy matrix, 44
Multiple Scales, 83
Multiple synchronization, 235
Multiple synchronization of vibro-exciters, 236
Multiplied by the factor, 346

Nondimensional variables, 82
Nonlinear dynamics, 74, 104, 118, 132
Nonlinear interaction, 76
Numerical integration, 75, 99, 102, 103, 106, 108, 110, 118, 119, 124, 127, 130, 131
Numerical simulation, 75, 76, 91, 92

Observer **O**, 6, 18
Observer **V**, 6, 7, 18
Oscillatory coordinates, 216, 237
Outfolding parameter, 105
Over-damped oscillator, 162
Overload parameter, 325

Parametric excitation, 333, 348, 351, 354
Parametric resonances, 151
Partially anglular velocities of rotation, 237
Perturbation analysis, 76
Phase synchronization, 183
Pitchfork bifurcations, 102, 107, 108, 109
Poincare and Lyapunov small parameter methods, 221
Poincare section, 183
Positions of quasi-equilibrium, 23
Post-critical behavior, 127, 131
Postcritical frequency range, 205
Post-resonance region, 224
Potential function, 32, 198, 204, 217, 218, 236
Potential of the generalized nonconservative forces, 241
Potential on the average dynamic systems, 56
Potential on the average systems, 32
Potentiality of the system on the average, 27
Precritical frequency range, 205
Pre-resonance region, 224
Problem of closure, 313
Pulsating flow, 120, 125, 130, 132
Purely inertial approximation, 21, 28

Quasi and "non-quasiconservative" idealization of the objects, 215
Quasiconservative objects, 218
Quasi-equilibrium, 88, 90, 168
Quasi-self-excited oscillations, 175
Quasi-static solutions, 75

Resonance function, 86
Rotational coordinates, 216, 237

Saddle-node bifurcations, 107
Sandwich beam, 261
Secular terms, 129
Selfsynchronization, 179
Semi-empirical theories of turbulence, 34
Separation of motions, 334, 354
Simple synchronization, 235
Skew-symmetric 'fast' motions, 272
Sliding disk, 74, 75, 76, 80, 91, 94
Slipping, 357, 360, 367, 394
"Slow" component, 6, 10, 170
Slow force, 6
Slow motions, 334, 346, 348, 354
Slow time, 6, 10
Slow time scale, 99, 128
Sources of discontinuities, 357
Speed-gradient algorithms, 186
Static instability, 97, 102
Statically and dynamically unbalanced rotors, 204
Stationary solution, 85
Statistical resonance, 152, 156, 164
Stereo-mechanical impact theory, 360
Sticking, 357, 358, 360, 361, 364, 367, 394
Stick-slip vibrations, 358
Stiffness matrices, 191
Stiffness modulation, 267
Strong nonlinear resonance, 362, 363, 370, 371
Strong response, 341, 353, 354
Subcritical, 106, 107, 108, 109, 110, 111, 118, 119, 130, 131, 133
Supercritical, 106, 108, 109, 110, 118, 119, 130, 131, 133
Support-excitation, 98, 101, 102, 103, 104, 106, 107, 108, 109, 110, 111, 112, 113, 119
Symmetric 'slow' motions, 272
Symmetry laws of the harmonic influence coefficients, 196
Synchronization, 179
Synchronization in the average, 182
Synchronization index, 181, 182
Synphase solution, 226

Systems with almost uniform rotations, 215
Systems with strong excitation, 333, 335, 336, 344, 354
Systems with very strong excitation, 333, 339, 341, 353, 354

Tangent-space approximation, 105
"Technical" and planetary-satellite version of the system, 212
"Technical" version of the system, 210
Tendency to the synchronization, 236
Theorem of Earnshaw, 168

Variable order discontinuous systems, 361, 363
Vibration damping, 76
Vibration induced displacement, 359
Vibrational closure of the holes, 317
Vibrational convection, 314
Vibrational force, 6, 18, 26, 170
Vibrational injection of gas into liquid, 317
Vibrational injection of the fluid into the fluid, 318
Vibrational mechanics, 7
Vibrational moment, 52
Vibrational resonances, 152
Vibrational rheology, 260
Vibro-genous force, 26
Vibro-jet effect, 317
Vibrorheological equation, 7
Vibrorheology, 7
Viscous damping, 82, 97, 120, 127

Weak response, 341, 352, 353, 354

Table of Contents of the Book *Vibrational Mechanics* (*Nonlinear Dynamic Effects, General Approach, Applications*) Iliya I. Blekhman, World Scientific, 2000

Part I. Fundamentals of Theory of Vibrational Mechanics 1

Chapter 1. Introduction. Subject--Matter of Vibrational Mechanics 3
1.1 Oscillatory Processes in Nature and Engineering 3
1.2 Damaging Vibration and Useful Vibration. Vibrational Engineering 3
1.3 The Theory of Vibrational Processes and Devices – a New Section of the Applied Theory of Oscillations 5
1.4 On the Effects Caused by the Action of Vibration in Non-linear Oscillatory Systems 6
1.5 Vibrational Mechanics and Vibrational Rheology. Observer **O** and Observer **V** 8
1.6 Vibrational Mechanics as a Section of the Mechanics of Systems with Hidden Motions 13
1.7 Mistakes and Paradoxes, connected with the Interpretation of the Phenomena under Consideration. The Observer **O** 14

Chapter 2. On the Mechanics of Systems with Hidden Motions 21
2.1 General Statements and Main Equations; Theorem 1 21
2.2 Special Cases, Examples 26

Chapter 3. Basic Statements and Mathematical Apparatus of Vibrational Mechanics 37
3.1 Vibrational Mechanics as the Mechanics of Systems with Hidden Motions 37
3.2 Method of a Direct Separation of Motions as an Effective General Method of Solving Problems of Vibrational Mechanics 38
3.3 On the Main Peculiarities and Advantages of the Approaches of Vibrational Mechanics and of the Method of Direct Separation of Motions as Compared to Other Similar Methods 63
3.4 On Other Methods of Obtaining Expressions for the Vibrational Forces and the Main Equations of Vibrational Mechanics 66

Chapter 4. Potential on the Average Dynamic Systems and Extremal Signs of Stability of Certain Motions 69
4.1 On Potentiality on the Average and on the Extremal Signs of Stability 69
4.2 Systems with Synchronized Objects. Integral Signs of Stability (Extremal Properties) of Synchronous Motions 75
4.3 Systems with Kinematic Excitation of Vibration (the Minimax Sign of Stability); Theorem 11 91
4.4 Systems with Dynamic Excitation of Vibration ; Theorem 12 96

Part II. Vibrational Mechanics of Machines, Mechanisms and Pendulum Devices 99

Chapter 5. Devices of Pendulum Type 101
    5.1 Pendulum with a Vibrating Axis of Suspension 101
    5.2 Chelomei`s Pendulum with a Vibrating Axis of Suspension 116
    5.3 The Follower-loaded Double Pendulum with a Vibrating Axis of Suspension 128
    5.4 Some Generalizations and Applications. On Vibrational Stabilization and Destabilization, Vibrational Shift (Drift) 130

Chapter 6. Rotor Mechanisms. Machine Aggregates 137
    6.1 Unbalanced Rotor on a Vibrating Base – Vibrational Maintenance of Rotation 137
    6.2 Devices "A Roller (Ball) in a Vibrating Cavity" and "A Ring on a Vibrating Rod (Hulla-Hoop)". Vibrational Maintenance of the Planetary Motion 152
    6.3 Unbalanced Rotor (Mechanical Unbalanced Vibro-Exciter) in the Oscillatory System – Vibrational Retardation of Rotation, Zommerfield`s Effect 158
    6.4 Machine Aggregates 168

Chapter 7. Self-synchronization of Mechanical Vibro-exciters 175
    7.1. On the Phenomenon of Synchronization of Unbalanced Rotors (Mechanical Vibro-exciters). A Brief Review of Investigations 175
    7.2. The Simplest Case: Self-synchronization of Vibro-exciters in the Linear Oscillatory System with One Degree of Freedom 180
    7.3. The General Case. A Short Review of the Results of Investigations 197
    7.4 Steadiness of Phasing the Vibro-exciters and the Adaptive Property of Vibrational Machines in case of Self-synchronization 202
    7.5 On Theoretic Investigation of the Devices with Self-synchronized Vibro-exciters. An Example of Using the Integral Criterion of Stability 206
    7.6 On the Synthesis of the Devices with Self-synchronized Vibro-exciters 209
    7.7 Regularities and Paradoxes of Self-synchronization of Unbalanced Vibro-exciters 210
    7.8 Supplement: On the Phenomena of Synchronization of the Oscillatory and Rotational Motions in Nature and in Technology. The modern state of the problem. General Definition of synchronization 214

Chapter 8. Generalized Principle of Auto-Balancing 219
    8.1 Laval's Principle and its Generalization, Ensuing from the Theory of Self-synchronization of Mechanical Vibro-exciters and from Some Other Investigations 219
    8.2 Application to the Theory of Group Foundations for the Unbalanced Machines 224
    8.3 Supplement to the Theory of Auto-balance Devices 227

Part 3. Vibrational Mechanics of Processes. (Vibrational Displacement and Shift) 231

Chapter 9. The Main Models and General Regularities of Processes of Vibrational Displacement from the Position of Vibrational Mechanics 233
9.1 On the Effect of Vibrational Displacement, its Theory and Applications 233
9.2 Simplest Model of the Process of Vibrational Displacement 235
9.3 Physical Mechanisms and Main Types of Asymmetry of the System, Causing Vibrational Displacement 240
9.4 More Complicated Main Models and Problems of the Theory of Vibrational Displacement 244

Chapter 10. Effects of Vibrational Displacement in Technique, Technology and in Nature 261
10.1 Vibrational Transportation 261
10.2 Vibrational Separation of the Components of Granular Mixtures 267
10.3 Vibrational Sinking and Intrusion, Vibrational Cutting 292
10.4 Vibrational Transformation of Motion; Vibro-engines 298
10.5 Vibrational Movement, Vibrational Coaches 303
10.6 The Vibro-jet Effect, Vibrational Pumps; Some More about Vibrational Coaches 313

Chapter 11. Vibrational Shift (Drift) 325
11.1 On the Notion of Vibrational Shift (Drift) 325
11.2 Effect of Vibrational Shift in Applications; Special Features of the Effect in Systems with Dry Friction 325

Part 4. Vibrorheology 329

Chapter 12. On Rheology and Vibrorheology 331
12.1 Rheology as a Section of Mechanics 331
12.2 Definition of Vibrorheology. Macro- and Microvibrorheology 332
12.3 Vibrorheological Equations, Vibrorheological Properties and Effective Vibrorheological Characteristics 333

Chapter 13. Effective Rheological Characteristics under the Action of Vibration 335
13.1 Effective Coefficients of Dry Friction under the Action of Vibration or Shock; Some Applications 335
13.2 Effective Friction when the System with a Positional-Viscous Resistance is acted upon by Vibration 345
13.3 Equation of Reinolds' as a Vibrorheological Equation. Effective Viscosity of Fluid in Turbulent Motion; Eeffect of the External Vibrational Action 350
13.4 On Other Cases of Using Concepts of Effective Viscosity under Vibration 352

Chapter 14. Vibrorheological Transformation of Nonlinear Mechanical Systems with Discontinuos Characteristics into Systems with Viscous Friction 353
14.1 Vibrorheology of Systems with Dry Friction 353
14.2 Vibrorheology of Systems with Periodic Collisions 355

Chapter 15. Vibrorheology of Granular Materials 363
    15.1 Vibrorheological Models of a Layer of Granular Medium 363
    15.2 Some Applications 372
    15.3 On the Behavior of the Granular Material in the Communicating
        Vibrating Vessels 383

Chapter 16. Penetration of Vibration into Certain Media 387
    16.1 Preliminary Remarks 387
    16.2 Penetration of Vibration into a Viscous Fluid 387
    16.3 On the Penetration of Vibration into Suspensions 393
    16.4 Penetration of Vibration into Granular Material 394
    16.5 On Penetration of Vibration into Concrete Mixes 399
    16.6 On the Theory of Vibro-conduction 399

**Chapter 17. Microvibrorheology: the Behavior of Suspension under
Vibration, Effective Viscosity and Effective Density of Suspension** 403
    17.1 Preliminary Remarks 403
    17.2 Structureless Suspensions – Solid Particles in a Viscous Fluid 403
    17.3 Structurized Suspensions – Particles in the Medium with the
        Resistance of the Type of Dry Friction 407

**Chapter 18. The Problem of the Control of Vibro-rheological Properties
of Mechanical Systems. The Idea of Creating Dynamic Materials** 411
    18.1 On the Problem of Forming the Properties of Nonlinear Mechanical
        Systems by Means of Vibration 411
    18.2 On Dynamics Materials 412

Chapter 19. Supplements 421
    19.1 Vibrorheological Effects in Macroscopic Homogeneous Media
        (Turbulent Viscosity, Vibrocreep, Vibrorelaxation, Vibroplasticity,
        Fatigue of Materials) 421
    19.2 Certain General Vibrorheological Regularities 423

Part 5. Some Other Problems 425

**Chapter 20. The Motion of the Particle in a Fast Oscillating
Nonuniform Field** 427
    20.1 The simplest Case: Rectilinear Motion in a Field of a Harmonic
        Standing Wave 427
    20.2 Special Case: a Pendulum with a Vertically Vibrating Axis
        of Suspension 432
    20.3 Generalization of the Problem 433

**Chapter 21. Resonance (Synchronization) in Orbital Motions of
Celestial Bodies** 435
    21.1 Preliminary Remarks 435
    21.2 Main Results of the General Theory of Synchronization and
        of the Theory of Potential on the Average Dynamic Systems
        as Applied to the Problem under Consideration 437
    21.3 The Case when the Orbits of Bodies Lie in the Same Plane or in
        Proximate Planes and Have Small Eccentricities.
        On classification of resonances 442
    21.4 The Case of Two Revolving Bodies: Comparison with the
        Results of the Direct Analytical Investigation and with
        the Observed Data 445

21.5 On Resonances in the Solar System. Hypothesis about
    Simple Resonancity 450
21.6 On Other Problems (Motion of Celestial Bodies with the Internal
    Degrees of Freedom) 452